George Francis Scott Elliot, James McAndrew

The flora of Dumfriesshire, including part of the Stewartry of Kirkcudbright

George Francis Scott Elliot, James McAndrew

The flora of Dumfriesshire, including part of the Stewartry of Kirkcudbright

ISBN/EAN: 9783337175191

Printed in Europe, USA, Canada, Australia, Japan

Cover: Foto ©berggeist007 / pixelio.de

More available books at **www.hansebooks.com**

THE FLORA OF DUMFRIESSHIRE.

THE

FLORA OF DUMFRIESSHIRE,

INCLUDING PART OF THE

STEWARTRY OF KIRKCUDBRIGHT.

BY

G. F. SCOTT-ELLIOT, M.A., F.L.S., F.R.G.S.

ASSISTED BY

J. M'ANDREW, J. T. JOHNSTONE, THE MISSES HANNAY,
G. BELL, R. SERVICE, REV. W. ANDSON,
B. N. PEACH, T. HORNE.

Dumfries:
J. MAXWELL & SON, 95, 97, AND 99 HIGH STREET.
1896.

PREFACE.

In writing a book of this kind, it is quite impossible to avoid all occasion of censure. There is not a single part of the scheme which I have followed that could not be severely criticised, either in one way or in another. I have followed, in the general arrangement of the families and in the case of the more critical genera, the classical work of Bentham and Hooker, but I have attempted to quote all the species mentioned in the Ninth Edition of the *London Catalogue*. I am quite aware that, in classing many of the "species" of this work as subspecies, I am probably exposing myself to a great deal of perhaps avoidable blame, but my reason for this course is perfectly conclusive to my own mind.

No one can doubt that the idea of a "species" depends entirely on the personal experience of the botanist. None who have worked on English Botany have had such a wide experience of plants of all nations as the authors of the Genera Plantarum, and I prefer to take their view as being in the main more serviceable and orderly than that of those whose experience, though profound, is confined chiefly to European plants.

The Record List has involved a vast amount of labour. All the MSS. of Watson's Topographical Botany in the Natural History Museum have been consulted, and the records are quoted under the names there mentioned. I have to thank Mr Carruthers for his kind permission in this respect. In the case of Dr Davidson's records, I have been put in a difficult position. His plants are now in California, and I cannot obtain any information as to who named them. With this exception it may, I think, be considered certain that the records are entirely correct, for I have taken the best advice as to all critical species, and spared no trouble in verification from all sources.

A great difficulty has been to know how to deal with those forms which have been in some manner introduced. I have been severely criticised on this point already. *First*, if I had taken only absolutely native or indigenous plants, I should have departed from the course pursued by every other author of a local Flora, and been very severely blamed for carelessness in not doing the work thoroughly. *Secondly*, as I have admitted every species now found established in a wild condition, however introduced, I shall be of course exposed to those who say that only well established plants should be admitted in a local Flora. The criterion I have taken is the establishment in a healthy condition of self-sown plants. None other is really of value, and certain interesting problems could not have been studied, if these doubtful forms had not been included. The climate of the county is so genial, that these introduced plants are exceedingly abundant, and exceedingly difficult to tell from plants undoubtedly native to Scotland. I do not know of any other method of treating this question, which does not either on the one side or on the other involve serious error.

In citing localities I have tried to pass, in the case of *Maritime Plants*, from the Mull of Galloway, along the coast of Glencaple, and thence to the English Border. The three great valleys are then taken in order, and the reader is supposed to pass up the Western and down the Eastern bank. Further subdivision than that here given did not seem to me of any practical use. The Cargen, Cairn, etc., are indubitably part of Nithsdale, just as the Æ and Kirtle are quite characteristically Annandale.

Of course no work of this nature can ever be complete. The *London Catalogue* shows a considerable growth in two years, and is still growing. I have chosen to produce this work as it stands, partly because the information is quite correct and will undoubtedly be of assistance, and partly because six years of African travel has greatly told on pedestrian efficiency, and I know that it could not be appreciably improved by my own exertions.

The amount of assistance received from my local friends will be visible to all those who read the Flora. I must thank all heartily for their kind and valuable help. Perhaps one of the pleasantest remembrances at the close of such a work is that of the many happy days spent in congenial companionship with fellow-students in botany. Mr M'Andrew, Mr Johnstone, the Misses Hannay, and Mr Bell are best described as part authors of the work; but I have had the greatest delight in finding that the older Botanists of Dumfriesshire were far more thorough and correct than is perhaps agreeable to some of those amongst us now. It is a pleasure to rescue their names from oblivion, and to put their notes in a form accessible to all. The Rev. E. F. and W. R. Linton have helped me very greatly with the critical Hawkweeds, and other forms. Mr Brunetti kindly named all the Diptera (I had asked Mr Verrall, who was, however, unable to spare the time), and Mr R. Service the Hymenoptera and Coleoptera. The plants collected by most of those named below are now in the Herbarium of the Natural History and Antiquarian Society of Dumfries, and open to inspection. I have also to thank Mr A. Bennett for much kind assistance.

The introductions by Mr Robert Service on the Hymenoptera, and the Geology by Messrs Peach and Horne, speak for themselves. I am very conscious myself of the value which they give to the following pages; and my thanks are also due to Mr Andson for the data, gathered by many years' careful observation, which we have brought together as an account of the Meteorology. The map has been specially obtained at great trouble, and it is to be hoped that it will be found satisfactory.

With this Herbarium and this Flora, I think there is scarcely any county in Britain so easily studied as mine; and it is with the most perfect confidence that I look to future Dumfriesians to carry out and prove some of the problems suggested herein.

G. F. SCOTT-ELLIOT.

November 25th, 1895.

CONTENTS.

	Page.
PREFACE	iii.
LIST OF ABBREVIATIONS, AUTHORITIES, AND ASSISTANTS	xxxvii.
TOPOGRAPHY, by G. F. Scott-Elliot	ix.
HABITAT ,, ...	x.
FLOWERING PERIOD ,,	xiii.
INSECT VISITORS ,, ...	xiii.
HYMENOPTERA OF MID-SOLWAY, by R. Service	xiv.
METEOROLOGY, by Rev. W. Andson and G. F. Scott-Elliot ...	xxii.
GEOLOGY, by B. N. Peach and T. Horne	xxvii.
FLORA ...	1
INDEX TO GENERA OF PLANTS ...	217
INDEX TO HOST PLANTS OF THE INSECTS MENTIONED ...	210

INTRODUCTORY.

TOPOGRAPHY.

The County consists roughly of the major part of the three great river valleys—Nithsdale, Annandale, and Eskdale; a certain amount of the first is included in the Stewartry of Kirkcudbright, and another portion of the upper part of Nithsdale belongs to Lanarkshire. A glance at the map will show that the lower portion of Eskdale belongs to Cumberland, and most of the Liddel is outside Dumfriesshire. Hence the County will be seen to be an extremely unnatural one, and for convenience sake I have attempted to include in the Flora the whole drainage area of the Nith and Annan, and only followed its regular boundary in the Southern and Eastern parts, as those parts of the Esk and Liddel drainage areas which are beyond Dumfriesshire would be most conveniently studied from Carlisle as a base.

The seaside part of Dumfriesshire is almost entirely composed of estuarine mud, on which grows an abundance of Armeria, Triglochin, and Plantago maritima. This is occasionally broken by sand and shingle or by a low cliff where the tide has reached, as *e.g.*, near Annan, a considerable hillock of boulder clay. Rocks are only represented by the concrete at Torduff and the sandstone quays at Annan and Glencaple. Hence it is surprising to find that most of the ordinary maritime plants have been discovered, though they occur usually in a very scattered and local manner.

It is not unusual in works of this kind to produce a full description of the general features of the County under review, giving the river names, general appearance, and other knowledge usually sought for in guide books. Most of this information is placed under the heads following, and much will be best found by a careful study of the map. The valley of the Nith is interesting in every sense. If we include the Cargen as a part of its drainage area, then the western watershed will be found a remarkable natural boundary, not only of plants, but of ethnological and geological value. Civilisation and human imigrants have followed the line of the Glasgow and South-Western Railway from the earliest prehistoric times. This is due probably to the depth and great inland extent of the valley. The 200 feet contour is not reached before Drumlanrig, and a careful perusal of the Geology will show that the depression occurred in the early Silurian age. All the valleys are of extraordinary beauty, but this is only understood by those who have the patience to go off the high roads and explore. In Nithsdale, the Glen, Blackwood Linn, the Scaur, and Craighope Linn may be recommended as the best examples of lower wood glens, but almost every burn has a beauty of its own from one end of the dale to the other. The western watershed begins with Criffel, over 1800 feet high, but its continuation is insignificant until the sources of Scaur, Kello, and Euchan. Queensberry (2285 feet) is the most conspicuous point of the eastern watershed. Annandale and the Caledonian line is a natural road of the same kind. Moffat is perhaps the best botanical centre in the County, as from it all the characteristic habitats of the district can be easily reached.

These include the Beld Craig Linn and Garpol, both typical woodglens, and the beautiful corries of Black's Hope and Grey Mare's Tail, above which are the highest elevations in the county.

Eskdale, between Langholm and Canobie Bridge, is the most beautiful wooded valley that the writer has seen in any part of the world. Higher up in the Eskdalemuirs, the scenery is desolate and wild in the extreme. Meikledale is an interesting corrie, not unlike those on Moffat water, but such ravines are rare in this valley.

Originally the County probably consisted of deciduous forest, broken along the river sides by stretches of marshy soil or peat moss. I believe this forest probably continued from nearly sea level to about 800 feet, from which level to that of the present peat haggs there may have been, either after a belt of conifers or throughout, rough grass and heather. The succession found in most parts of the world of deciduous forest, conifers, and heather or moss may be traced therefore without much difficulty. Arable land now replaces the deciduous forest, which still persists along the rivers and burns up to 2200 feet.

The hill farms and permanent pasture represent probably what was once rough grass and heather, but man has not yet been able to bring the peat-haggs into cultivation. These are laid down in a capping of peat some ten to twelve feet thick, though varying enormously in depth over all the higher hills in Dumfriesshire. Peat-mosses are found at all elevations, from the Solway to that of Loch Skene and higher, but I can find nowhere a satisfactory explanation of this formation.

I must refer the reader to the map for the source and tributaries of the three great rivers, Nith, Annan, and Esk, as the impression of these three large valleys is by far the most valuable for all practical purposes.

THE HABITAT.

It was chiefly on account of the great importance, to my mind, of a study of the suitability of plants to their habitat, climate, environment, milieu or monde ambiant (for all these terms really express the same thing), that I undertook this Flora. I soon found that in this country it was not possible to obtain a very clear idea of those tendencies to variation which are produced by climate, mainly because the climate itself does not vary in a sufficiently marked manner, even in such a diversified area as one finds in Dumfriesshire. My original scheme has been therefore modified more than once, and in its final form consists in pointing out the following factors for every species—soil, exposure to wind, and exposure to sun. Insect visitors, really an essential part of the environment, are treated of by themselves.

These three leading factors of the habitat are found inextricably mixed in practice. The most usual combinations in Dumfriesshire are the following:—

1. *The Seashore* (sand, shingle, estuarine mud flats, and rarely cliffs of boulder clay; or, in Galloway, of rock).
2. *Holms* (flat river-alluvium)—roadsides and arable fields on boulder clay.
3. *Ordinary Arable Land* extending from sea level to about 800, or rarely 900 feet.
4. *Permanent Pasture* or sheep farms, chiefly on Silurian rock or drift, and extending from about 800 feet to the lower level of the peat-haggs.
5. *Woods and Linns*, from about sea level to nearly 800 feet.
6. *Corries and Glens*, or mountain ravines, from 800 to 2200 feet.
7. *Peat Mosses and Haggs*, at almost all elevations from nearly the sea level at Lochar Moss, and forming a thick capping over all the highest hills of the district.
8. *Craigs and Scaurs* of bare whinstone rock or mudstones; occasionally screes and accumulations of broken boulders and stones.
9. *Railway Stations and Tracks*, as well as waste ground generally.
10. *Lochs and Rivers*.

A large proportion of the Flora is rigorously confined to one or other of these divisions. Unfortunately the meteorological and geological factors are so insufficiently known that one cannot in any manner obtain data of the degree of moisture of the soil or atmosphere, the wind exposure, the amount of sunlight received, the porosity and fertility of the soil, or even its geological character in any one of them. Without distinct data on these points it is useless to attempt to draw up statistics of the general characters of leaf, branching, or inflorescence found in these particular habitats which would be, if possible, of enormous importance.

As regards plants, the different effect of light, heat, and wind exposure can scarcely be distinguished in the field, because, *e.g.*, shelter from wind brings in its consequence rank vegetation, whether as a wood or as in a corn field, and rank vegetation means a certain amount of shade and a moist atmosphere. Shade in the same way involves wind-shelter and a lower temperature. It is for this reason that exposure is the most convenient term available.

The effect of exposure is most marked in the divisions one and eight. Some characteristic desert adaptations, such as the rosette type of plant, dense twiggy, cushion-like shrublets (which are really, I think, to be considered as a bunch of rosettes supported on small branches), fleshy leaves, and a coat of woolly hairs, are found both on the seashore and on exposed rock ledges, as *e.g.*, Raven craigs in Black's Hope.

Spines and thorns, on the other hand, as well as the nameless type represented by the seaside Euphorbias and Chenopodiaceæ, do not, to my knowledge, occur on the high exposed mountain ledges.

This is probably because the last two are special adaptations against or effects of strong sunlight, which may injure the tissues or chlorophyll, while the others are protections against transpiration generally, whether due to the wind or sun.

The actual humidity of the atmosphere in the immediate neighbourhood of the sea must, unless I am greatly mistaken, be at some period of the 24 hours far below that of the low-lying country and hills, and it is not till the exposed rock summits of about 1500 to 2200 feet are reached that there is the same amount of transpiration.

The same condition of exposure, though it is not quite so clearly marked, is obvious on (division nine) railway lines and all waste ground which is kept clear of plants. It is obvious that in such places the conditions of temperature and radiation of heat are extreme, and possibly this explains the presence of such plants as Linaria minor, Tragopogon, and Hieracium aurantiacum, which become yearly more abundant along the different systems.

The opposite extreme, that is of shade and shelter, is most marked in number five, but in parts of number six there is considerably more moisture even than in the lower woods, and I am rather doubtful whether the separation is entirely justifiable.

Probably the typical peat plants are more thoroughly represented in Dumfriesshire than in any other county. Whether their characteristics are due to the spongy water-holding nature of peat, to its antiseptic property, or to the large amount of decaying organic substance, is quite unknown to me. The Droseras, Utricularias, and Pinguicula seem to depend on the latter quality, and possibly the antiseptic property is responsible for the limited flora, but it is not safe to say more on the subject.

The effects of a clay and a sandy soil are quite distinct in their extreme forms, but as one finds them under ordinary circumstances it becomes excessively difficult to trace the effect on vegetation in any one field; one may discover typical sand-loving plants in the stony places of an alluvial holm and along the roadsides, where dust and road sweepings are perpetually heaped against the bank, one may find all the clay-loving plants in a sandy district.

The only habitat which appears to me to be explained by soil is that of Helianthemum vulgare. Its curiously isolated position may conceivably be due to the drift, after passing Lanarkshire, having left along the left bank of the Annan Valley and the Beeftub considerable quantities of limestone fragments; but I am not sure if this explanation is correct. (See Geology). The present condition of the study of botanical environment is so chaotic and speculative that it is not advisable to point out more than these obvious factors.

FLOWERING PERIOD.

My data for this are chiefly due to the excellent observations made for me by Messrs Johnstone, G. Bell, and J. Shaw. These are only intended to be approximate, and in cases where the two former have observed the same plant there is sometimes a difference of six weeks, which is quite what one would expect. I did not realise at first the importance of this question, both as proving that a definite "thermal constant" is required before any given flower appears, and also in practically isolating two varieties of the same species which grow under different conditions of exposure. Observations are much required on all these points, and both the hour of the day at which flowers open and close, as well as the commencement and end of the flowering season, must sometimes contribute to complete isolation of incipient varieties and subspecies.

INSECT VISITORS.

The catching of insect visitors has probably occupied more than nine-tenths of the time which I have spent on the work. I soon found that no human being could expect to go through the whole Flora, species by species, as it would scarcely be possible to do six species thoroughly in one season, while the number here alluded to is nearly 900. I have therefore attempted to give an idea of the more common visitors of about 270 species. Miss Hannay, Mr Armstrong, and Miss Ethel Taylor have given me most valuable assistance, and a very important paper by Mr J. C. Willis* has greatly pleased me, because my own observations are very markedly supported by his, which were taken independently at Auchencairn.

The result is to leave me more impressed by the importance of this study, for it is obvious that for the distribution of many plants, certain insect visitors are absolutely essential. They are also to my mind important agents in the isolation of varieties, for though the same plant may grow inside a wood and in the neighbouring field, it is quite certain that the same insects will not visit its flowers in each case, and hence these plants are absolutely isolated.

* Annals of Botany, vol. ix., June, 1895.

I have much doubt as to wind-fertilised flowers. I have seen insects on Grasses, Sedges, Plantago, Thalictrum, and of course on Willows. It is obvious that if two flowers of Thalictrum are separated by five feet, the chance is one to thirty that a pollen grain should go within a foot of the right direction, and the wind and gravitation probably diminish the probability to an indefinitely small extent. If an insect is, for any reason, on one, the chances are ten to one that it will go to the next visible flower.

I should, if I had drawn this introduction out at the beginning of my work, have added tables showing the proportions of Hymenoptera, Diptera, etc., but now this does not seem to be advisable. No one would have supposed that a plant which is visited by one butterfly, four species of humble bee, and two Syrphids was anything but a red or purple Corolliflor, whereas these were all caught on the common bramble, Rubus fruticosus * I am, however (*vide* Labiatæ and Caryophylleæ), firmly convinced that the shape of every flower as well as the modes of dehiscence, general arrangement, and so on, are entirely suited to the average insect visitor of the particular species. In most cases the corolla is nearly an exact mould of the shape and motions of the head and proboscis of the average visitor; it is very rarely the shape of its head at rest. In fact, granted growth of the corolla and the obvious modifications due to mechanical strains and stresses, most flowers seem to have suited themselves exactly to these shapes and motions, like a foxglove to a humble bee or an old glove to its wearer's thumb.

Much as I should like to uphold Professor Henslow's theories as to the exact effect of probing in stimulating the flow of honey and hairs, the subject does not seem to me proved, and scarcely to be proved, without exceedingly delicate and difficult experiments.

It is not too much to say that the knowledge of British Hymenoptera, and particularly Diptera, is disgracefully behind that of British Botany. I have been most fortunate, however, in obtaining the invaluable assistance of Mr Robert Service for the former and of Mr E. Brunetti for the latter. These gentlemen have named all the insects cited under my name, that of Misses Hannay, Miss Taylor, or Mr R. Armstrong, and I have to sincerely thank them for this.

THE ACULEATE HYMENOPTERA OF MID-SOLWAY.

By ROBERT SERVICE.

Amongst all the vast multitudes of the Insect tribes, I think it will be generally conceded that in variety of habits, in the high intelligence displayed in their social intercourse, and in beauty of form and colouring, the Aculeate Hymenoptera stand unrivalled. Unfortunately, with

* This flowers when there are practically no competitors.

but few exceptions, this district is not at all rich in species, and still less in the numbers in which the species that do occur are represented. Of course, owing to the very insufficient way in which the very few collectors who have studied these insects have of necessity examined the district at large, it is certain that many species have escaped observation; but enough is known to prove that we are comparatively poor in the classes under notice. This is doubtless caused by the general absence of suitable sandy soil. The district of whose Aculeates we propose to give an account may be stated broadly as the country that lies betwixt Annandale on the east and the valley of the Dee on the west. It is thus the mid portion of the faunal territory, known to all Scottish Naturalists as "Solway," being one of the zoological divisions into which the late Dr Buchanan White so admirably mapped out Scotland for faunistic purposes. In the meantime, owing to the almost complete absence of materials, there is no use in extending our present remarks outside the limits of Mid-Solway. The nomenclature employed is that of Mr Edward Saunders—*Catalogue of British Hymenoptera* (Aculeata), 1883.

The Ants stand *par excellence*, not only at the head of the Aculeates, but, in the stage of evolution to which they have attained in respect to their well-ordered communities, and in the high degree of intelligence reached by the individual, they are unquestionably in advance of all other insects whatsoever.

Formica cunicularia is the first species on our list, and is fairly common. It varies much in size, as do so many other species of ants, and some of the "races," as they are termed, have received names, having been considered at one time to be distinct species. Lasius niger is the small ant so abundant everywhere, and sometimes so troublesome to gardeners when it takes up its abode in glass-houses. Tapinoma erratica is in general appearance similar to the last named, but is much darker, and it has also some structural peculiarities. It is a very scarce species here. Myrmica is a genus of small ants that is represented here, as elsewhere in Britain, by a single extremely variable species. No less than five very distinct "races" are recognised by Mr Edward Saunders, the great authority on the Aculeates. Of these I have very commonly taken—indeed they are the commonest of all our ants— Ruginodis, Scabrinodis, and Lævinodis. Sulcinodis and Lobicornus have not been detected as yet. Leptothorax acervorum is a small species sometimes found in little communities under bark or rotten stumps in the Mabie woods. I have not found it elsewhere. The last ant amongst our local species is a very interesting one, inasmuch as it has only been known to exist here within the last ten years or so. It is the Monomorium Pharaonis—rather a startlingly long name for such a minute creature. It is confined to places where it finds sufficient heat, such as bake-houses, hot-houses, and dwelling-houses kept at a sufficiently high temperature. Lately I was called in to see this ant at home in a house which it had over-run from cellar to attic. Its myriads were past comprehension, and in some places it distinctly coloured the white wall with its hosts.

The Fossorial portion of the Aculeates next claims our attention. Tiphia minuta, a very small dark species, has been taken several times on the flowers of goutweed (Ægopodium podagraria). Tiphia femorata, the other member of the genus, has been taken near Moffat frequenting the flowers of various Umbellifers. Amongst the Pompilidæ we have a few local species, these being Pompilus plumbeus, of which an occasional specimen is to be seen on Umbelliferous flowers; P. viaticus, a few in August near Mabie, and P. fuscus, also captured at Mabie. Two other species of Pompilidæ were originally described by the late Frederick Smith from specimens captured here. One of these is P. acuminatus, found in Kirkpatrick-Juxta many years ago by the late Rev. W. Little; the other is P. approximatus, and was found at Eccles by Dr Sharp during his residence there. Neither species has been met with since by any other collector. Passing over a large number of Aculeates that are not found in these counties, we next come to the genus Pemphredon. All of the three British species, viz., lugubris, unicolor, and lethifer are found, the last named being quite common. It can be most easily collected by looking for broken bramble stems, in which the exposed end has been burrowed in, or perforated. These are almost certain to contain larvæ of this species, and if the bramble stems are cut off and kept, the perfect insects will duly make their appearance at the end of June or early in July. Mimesa bicolor is found not infrequently. Mimesa Dahlbomi was taken by Mr Scott-Elliot at the Mill Loch, Lochmaben, in June on flowers of Cicuta virosa. One or two specimens only of Harpactus tumidus have been taken. Nysson dimidiatus, though the rarest of the genus, is the only one I have taken. I found a solitary individual on Ragwort. The others ought to be found here, but so far have not been recorded.

A very pretty species is Gorytes mystaceus, and it is common in most seasons. It provisions its nest with the green larvae of the Cuckoo spit, or "Gowkspittle" (Anthrophora spumaria), so abundant on herbage throughout the summer months. Mellinus arvensis is another very common insect. It burrows regularly in the sides of potato furrows in dry sandy districts, and is, like the previous species, a carnivorous insect. M. sabulosus has only been taken on one occasion. Only a few representatives of the great genus Crabro are to be met with, but where they do occur they are tolerably plentiful. C. palmipes is a minute species that may be collected in quantity, skipping about on the leaves of brambles, or on similar surfaces in the tangled growths about sunny lanes and hedgerows. They may often be seen preying upon the tiniest Diptera. C. varius has been taken, but is scarce. It occurs about flowers on railway banks. A few specimens of C. elongatulus were found on one occasion on the moss road near Mabie. C. dimidiatus has only occurred in single specimens as yet. C. cribrarius is a most handsome species; it is very common on composite and umbelliferous flowers. A few individuals of C. peltarius have been got in the Mabie plantations some years ago, and the same may be said of C. vagus, although it is perhaps rather oftener met with. C. chrysostoma, a species with a splendidly golden coloured fringe of hairs on the upper lip, is frequent, and seems to be attached to rotten willow trunks. Oxybelus uniglumis used to be taken near Moffat by the late Rev. W. Little.

Next in order are the Wasps, Social and Solitary. The Social Wasps are very fully represented indeed. Vespa vulgaris, V. rufa, V. germanica are all equally common, and of general distribution, appearing in some seasons in enormous swarms, and at such periods doing considerable damage to ripe fruit. The good they always do in destroying vast quantities of noxious flies and larvæ, as well as in acting as general scavengers, is apt to be over-looked. There can be no question that the good qualities of Wasps far out-balance any evil they do, but this is very often lost sight of. V. norvegica is also an abundant species. Its small hanging nests are often hung on low bushes and similar situations, and are a subject of dread to those ignorant of wasp habits. V. sylvestris is rather a rare species, while V. arborea does not seem to be found here at all. It might be as well to correct the rather common error that V. crabro, the Hornet, is found in this region. It has never been seen here, nor is it ever likely to be, as it is strictly a southern species.

Of the solitary species we have Odynerus spinipes, O. parietum, O. trimarginatus, O. parietinus, and O. gracilis. The last named seems to be plentiful near Rockcliffe in Colvend, but none of the others are really common except O. parietinus. This species concludes our meagre list of Solitary Wasps.

The Bees proper come next in order for consideration, and although neither in species nor in the number of individuals with which they are represented are these at all abundant, still they yield to no other division of the insecta in the interest attached to their varied and sometimes very curious life histories. Of the genus Colletes, we have C. fodiens and C. succincta. The former is usually taken on ragwort bloom, and the latter has been caught in Lochar Moss on the flowers of the heather. Of Prosopis, which comprises no less than nine British species, I have never taken or seen a single individual; but Mr Scott-Elliot has had the good fortune to capture a male and female of Prosopis hyalinata on flowers. Sphecodes is a very pretty genus of small red and black bees found burrowing in garden walks. S. gibbus is fairly common; a few specimens of S. pilifrons have been taken at intervals; while S. ephippium and S. subquadratus are seldom met with.

The great genus Halictus is very well represented throughout this district, and the active little bees that belong to it are very numerous and conspicuous on many kinds of flowers. On the flowers of the dandelion, very early in spring, the females of H. rubicundus are pretty sure to be seen. The males, as is the case with this sex throughout the genus, do not put in an appearance till mid-summer, and on fine autumnal days they are very abundant on ragwort and other late flowering plants. H. leucozonius has been taken near Moffat, and so also has H. quadrinotata, both by the late Rev. W. Little. H. lævigatus is a very scarce member of the genus; it has been taken at several places in Troqueer. H. cylindricus and H. albipes are very closely related; both species are very abundant, and their colonies may be seen on almost every dry pathway. H. subfasciatus seems rather scarce; it has been taken on ragwort. H. villosulus is very commonly found on dandelion and hawkweed flowers. H. nitidiusculus is a common species, generally found at the flowers of weeds in waste places in company with H. Smeathmanellus. H. minutus, H. leucopus, and H. morio are very small species found in some abundance and of general distribution. Andrena is the largest genus of British bees, numbering, according to the catalogue of Mr E. Saunders, no less than forty-eight native species. Many of them are as "like as two peas," and it is in such cases almost hopeless for the novice to attempt to name his captures. By and bye, however, he will begin to see and appreciate their differences, and when the collector once learns to discriminate between the more closely allied species he will find their study a very fascinating pursuit. Andrena albicans is common everywhere, frequenting dandelion and chickweed flowers. The males have a special preference for the flowers of Mahonia aquifolium in shrubberies. A. trimmerana is an interesting species that builds its nests in old walls, especially where the lime is crumbling out from betwixt the stones. In such places it is not infrequent in May. A. Clarkella is a particularly pretty species found early in April as a rule, but occasionally even in March. I have taken it in several localities, making its burrows on sandy pathways. A. nigro-aenea has been captured only once, when I took a specimen at Mabie on willow catkins in April. Both forms of A. gwynana occur very freely, the var. bicolor being very partial to flowers of the harebell. I have taken one or two specimens of A. lapponica. This species was first captured in Britain, at Kirkpatrick-Juxta, by the late Rev. W. Little. A. varians seems to be rather scarce. A. fucata is common in Lochar Moss, and some other localities. Only a few individuals of A. nigriceps have been captured at some localities on the Galloway side of the Nith. A. denticulata, though apparently a rare species in Britain, is very common here. One specimen of A. tridentata has been taken. At certain spots there are immense colonies of A. albicrus. I know no other of the wild bees that is found in our district in such multitudes. One such place is on the Moss, just behind Douievale House. A. coitana is a small species that is found everywhere. A. minutula is another species equally common, its var. parvula being found in about equal numbers. A. nana is another of the small species that may be captured in some seasons in great abundance. A. afzeliella occurs freely near Craigs, and the last on our list of local species, A. Wilkella, is rare, only one or two specimens having been taken on dandelions near Dalscairth.

The next division that comes in for enumeration is that of the Cuckoo Bees, forming the genus Nomada. Superficially they resemble slender wasps, being mostly more or less banded with black and yellow. Their economy is most interesting. Stated generally, these bees attach themselves to certain other species of wild bees, depositing their eggs in their cells, and leaving the eggs and young larvæ to the care of other species, so that their life history is not unlike that of the cuckoo amongst birds. Hence their name of Cuckoo Bees. Nomada solidaginis, N. alternata and N. obtusifrons are each very commonly found. N. lateralis, N. fabriciana, and N. flavoguttata are frequent, though not numerous. A few specimens of N. ruficornis and N. furva have been caught, while of N. roberjeottiana only one specimen has been taken.

The genus Caelioxys comes next on our local list, and of its representatives we have C. vectis, C. rufescens, and C. elongata, all occurring with tolerable frequency.

The fine genus Megachile is well represented here in numbers of individuals, if not in species. M. Willughbiella is very common everywhere; this is the bee which is seen so frequently making its burrows in rotten trees and stumps. M. centuncularis is the familiar "leaf-cutter" bee; it makes its burrows in wood or in old walls, but occasionally in the ground; it lines its cells in the most artistic way with little round pieces neatly clipped out of the leaves of bushes, those of the rose by preference.

Anthidium manicatum is to me personally the most interesting of insects, for it was probably the first that attracted my attention. Its strange mode of courtship, much akin to what is related of the lovemaking (?) of the Australian aborigine, I used often to watch when I was a very small boy indeed. But at that time I had no idea when I saw the headlong rush of the male, and saw him clutch his partner in such a vicious looking embrace and observed him bear her aloft into the air, that his intention was anything else than to devour her forthwith. A favourite occupation of this species is that of scraping the tomentum off such plants as supply this material, making it up into little bundles, and then carrying it off for lining its nests with. The species is common in most situations suitable for it. I have nowhere seen it in such abundance as upon the bramble flowers in the loaning that leads up from Rockcliffe past the avenue to Baron's Craig.

It is somewhat remarkable that the genus Osmia should be unrepresented in the district, at anyrate so far as my experience goes. There are certainly specimens of Osmia fulviventris in the collection of the Rev. W. Little, now in my possession, but they have no data attached, and so it is uncertain whether they are of local origin or not.

Of the Anthophoræ I have only taken Anthophora retusa, of which I took specimens on one occasion near Threave Castle.

There is no more conspicuous tribe of insects at all times, from the opening of the leaves in spring until the chill airs of October have turned the foliage to the brillance of autumn hues, than the Humble Bees. With the flowering of the willow catkins in March, and often much earlier, the large fine coloured females of Bombus terrestris are out in force from their hibernating quarters. Their noisy hum, as they buzz through amongst the willow branches to sip the fragrant greenish-coloured nectar from the catkins, is the most welcome music to the naturalist's ear. This species is usually considerably in advance of the others, but if the weather continues mild, B. hortorum, so readily distinguished by the length of its face, soon follows. And then one species after another puts in an appearance as the spring advances, till when we have seen the beautiful B. distinguendus drinking at the bottom of the rhododendron chalices, we may conclude that all our old friends amongst the Humble Bees are once more in full flight. As yet there is, however, of the different sexual forms none, except the old females to be found. These, fertilised in autumn, have been hibernating all through the cold season, till a suitable temperature and other conditions call them forth in spring. Then they flit about looking for a suitable site for a domicile, and prepare for setting up house. Their nidus may be in an old mouse hole, in a dyke, or in a bank or slope, or under a stump, or it may even be on quite open level ground if a suitable aperture into and underneath the surface soil can be discovered. That is in the case of the ground builders. Other species build on the surface amongst the herbage, and make a very neat and snug domed dwelling of felted grass and vegetable fibres, very often adopting to this purpose the nest of the Short-tailed Field Vole. Many times I have seen nests of B. muscorum built inside those of birds. The nests of the common wren, yellow hammer, titlark, robin and willow warbler, I have found filled up with a nest of the last named species of Bombus, and more than once I have seen where the intruder had built her own nest over the eggs of the rightful owner. In such cases the birds had in all probability deserted their nests before the bees took possession, for one can hardly think the birds would give way to such a tiny burglar, and one, moreover, that would furnish such a nice little eatable morsel. Nests being fairly into shape, cell-building and egg laying proceed in due course, then the young grubs hatch and go through their interesting series of transformations, and by the middle of May, the workers of an early species, like B. pratorum, may be noted enjoying their first flight. Within a fortnight or three weeks later, workers of all the species found here may be observed, with perhaps the exception of B. distinguendus, which is a somewhat late species. Males are not developed till far on in the summer, indeed autumn has fairly arrived before there is any great show of the male Humble Bees. This sex is in all the species, as it is in so many other families of insects, very much smaller than the female sex, and is even smaller, as a rule, than the workers or neuters (which are really abortive females). To compensate for their lesser bulk, most of the species have males of more diversified and fully brighter colours than those which adorn the females. Thus the males of Bombus terrestris, pratorum, soroensis, Derhamellus, and lapidarius, are much prettier than their respective partners. All the species already mentioned are plentiful and generally distributed, with the exception of B. distinguendus, a species that is by no means a common one, although it is almost always present wherever rhododendrons are in bloom. Then we have, in addition to the foregoing Bombi, B. cognatus, a very beautiful species, of which I have seen only a very few examples. B. latreillellus is another species of great rarity hereabouts; while B. schrimshiranus, though scarce, may be found by diligent search. There is thus a total of eleven species that occur in this district out of the entire number of sixteen that are known as British species. Doubtless one or two more species may yet be added to our local list by an assiduous collector.

Nearly allied to the Humble Bees, and so similar in general aspect to them that most folks never regard them as anything else, are those curious parasites known to systematists by the generic name of Psithyrus (Apathus). They have quite a strange history. They make no nests for themselves, each of the five British species—all of which are found in this district—being attached to certain species of Humble Bees. In appearance they are exactly similar to Bombi, one main distinction being that they are destitute of the corbicula, or pollen basket along the outer portion of the hind tibiæ, which when filled is so conspicuous in the females and workers of the Humble Bees; and not making any nests for themselves and also leaving their young, like the cuckoos, to be brought up by others, no workers are therefore required, so the sexes consist of males and females only. They are much later in making their appearance in spring than their hosts, for a wise provision of Nature keeps them slumbering in their hybernacula until the nests of the Humble Bees are fairly set going for the season. Then the females of Psithyrus come forth and search for the nests of the particular species of Bombus to which they are assigned. Having found what is required they have no difficulty in entering, for they are so similar to the rightful inmates of the unsuspecting household that they enter it as freely as if it was all their own. Once in, eggs are laid in the cells prepared for those of the Bombi, and the latter, never guessing at the presence of foster children, bring up the young larvæ with all the care lavished on their own brood. Towards the latter end of the summer, the old females of Psithyrus give place to a new brood that consists of both females and males. They occur in great abundance on all the wild flowers during the autumn months, the females that have been paired at once proceeding into winter quarters, while the males die off gradually till they finally disappear with the setting in of cold weather. As to their relative abundance and distribution:—Psithyrus rupestris, which is said to be solely attached to Bombus lapidarius, seems to be very scarce throughout this district. P. vestalis, which lives with B. terrestris, is very common on thistles and other flowers in autumn. P. barbutellus is rather a rare species, so far as my experience goes; it frequents the nests of B. pratorum, B. schrimshiranus, and B. Derhamellus, according to some authorities. P. Campestris is common and is found wherever its host, B. hortorum, dwells. P. quadricolor lives with B. pratorum and B. schrimshiranus, and is fairly common in some localities.

Such is our review of the genera and species of Aculeate Hymenoptera that have been taken by a very few collectors and observers in the district betwixt Annan and the Dee. Many more species remain to reward the labours of those who diligently search out the haunts of the wild bee. A rich field, almost untrodden, lies ready for exploration by any student who cares to take up one of the most fascinating branches of entomological study.

METEOROLOGY.

By the Rev. Wm. ANDSON and G. F. SCOTT-ELLIOT.

The Meteorology of Dumfriesshire, as a whole, cannot be easily described, for there are no records from, *e.g.*, the Moffat Hills or the Merrick, and without these, it is scarcely possible to understand the bearings of the peculiarities shown in the county.

The following tables show the results which are at present available. The death of Mr Dudgeon of Cargen has been a severe loss to meteorology, as well as to many other sciences, and the first table which gives the mean of no less than thirty years' observations (1860-1890) taken by him, is one of extreme value.

CARGEN (Elevation, 90 Feet).

	Jan.	Feb.	Mch.	Apl.	May.	Jun.	Jul.	Aug.	Sep.	Oct.	Nov.	Dec.	Year.
Average mean Temp.	38·1	39·6	40·7	45·9	50·7	56·5	59·3	58·4	54·9	48	41·5	38·6	47·7
Inches of Rainfall......	5·07	3·77	3·13	2·22	2·62	2·56	3·19	3·77	4·16	4·58	4·05	4·70	43·82 in.
Hours of Sunshine.....	59	88	128	165	212	236	239	210	174	120	85	57	1773 hrs.

As a result of five years' observation obtained at

METEOROLOGY.

DRUMLANRIG (Elevation, 191 Feet).

	Jan.	Feb.	Mch.	Apl.	May.	Jun.	Jul.	Aug.	Sep.	Oct.	Nov.	Dec.	Year.
Average mean Temp.	34·8	36·2	36·1	42·3	48·4	54·9	56·7	55·1	51·1	45·1	41·1	35·1	44·7
Inches of Rainfall	3·84	1·18	2·22	1·28	2·12	1·06	3·20	4·22	3·50	4·12	5·16	4·54	36·44 in.

The observations at Dumfries are obtained from nine years' observation, and the instruments have been carefully examined by Dr Buchan and pronounced satisfactory.

DUMFRIES (Elevation, 60 feet).

	Jan.	Feb.	Mch.	Apl.	May.	Jun.	Jul.	Aug.	Sep.	Oct.	Nov.	Dec.	Year.
Average mean Temp.	37·3	38	40·7	45·8	52·2	57·6	58·6	57·6	54·2	46·8	42·3	37·8	47·4
Inches of Rainfall	3·38	2·44	2·08	1·60	2·90	2·01	3·58	4·06	2·85	3·92	4·03	3·98	36·83
Relative Humidity (Sat. =100)	91	89	87	79	78	78	82	85	87	88	90	91	85

From other stations there are no observations which can give thoroughly trustworthy averages extending over a large number of years. The following table is, however, of interest as showing the comparative differences at different elevations:—

METEOROLOGICAL REPORT FROM 1ST DECEMBER, 1887.

Stations and Elevations.	Winter Quarter. Dec., Jan., and Feb.		Spring Quarter. Mar., April, and May		Summer Quarter. June, July, and Aug.		Autumn Quarter. Sept., Oct., and Nov.		Per Year.	
	Mean. Temp.	Rainfall.	Mean. Temp.	Rainfall.	Mean. Temp.	Rainfall.	Mean. Temp.	Rainfall.	Mean. Temp.	Total Rainfall.
Dumfries, 60 feet	37·3	8·80	43·3	6·68	55·8	11·25	48	10·29	46·1	37·02 inches.
Cargen, 90 feet	37	9·36	42·9	7·06	54·6	9·50	47·1	11·01	45·4	36·93 ,,
Drumlanrig, 191 feet	35·7	9·30	42·6	6·60	54·2	11·10	46·5	10·90	44·7	37·90 ,,
Cally, 140 feet	39	10·12	42·8	8·62	54·1	10·42	46·7	10·98	45·6	40·14 ,,
Glenlee, 206 feet	35·8	13·0	42·3	8·47	54	14·51	45·9	15·53	44·5	51·51 ,,
Wanlockhead, 1334 feet	31·7	10·57	36·8	17·10	48·8	11·32		*	(38 ?)	(50? or more)
Dalshangan, Carsphairn, 500 feet (for 1889)	36·6	13·03	44·8	9·89	56 ?	9·77	45·6	14·43†	45·9	44·50 inches.

* No report of rainfall for August, September, October, and November. † Accuracy doubtful. Therm. probably unprotected.

The heavy rainfall at Glenlee is probably due to its situation on the eastern side of a hilly region where many of the summits are over 2000 feet in altitude.

The westerly and south-westerly winds laden with moisture from the Atlantic meet these hills, and are forced up into a higher stratum of the atmosphere, where they expand under the lower pressure there existing. Consequently the aqueous vapour is condensed, and heavy rain falls on the hills, and particularly their eastern slopes. There was a good example of this on a smaller scale in 1888 at Drumpark, where 47·20 inches fell, as compared with only 37 inches at Dumfries.

It is also commonly observed that across the watershed of the Nith valley, for example at Colvend, there is frequently no rain, while there is continuous wet weather about Dumfries.

The direction of the wind seems to be chiefly westerly. Thus the mean of nine years gives as below :—

	S.W.	W.	N.W.	N.E.	E.	S.	S.E.	N.	Calm or Variable.
Days,	89	62	42	39	36	32	26	23	17

The following table is also interesting :—

WIND.

	N.	N.E.	E.	S.E.	S.	S.W.	W.	N.W.	Calm or Var.	Mean pressure on Square Foot.
Dumfries	30	42	44	19	26	71	64	51	19	Lbs. 2·00
Cargen	28	44	100	22	43	45	55	33	...	1·70
Drumlanrig	66	60	15	27	52	57	23	66	...	0·4
Cally	84	28	44	12	54	24	63	23	34	1·55
Glenlee	15	28	26	21	38	49	48	40	70	1·83
Wanlockhead	41	26	59	22	44	8	85	19	...*	1·70

Cyclones passing over the county almost invariably pass from S.W. or S.S.W., to N.W. and N. The wind veers from the former to the latter direction as the depression passes. It is possible this may account for the curious difference betwixt the wind observations, Dumfries having 71 south-westerly wind days, Drumlanrig 57, and Wanlockhead only 8.

* No reports of wind at Wanlockhead for months of April, July, August, and September.

All inhabitants of the district are, of course, aware that the climate is peculiarly mild and genial. Unfortunately sunshine-recorders are neither common nor in every way satisfactory, but the results given in a paper by Mr H. N. Dickson in the *Scottish Geographical Magazine*, August, 1893, are well worth considering as exemplifying the mildness of our climate. This paper contains a map in which curves represent the amount of sunshine (by the Campbell Stokes burning recorder) received at various points in the British Islands. The three counties receive between 1300 and 1400 hours of sunshine out of 4400 hours "possible sunshine." In this respect they are only approached in Scotland by Banff, Elgin, and Aberdeen. If we follow the lines representing the same amount of sunshine through England, the result is astonishing. Salisbury Plain, the Weald of Sussex, and Greenwich are practically in the same position as Dumfriesshire, while no part of the east coast between Montrose and Skegness in Lincolnshire is so fortunate in sunshine as the Burrow head. Nottingham has less than, *e.g.*, Drumlanrig and Dumfriesshire generally, and this also seems to be the case with the whole of Yorkshire, Durham, and Northumberland. In fact, so far as sunshine goes, Wigtownshire particularly should have as genial a climate as such health resorts as Aberystwith and the Malvern Hills.

This probably explains why an enormous number of escapes, such as Datura Stramonium, Gagea lutea, and Tragopogon porrifolius are found in the Flora. I should not wonder if observation were to show that in the neighbourhood of Whitcoombe and Loch Skene, a small tract of country exists in which not more than 1200 hours' sunshine is received annually, and if this is found to be the case, then the three counties contain every variation found in Scotland.

On the whole, the Solway, resulting from a great depression which separates the granites and whinstones of Dalbeattie and Kirkcudbright from those of Scafell and Cumberland, acts in a double manner. The clouds keep to the course of the mountains from the Cumberland Hills round by Whitecoombe and the Lead Hills, to Carsphairn and the Merrick, and hence give us more than our correct allowance of sunshine; and this explains also the abundance of westerly winds and the resulting humidity.

On the other hand, the Gulf Stream probably assists in the production of the complex currents round the Mull of Galloway. Some of these currents, to judge by the Flora and garden experiments, must bring a very warm climate with them round the shore of Galloway. Observations at Kingholm and Dumfries on the temperature of the river, as contrasted with similar observations at the Island of Little Ross, show in fact that in autumn and winter the temperature of the Solway is considerably higher than that of the rivers which flow into it. The seasonal variation of the rivers and estuary is perhaps best shown by the following table:—

	Spring Months.	Summer Months.	Autumn Months.	Winter Months.
Nith	47·8	60·2	47·1	38·9
Dee	50·9	61·1	49·8	40·2
Estuary (Little Ross)	47	57·5	53·1	43·5

The width of the estuary being about two miles at Little Ross, it is probable that these means may be considered applicable to the Solway Firth rather than the estuary.

At Little Ross the water temperature was higher than that of the air in seven months out of twelve, and the difference ranged to from 2·2° to 2·9°. The mean monthly temperature of the rivers, on the other hand, was never higher than that of the air, though there were occasional days in which the water was warmer.

	Air.	Water.	Difference.
Nith	52·8	48·5	4·3
Dee	54·5	50·5	4

The difference between the Nith and Dee observations are probably due to two circumstances. First, those of the Nith were taken in the year 1889, and those of the Dee in 1890; and secondly, the hour of observation was as a rule later, and nearer the maximum heat of the day on the Dee than in the case of the Nith. The observations of the temperature of the Dee were taken at Langland by the Rev. W. T. Gordon, minister of the parish; and those of the estuary or Solway at the island of Little Ross, by Wm. Macdonald, the lighthouse keeper there, and in both cases with great regularity.

THE GEOLOGY OF DUMFRIESSHIRE.

By B. N. PEACH, F.R.S., F.G.S., AND T. HORNE, F.R.S.E., F.G.S.

With the exception of certain limited areas in Nithsdale, Annandale, and along the fertile region bordering the Solway, the County of Dumfries is composed of Silurian strata, forming part of the ancient Silurian table-land that stretches from St. Abb's Head to Portpatrick. There is little variation in the lithological characters of the rocks, as they consist mainly of massive grits, greywackes, flags and shales singularly destitute of fossils. Fortunately, however, there are certain bands of black shales richly charged with graptolites, by means of which Professor Lapworth demonstrated the true order of succession of the beds. The strata have been thrown into innumerable folds, frequently inverted, whereby certain zones, not exceeding several hundred feet in thickness, have been made to cover areas several miles in width. Further, the members of the black shale series, which are typically developed in the Moffat region, undergo important modifications when followed north-westwards to Wanlockhead. The higher fossiliferous zones gradually disappear, and they are represented by coarser sediments. For these reasons the stratigraphical relations of the strata are extremely complicated, but with the aid of the graptolites it is possible to determine the age of the various rock groups, and to correlate them with the subdivisions of the Silurian system in other regions.

The lowest zones of the Moffat black shale series are accompanied by cherts containing Radiolaria which were detected in the course of the geological survey of the Abington area. The latter form an important horizon from their great horizontal extension and their constant association with volcanic rocks. From these cherts Dr Hinde has described twenty-three new species of Radiolaria belonging to twelve genera, of which half are new. He concludes that these Silurian cherts from the south of Scotland are due to the accumulation of the tests of Radiolaria, forming a pure Radiolarian rock resembling the Teritary beds of Barbados and the Nicobar Islands, which, according to Hæckel, correspond to the Radiolarian ooze of existing seas. There can be no doubt, therefore, that these cherts are true deep-sea deposits, which must have accumulated beyond the limit of sedimentation. The horizon of this important zone is accurately defined in the Ballantrae area, Ayrshire, where the cherts are underlain by a band of black shales containing Avenig graptolites. Throughout the southern uplands they are always overlain by the Glenkiln black shales yielding graptolites of Upper Llandeilo age. Hence they belong partly to Avenig and partly to Llandeilo time.

The following table gives in descending order the subdivisions of the black shale series, which have been established by Professor Lapworth in the typical Moffat region :—

LLANDOVERY...	Upper Birkhill or Grey Shale Group.	Zone of Rastrites maximus, Carr. Zone of Monograptus spinigerus, Nich. Zone of Cephalograptus cometa, Gein.
	Lower Birkhill or Black Shale Group.	Zone of Monograptus gregarius, Lapw. Zone of Diplograptus vesiculosus, Nich. Zone of Diplograptus acuminatus, Nich.

		Upper Hartfell or Barren Mudstones.	Zone of Dicellograptus anceps, Nich. Zone of Barren Mudstones.
CARADOC	...	Lower Hartfell or Black Shale.	Zone of Pleurograptus linearis, Carr. Zone of Dicranograptus Clingani, Carr. Zone of Climacograptus Wilsoni, Lapw.
LLANDEILO	..	Glenkiln Black Shales.	Comograptus gracilis, Hall. Thamnograptus typus, Hall. Didymograptus superstes, Lapw.
		Avenig.	Radiolarian cherts, mudstones, and volcanic tuffs.

The typical sections, where the various divisions of the black shale series are displayed, occur within the limits of the county, at the localities from which they take their name. The members of the lowest division, corresponding to the Upper Llandeilo rocks of Wales, are met with in the Glenkiln burn, near Rachills, to the south-west of Moffat; those of the middle division, representing the Caradoc rocks, occur on Hartfell at the spa, north-west of Moffat; while the members of the Upper division of Llandovery age, are seen to advantage in the famous section in Dobb's Linn, near Birkhill, at the head of Moffatdale.

The representatives of the black shale series, in the Moffat region, are exposed along axial folds in the midst of younger strata. They are arranged in parallel bands running in a north-east and south-west direction, which can be readily distinguished from the surrounding greywackes and shales by their colour and composition. Where the folds are normal, the lowest beds occur in the centre, and the higher zones follow in regular order on either side. Frequently the arches are inverted, and both limbs dip in one direction, while the beds are traversed by normal and reversed faults. In order to study the succession of the strata with advantage, the observer must visit the typical sections of the Moffat region, where the various lithological and palaeontological zones are admirably displayed.

One of the remarkable features of these Silurian rocks, ranging from the horizon of the Radiolarian cherts to the top of the Birkhill shales, is the variation in the character of the deposits when followed across the strike of the beds, that is from south-east to north-west. This variation, together with the complicated system of folding, has led to much of the difficulty in interpreting the succession. Excluding the Radiolarian cherts, which are evidently true oceanic deposits, the strata in the Moffat region consist of black and grey shales, clays and mudstones indicating deposition in comparatively deep water. The whole series does not exceed three hundred feet in thickness, and yet on palaeontological grounds it is evident that the beds represent the Llandeilo, Caradoc, and part of the Llandovery formations of Wales, which in the latter territory are measured by thousands of feet. When, however, we examine the successive reappearances of the black shale series between Moffat and the Dalveen Pass, the Birkhill shales gradually disappear, being represented by thin dark blue shales with a few characteristic Birkhill graptolites. Passing still further to the north-west, beyond the limits of the Llandovery area, the Barren Mudstones are represented by grey sandy shales with lenticular bands and nodules of limestone. Greywackes, grits and conglomerates, yielding Caradoc fossils, appear; and, in the northern part of the county, overlie the Glenkiln-Hartfell black shales. Proceeding still further to the north-west, towards the county boundary, west of Sanquhar, the higher zones of the Hartfell black shales disappear, and the lower zones of this division occur as dark seams in grey sandy shale. Taking these facts into consideration, it is evident that the old land surface from which the sediment was derived lay to the north-west. This conclusion receives support from the occurrence of pebbles of crystalline schists from the Highlands, in some of the Caradoc and Llandovery conglomerates, along the northern portion of the Silurian tableland.

The volcanic rocks underlying the Radiolarian cherts consist mainly of contemporaneous lavas and agglomerates, with occasional masses of intrusive dolerite and gabbro. They come to the surface along sharp anticlines in the northern part of the county; the extent of the exposures depending on the breadth of the anticline and the depth to which they have been cut by denudation. They are visible near Wanlockhead, in the tributaries of the Euchan water, and on Bail Hill, north of Sanquhar. The volcanic rocks underlying the cherts are not met with in the Moffat area, because the various folds have not been denuded deep enough to expose this horizon. Volcanic tuffs are, however, associated with the cherts and barren mudstones at certain localities.

The various bands of black shales in the Moffat area are succeeded by greywackes, grits, conglomerates, and shales of Llandovery age (Queensberry grits and Hawick rocks), occupying a broad belt of territory. A line drawn from the Dalveen Pass south-westwards by Moniaive to the county boundary near Castlefearn marks the northern limit of this formation. To the north of this line they are underlain by the Caradoc and Llandeilo rocks, while towards the south of Eskdalemuir they pass conformably upwards into the Wenlock beds. In the Moffat region there is no difficulty in drawing the base line of the Llandovery formation, as it is everywhere indicated by the Diplograptus acuminatus zone of the Lower Birkhill shales. The northern boundary line from Dalveen Pass to Moniaive is rather uncertain, owing to the disappearance of most of the well defined Birkhill zones. Though Llandovery rocks occupy a wide area, it is evident that their thickness is not excessive, because the same beds are constantly repeated by folding. In the central area, the dominant types, consisting of massive grits, occasionally conglomeratic, and shales are extremly barren. Only a few forms have been obtained from some of the shales, including Monograptus exiguus, Dexolites gracilis, and Crossopodia Scotica. The Hawick type of the Llandovery rocks consists of grey sandy shales, flags and greywackes, associated with pale-coloured clayey shales, and red shales. The shales have yielded Protovirgularia Harknessi.

The foregoing strata are succeeded southwards by the representatives of the Wenlock formation. A line drawn from the head of Ewes Water in Eskdale, south-westwards by Lockerbie to Mouswald, marks the boundary between the Llandovery and Wenlock rocks. Consisting of brown crusted greywackes, flags and shales, resembling to some extent the Hawick series, they are readily distinguished by certain bands of dark shales yielding graptolites characteristic of the Wenlock beds of other countries, as, for example, Cyrtograptus Murchisoni, Monograptus vomerinus, Monograptus colonus, Monograptus priodon, etc. Occasionally thick zones of olive-coloured shales are met with, resembling the typical Wenlock shales of Wales. Along the fertile tract stretching from Langholm to Ruthwell, these Upper Silurian strata are covered unconformably by the Old Red Sandstone and Carboniferous rocks.

At the northern boundary of the county in the the basin of the Spango Water, the Silurian rocks are invaded by a mass of granite, probably of the same age as the granite masses of Galloway, which are later than the Upper Silurian time, and older than the Upper Old Red Sandstone. Excellent examples of the contact metamorphism produced by this igneous intrusion are to be found in the zone of altered Silurian rocks surrounding the granite. Most of the dykes of felsite, diorite, and other igneous rocks traversing the Silurian area are probably of the same age.

Towards the close of the Silurian period the marine deposits, which had accumulated during that long interval of time, were elevated so as to form a prominent barrier of land. In the hollows worn out of this ancient table-land the strata belonging to the Old Red Sandstone, Carboniferous, Permian, and Triassic periods, were deposited. These newer formations have, however, been so much denuded that only isolated fragments remain of what were once more extensive deposits.

The representatives of the Old Red Sandstone belong to the lower and upper divisions of that system. Along the County boundary in Upper Nithsdale the members of the lower division occupy a limited area, stretching north-eastwards from the northern margin of the Sanquhar coal field. They consist of sandstones and conglomerates, prominently developed on the slopes of Corsoncone beyond the County boundary, where they are associated with contemporaneous volcanic rocks. They form part of the great belt of lower old red strata, stretching from the Braid Hills, near Edinburgh, into Ayrshire. The upper Old Red Sandstone, on the other hand, forms a narrow fringe underlying the carboniferous rocks from the County boundary east of the Ewes Water, south-westwards by Langholm to Birrenswark. At the base they consist of conglomeratic sandstones, the included pebbles having been derived from the erosion of the Silurian flagstones and shales; these are overlaid by friable red sandstones and marls.

The Old Red Sandstone strata, within the limits of the County, have not proved fossiliferous, but elsewhere in Scotland they have yielded land plants and ganoid fishes. From the lithological characters of the strata and the nature of the organic remains, it was long ago suggested by Fleming, Godwin-Austin, and Ramsay that they had been deposited in lakes or inland seas—an opinion which has been generally adopted by geologists.

Towards the close of the upper Old Red Sandstone period there was a remarkable outburst of volcanic activity on the south slopes of the Silurian table-land, giving rise to an interesting series of igneous rocks that always intervene between the red sandstones and carboniferous strata. They consist mainly of slaggy and amygdaloidal andesites, which were spread over the sea-floor as regular lava flows. They can be traced more or less continuously from the Tarras Water by Langholm to Birrenswark Hill. This picturesque hill is formed of an isolated mass of lava, surrounded by a narrow fringe of upper Old Red Sandstone. Some of the volcanic orifices from which these igneous materials were discharged are still to be found on the watershed between Tarras Water and Liddlesdale, and another of considerable size on the east bank of the Ewes Water, about a mile north of Langholm.

The representatives of the carboniferous system—by far the most important from an economic point of view—occur in three separate areas: (1) in the district extending from Langholm to Ruthwell; (2) at Closeburn, near Thornhill; (3) in the neighbourhood of Sanquhar. The first of these is the most extensive, measuring about twenty-two miles in length and varying in breadth from two to seven miles. The following zones, given in descending order, were established in the course of the geological survey of the district :—(8) Reddened shales with plants belonging to the true coal measures; (7) Canobie coals and associated strata; (6) marine limestone series of the Esk, Penton, Ecclefechan, and Kelhead; (5) volcanic zone of fine tuff and andesite, including about 50 feet of fine shales; (4) Woodcock air sandstones; (3) Tarras Waterfoot cementstone series; (2) white sandstones underlain by (1) the andesite lavas of Birrenswark and Ward Law.

In the course of the geological survey an important discovery of a large number of new organisms was made in the beds of zone 5 and partly in zone 3. The most celebrated locality occurs on the banks of the River Esk, north of Canobie, in a particular band of shale associated with volcanic tuff in zone 5. The fossils are in a splendid state of preservation; in some instances they have been so protected by their matrix of fine clay as to retain structures which have never before been recognised in a fossil state. Upwards of twenty new species of ganoid fishes were obtained from these beds, and out of the sixteen genera to which these species belong five are new to science. Few of the species are common to the carboniferous rocks of the Lothians, which has an important bearing on the physical history of that period. Along with the fishes were found about an equal number of crustacea new to science, comprising about twenty species of the higher crustacea, together with Eurypterids and Limaloids. No less interesting is the discovery of several new species of scorpions, the occurrence of which in carboniferous rocks has been extremely rare. The specimens recently obtained are admirably preserved, and from a minute examination of them it is evident that they closely resemble their living representatives. Every structure of the recent form has been recognised in the fossil scorpions from this horizon, including the hairs and hooks on the feet. The sting alone has not been certainly observed, but that it existed may be inferred from the presence of the poison gland which has been detected in the fossil state. The remains of several new plants were also found in the fine shales from the river Esk.

The organic remains found in the different subdivisions of the carboniferous rocks bordering the Solway are of great value in correlating them with their representatives in the midland valley of Scotland. The cementstone bands of Tarras Water and Ecclefechan are largely composed of microzoa, chiefly Entomostraca; others are almost entirely made up of minute gasteropods.

Between the foregoing horizon and the true marine limestone series of Penton, the river Esk and Kelhead, there is an intervening group of thin limestones, which, from the presence of Lamellibranchs and gasteropods, indicate shore conditions during their deposition. The Lamellibranchs are represented by Myalina Crassa, M. lamellosa, Modiola modioloformis, M. Macadami, Aviculopecten dissimilis; the gasteropods by Natichopsis plicistria, Bellerophon Urei, Murchisonia Verneuilliana; and the brachiopods by Camerophoria Crumena, Athyris ambigua, Productus semireticulatus, etc.

The marine limestone series of Penton, the river Esk, etc., from the abundance of corals, point to deposition in clear water. Amongst the corals obtained from these beds may be mentioned Lonsdalea floriformis, Lithodendron junceum, L. irregulare, Lithostrotion Portlocki, L. basaltiforme, Laplorentis cylindrinca, Chaetetes tumidus. The brachiopods are represented by Productus semireticulatus, Athyris ambigua, and Rhynconella pleurodon; the Lamellibranchs by Edmundia sulcata, etc. From the foregoing assemblage of organic remains, it is highly probable that some of these marine limestones may be the equivalents of the marine zones of the carboniferous limestone series of the midland valley.

The recent researches of Mr Kidston have led him to the conclusion that the plants met with in the Millstone Grit, and the overlying true coal measures, are specifically distinct from those which preceded them in the carboniferous limestone and cementstone series. If this generalisation should prove to be correct, then it is clear that the vertical distribution of the fossil plants in the carboniferous system may be of the greatest service in determining the horizon of the beds. Amongst the plants obtained from the beds at the foot of Byre Burn, Archerbeck, the Rowan Burn, and the Canobie coal field, and determined by Mr Kidston, the following may be mentioned:—Sphenopteris L. and H., Sph. obtusiloba Brongt., Staphylopteris sp., Neuropteris flexuosa Brongt., Alethopteris lonchitica Schl, Pecopteris nervosa Brongt., Lepidophyllum lanceolatum. According to Mr Kidston, some of these forms are never found below the horizon of the true coal measures, and it is highly probable, therefore, that the Canobie coal field may represent the true coal measures of the central valley of Scotland. The reddened shales occurring to the south of the Canobie coal field yield Neuropteris flexuosa Brongt., a form confined to the true coal measures.

Within the Silurian area, carboniferous rocks are met with in the Thornhill and Sanquhar basins, filling hollows worn out of the old Silurian tableland. At Closeburn and Barjarg there are beds of marine limestone associated with sandstones and shales. Still further north, at the south-eastern limit of the Sanquhar coal field, there are small outliers of the carboniferous limestone series, consisting of sandstones, shales, and thin fossiliferous limestones. The latter rapidly thin out, and the true coal measures rest directly on the Old Silurian platform. From these data it would appear that in Upper Nithsdale the Silurian barrier did not sink beneath the sea level till the latter part of the Carboniferous period. The Sanquhar coal field is about nine miles in length, and from two to four miles in breadth. It contains several valuable coal seams, and from the general character of the strata it is probable that they are the southern prolongations of the Ayrshire coal measures.

Along the south-west part of the Sanquhar field, the strata are traversed by three narrow dolerite dykes, which send out intrusive sheets along the coal seams. The igneous rock is much decomposed, having the same character as the white trap so common in the Ayrshire coal fields. The coals are so altered by it as to be unworkable; indeed in some places they have been converted into columnar anthracite.

The strata next in order are of Permian age, which are invariably separated from all older rocks by a marked unconformability. They occur in four separate areas—(1) at Moffat; (2) at Lochmaben and Corncockle Moor; (3) the Dumfries basin; (4) the Thornhill basin. The beds consist of red sandstones and breccias, the latter being mainly derived from the denudation of the Silurian platform. Professor Harkness, however, long ago noted the occurrence of fragments of carboniferous limestone in these breccias in the basin of the Annan, south of Lochmaben. Recently, when cutting the road to the hydropathic establishment at Moffat, nodules of impure limestone were found in the breccia. From these nodules the following fossils were obtained: Natichopsis plicistria, Camerophoria crumena and Bellerophon, belonging to the cementstone series. It is clear, therefore, that patches of carboniferous strata must have existed in these valleys during the deposition of these breccias. In the red sandstones of Corncockle Moor and to the south of Dumfries, reptilian footprints have been detected. The breccias have probably accumulated in narrow fjords, and the red sandstones may have been deposited in enclosed basins. The strata are singularly destitute of organic remains, and though at present they are provisionally regarded as Permian, it is not improbable that they may be of younger date. An interesting feature connected with the Thornhill basin is the occurrence of contemporaneous volcanic rocks at the base of the Permian beds. They form a prominent belt, usually rising to the surface as a distinct ridge, between and underlying carboniferous strata and the over-lying Permian rocks. They consist of slaggy diabase lavas and tuffs, which are occasionally interstratified with the sandstones.

Beyond the limits of the Thornhill basin there is further evidence of the extension of these volcanic rocks at Lockerben, about ten miles to the east of Thornhill. Here there is a small isolated area of carboniferous and Permian strata. In the course of the Garroch Water, red and liver-coloured sandstones, probably of carboniferous age, are overlain by red ashy breccia, composed of Silurian fragments and blocks of amygdaloidal lava, followed by slaggy diabase lava. In the Sanquhar basin also there are several "necks" or volcanic rents filled with agglomerate, which in all likelihood mark the site from which lavas of Permian age were discharged. There is also a small out-lier of diabase lava, resting on the coal fields north of the town, which is regarded as of Permian age. Further evidence might be adduced of the original extension of Permian or Triassic rocks over areas from which they have been removed by denudation. Some of the carboniferous strata in the Sanquhar coal fields have been stained red by the infiltration of iron oxide, and the same feature is observable in the Silurian rocks round the margin of the Thornhill basin. The shales over-lying the Canobie coals have been so much reddened as to resemble externally Permian or Triassic rocks. In these cases the older rocks were formerly buried underneath red sandstones, from which the percolating water derived the iron oxide.

Between Canobie and Annan there is a strip of red sandstones and marls, which have been correlated with the Triassic sandstones on the south side of the Solway Firth. In the latter region, to the west of Carlisle, they are succeded by Liassic strata.

Several prominent basalt dykes traverse the Silurian and Permian rocks in a north-westerly direction, which may probably belong to the remarkable outburst of volcanic activity during the Tertiary period in the Western Isles. One example is worthy of notice, as it has been traced from Leadhills south-eastwards by Moffat, and across Eskdalemuir by Langholm to the English border. In texture it varies from a dolerite to tachylite, which is the glassy form of basalt. Another is traceable for a short distance through the Permian sandstones, about a mile to the north of Caerlaverock Castle.

Throughout the County there are abundant proofs of the intense glaciation which this region experienced during the glacial period, in common with the rest of Scotland. By means of the striae on the rocks and the transport of stones in the boulder clay, we can follow the path of the ice-sheet during the extreme glaciation. The general trend in Nithsdale and Annandale is towards the south-east, but on approaching the shores of the Solway the ice seems to have been deflected eastwards, across the belt of low ground towards Langholm and Carlisle, and thence across Northumberland and down the valley of the Tyne. This singular deflection was evidently due to the opposing mass of ice radiating from the northern part of the Lake District. This easterly movement is placed beyond doubt by the occurrence of boulders of Criffel granite in the boulder clay near Langholm and eastwards towards the valley of the Tyne. The wide-spread covering of boulder clay which is now found in the upland valleys, and on the low grounds, is the relic of this ancient glaciation. But in the valleys draining the larger masses of high ground, there are numerous moraines deposited by local glaciers, of which some of the finest examples are to be found round Loch Skene, at the head of Moffatdale. Moraine mounds, however, occur, far removed from any great mass of high ground, suggesting the probability that local or district ice sheets may have operated in producing them. During the disappearance of the ice sheet and valley glaciers, those fluvio-glacial gravels may have been formed which stretch along the basin of the Annan as far as Moffat and along the valley of the Nith. The high level terraces of the river valleys may have been laid down during the stage of the 100 feet sea beach, the remains of which are now to be found bordering the Solway. Even then arctic conditions seem to have prevailed in the South of Scotland. In the alluvia of old lakes occupying hollows in the boulder clay near Edinburgh, the remains of arctic plants have recently been found. Among them are dwarf birch (Betula nana), willows (Salix polaris, S. herbacea, S. Reticulata), Dryas octopetala, etc., together with a phyllopod Apus glacialis, now found only in fresh water lakes in Greenland and Spitzbergen. It is highly probable that some of these relics of an Arctic Flora and fauna may yet be found in the old lake deposits within the County. The latter stages in the geological history of the County are indicated by the growth of extensive peat mosses, by the fifty feet and twenty-five feet sea beaches fringing the Solway, and by the successive terraces of alluvium which tell us of the erosion of the land by existing streams.

LIST OF
AUTHORITIES AND ABBREVIATIONS EMPLOYED.

ab.	Abundant.
Ad. and S. D. J.	Miss E. G. Adams and Miss S. D. Johnstone. Herbarium of Cowhill seen.
Ak.	Miss Aitken. Specimens seen.
Arm.	G. Armitstead, Esq.
R. A.	R. Armstrong, Thornhill. Notes and Insects.
Arn.	Dr Walker Arnott. Visited Wigtownshire, 1848.
C. C. B.	C. C. Bailey. Records in Wigtownshire, 1883.
Dr Bl.	Dr Balfour. Transactions, Philos. Soc. of Glasgow, 1841-4. Visited Wigtownshire in 1843? Also, Moffat.
Bea.	T. Beattie, Esq., Langholm.
D. Bl.	D. Bell, Esq. Trans. Bot. Soc., Edin. Vol. i.
G. Bl.	George Bell, Esq., Lockerbie.
R. Bl.	R. Bell, Esq., Castle O'er.
T. Bl.	Rev. T. Bell. In M'Andrew's List.
A. Bn.	A. Bennett, Esq., F.L.S. Notes in *Scottish Naturalist*.
W. Bn.	Rev. W. Bennet, Moffat.
Bk.	Miss Black, Lochmaben. Specimens seen.
W. Br.	W. Brand. Trans. Bot. Soc., Edin. Vol. i.
J. A. Br.	J. A. Brown, Edinburgh Herbarium, about 1838.
Brem.	Rev. W. Bremner, Kirkmichael Manse.
T. Br.	T. Brown, Esq., Auchenessnane, Penpont. In Dumfries Herbarium and M'Andrew's List.
E. Br.	E. Brunetti, Esq., London. Authority for Diptera.
B. W.	J. Buchanan White, M. A. *Journal of Botany*, vol. ix.
Dr Br.	Rev. Dr. Burgess. Lightfoot (1789).
c. p.	Collecting pollen (of insects).
Carr.	Mrs Carr, Maryport.
W. Ca.	Dr Carruthers, F.R.S. *Notes on Moffat Flora*, 1882, and *Moffat Fern Album*, 1863. (Mrs Carruthers).
S. W. Ca.	Dr Carruthers. *Notes on Moffat Flora*, 188-?
C. Y.	Mrs Carthew-Yorstoun, Irvine House. Herbarium seen.
F. R. C.	F. R. Coles, Esq. In M'Andrew's List.

LIST OF ABBREVIATIONS.

E. M. C. .... Miss Copland, Newabbey. Specimens seen.
J. Cr. .... J. Corrie, Esq., Moniaive. Notes and Specimens seen.
C. C. .... Dr Craig Christie. Edinburgh Herbarium.
Cro. .... A. Croall, Esq. In Watson's MSS.
J. Cru. .... J. Cruickshank, Esq. Bot. Soc., Edin., Report, 1837. In Edinburgh Herbarium and Watson's MSS.
J. G. C. .... Sir J. G. Cullum. In Watson's MSS.
N. A. D. .... N. A. Dalzell, Esq. In Watson's MSS.
Dr Dv .... Dr Davidson. Trans. Nat. Hist. Soc. of Dumfries. (Naming of Specimens unknown). Brown's *History of Sanquhar*, 1891.
J. H. D. .... J. H. Dixon, Esq., Dabton. Notes.
R. Do. .... R. Doughty, Esq., Canobie. Notes.
G. C. Dr. .... G. C. Druce, Esq. Bot. Record Club, 1883.
Dfs. .... Dumfriesshire.
e. .... Efficient (of insect visitors).
Exc. .... Field Club Excursions. Trans. Nat. Hist. Soc. of Dumfries.
Fq. .... Rev. J. Farquharson. *Scottish Naturalist*, vol. ii., p. 80.
J. Fn. .... J. Fingland, Esq., Thornhill. Trans. Nat. Hist. Soc. of Dumfries.
J. Fr. .... Rev. J. Fraser, Colvend. Stat. Acct. of Kirkcudbright? Trans. Nat. Hist. Soc. of Dumfries. In M'Andrew's list.
Dr Gl. .... Dr Gilchrist. In M'Andrew's list.
G. C. .... Mrs Gilchrist-Clark. List of Ross Plants, 1867.
G. F. .... J. Gillon-Fergusson, Esq., Isle.
G. Go. .... G. Gordon. In Watson's MSS.
Dr Gr. .... Dr Graham. Excursion Wigtownshire, 1835, and Trans. Bot. Soc. Edin. Report, 1836.
P. Gr. .... P. Gray, Esq. Catalogue in Watson's MSS. Phytologist, vol. i., p. 257, 416; vol. iii., p. 348. New Stat. Acct., Dumfries, 1876.
Grev. .... Greville Herbarium at Edinburgh.
F.W.G. .... Dr F. W. Grierson. Herbarium now at Dumfries, and seen.
A.B.H. .... Allan B. Hall, Esq. Thirsk.
M.J.H. .... Miss Hamilton, Dumfries. Specimens seen.
Hn. .... Misses Hannay. Specimens seen.
Harv. .... Miss Harvey. In Watson's MSS.
Hg. .... W. S. Hogg, Esq. In M'Andrew's list.
Hook. .... Hooker and Arnott. *British and Student's Flora*.
F.A.H. .... Miss Hope. *Gardens and Woodlands*, 1881.
G. Ho. .... George Horn. In Watson's MSS.
Ht. .... Mr Hutton. In Watson's MSS.
Irv. .... General Irving. Herbarium, seen by J. M'Andrew.
Jack. .... Mr Jackson. Withering's Flora, 1796.

LIST OF ABBREVIATIONS.

Jard. ....Sir W. Jardine. Hooker's Flora.
J.T.J. ....J. T. Johnstone, Esq., Moffat. Trans. Bot. Soc., Edin., 1889, 1890, 1891, 1892; Nat. Hist. Soc., Dumfries; Guide to Moffat; Herbarium, etc.
Kcd. ... .Kirkcudbright.
Kd........................W. Keddie. *Moffat: its Walks and Wells*, 1854.
Lg. ....Lightfoot. *Flora Scotica*, 1777 and 1789.
Lind. ....Mr Lindsay. In Watson's MSS.
E.F.L. ....Rev. E. F. Linton. Authority for Hieracia.
W.R.L. ....Rev. W. R. Linton. *Journal of Botany*, 1890 and 1893.
Lt........................Rev. Dr W. Little.
G. N. Ll...............G. N. Lloyd. Watson's New Botanical Guide, 1837.
J. M'A.................J. M'Andrew, Esq., New-Galloway. Many papers Nat. Hist. Soc., Dumfries, and specimens in Herbarium. List Kirkcudbright plants, 1882. List Wigtownshire plants, 1893, etc.
G. M'C................Rev. G. M'Conachie. In M'Andrew's list.
M'K. ....J. M'Kay. In Hooker's *Flora Scotica*.
M'N. ....Dr G. M'Nab. Notes in Watson's MSS.
R. H. M...............R. H. Masterman, Esq. Notes.
J. Mt....................J. Matthewson, Esq., Dalbeattie. In M'Andrew's List.
Mau......................Mr Maughan. *Flora Scotica*.
M.-W. . ...Miss Maxwell-Witham. Notes.
A. M....................Alexander Menzies, Esq., Dabton. Notes.
C. E. M...............Miss Milligan. Specimens and Notes.
W. M. H. M........Rev. W. M. H. Milner, Lockerbie. Notes.
Oliver. ....Professor Oliver, F.R.S.
Pag. ....Rev. J. Pagan. In M'Andrew's List.
Pat......................Mr Patrick. *Flora Scotica*.
Pl........................Rev. P. Lyon Playfair, Glencairn. Notes.
r. ....Rare.
reg. ....Regular.
S. O. R. ....S. O. Ridley, Esq. Herbarium of British Museum.
R. R....................R. Rimmer, Esq., F.L.S., Dalawoodie.
Ro.......................Miss Robb. Notes.
J. Ru...................J. Rutherford, Esq., Jardinton. Notes.
s. ....Sucking (of insect visitors).
J. Sd...................J. Sadler. Moffat Register, 1858. *Rambles among Wild Flowers*, 1857.
S.-E.....................G. F. Scott-Elliot, Esq.
R. Se...................R. Service, Esq. Authority for names of Hymenoptera and some Diptera.
J. Sh...................J. Shaw, Esq., Tynron. In M'Andrew's List. Notes and Specimens.

LIST OF ABBREVIATIONS.

W. Sh..................W. Sheffield. *Flora Scotica*, 1789.
Sb.......................Sibbald's *Scotia Illustrata*, 1684.
Dr Sn................... Rev. Dr Singer. Authority Stat. Acct. of Dumfries 1843? Agricultural History of Dumfriesshire.
L. Sm..................Miss Lorraine Smith. Notes.
W. St..................W. Stevens. Phytologist. Vol. iii., p. 390.
St.......................Mrs Stewart, Shambellie. Notes.
J. T. S................J. T. Syme. In Watson's MSS.
E. Ty..................Miss Ethel Taylor, Kirkandrews. Notes, Specimens, and Insects.
Th......................Mrs Thomson. Notes and Specimens.
v. c.....................Very Common.
v. r.....................Very Rare.
Dr Wl.................Rev. Dr Walker.
Wall....................Professor Wallace, Edinburgh. In M'Andrew's List.
H. C. W.............Watson's MSS. for Topographical Botany, etc., in the Natural History Museum, consulted by me.
F. M. W.............F. M. Webb. Edinburgh Herbarium, 1877.
A. We.................Miss A. and Captain Wedderburn. Herbarium at Dumfries seen.
Wel....................Mrs Welsh, Moffat. Notes.
Wgt....................Wigtownshire.
J. Wn. and R. Bl...Miss J. Wilson and Mr R. Bell. List of Plants.
J. C. W...............J. C. Willis, F.L.S. Annals of Botany, vol. ix., Insect Visitors.
J. Wl...................J. Wilson, Esq. Notes.
Win....................Mr Winch. *Flora Scotica*.

FLORA OF DUMFRIESSHIRE.

Thalictrum alpinum. *Linn.*

RECORDS: *Dfs.*—W. Sheffield, 1789. *Kcd.*—G. N. Lloyd, 1837.

LOCALITIES: *Annandale*—Hartfell, W. Sh.; Saddleback, W. St.; Black's Hope, Whitecoombe (1750-2000 feet), J. H. Bl., J. T. S.

Appears May 28 to June 22, J. T. J. On slate or whin rocks; in sun, but wind-sheltered.

Thalictrum minus. *Linn.*

RECORDS: *b.* Montanum, *Dfs.*—E. F. Linton, 1889. *Kcd.*—G. N. Lloyd, 1837.
c. Flexuosum, *Dfs.*—Dr J. H. Balfour, 1856. *Kcd.*—F. R. Coles, 1882. *Wgt.*—G. C. Druce, 1883.

LOCALITIES: *Nithsdale*—Scaur, R.A.; Drumlanrig, var. *c.*, J. Fn., R. A., Dr Dv. *Annandale*—Black's Hope (800-1750 feet), J. T. J.; Grey Mare's Tail, vars. *b.* and *c.*, Dr Bl., P. Gr., Hg., J. T. J., S.E.

Appears May 28 to August, J. T. J. Moist ground near streams on whinstone soils or mud; in sun, or half-shaded and wind-sheltered in narrow ravines.

VISITOR: Hydrotea dentipes *c. p.*, S. E.

Thalictrum flavum. *Linn.*

RECORDS: *Kcd.*—J. Matthewson, 1882. *Wgt.*—Rev. J. Fraser, 1843.

LOCALITIES: *Along the shore*—Garliestown, J. Mt.; Newton-Stewart, J. Fr.; Kenmure Holms, J. M'A.; Auchencairn Bay, J. Mt.

Anemone nemorosa. *Linn.* (Wood Anemone).

RECORDS: *Dfs.*—J. Cruickshank, 1837. *Kcd.*—P. Gray, 1843. *Wgt.*—G. C. Druce, 1883.

LOCALITIES: Very common in all the valleys to 1000 feet.

Appears April 1, G. Bl.; April 5 to 27, J. T. J. Usually moist leaf mould, holms, clayey soils; usually shaded or half-shaded and wind-sheltered, though on ground bare of other plants.

VISITORS: Roctotrypes, Melanostoma mellinum, gracilis, Empis sp. Anthomyidae; S. E.

Ranunculus hederaceus. *Linn.*

RECORDS: 23. Lenormandi, *Dfs.*—W. Stevens, 1849. *Kcd.*—J. M'Andrew, 1882. *Wgt.*—G. C. Druce, 1883.
24. Hederaceus, *Dfs.*—F. M. Webbe, 1835 (?) *Kcd.*— J. M'Andrew, 1882. *Wgt.*—G. C. Druce, 1883.
b. Omiophyllus, *Dfs.*—J. Corrie, 1891.

LOCALITIES: *Nithsdale*—Carlaverock, W. St.; Glencaple, P. Gr.; Cargen, Hn.; Glencairn, Pl.; Moniaive, 300 feet (24, var. *b.*), J. Cr.; Holywood, Isle, S. E.; Thornhill, W. St., R. A.; Durrisdeer, J. Wl. *Annandale*—Lochmaben, F. M. W.; Corehead, J. T. J.; Andrew's Well, W. Ca. *Eskdale*—Glenzier, E. Ty.; Wauchope, S. E.

Appears end of January, J. T. J.; April 9, G. Bl. On wet mud of alluvium or peat springs, etc.

Ranunculus aquatilis. *Linn.* (Water Crowfoot).

RECORDS:

SUBSPECIES*—R. Petiveri Koch.
21. *b.* Confusus, *Dfs.*—Dr Davidson, 1886.

SUBSPECIES—R. Peltatus. *Schrank.*
18. (of 8th Ed.) Elongatus, *Dfs.*—Miss Hannay, 1891.
20. *b.* Truncatus, *Dfs.*—G. F. Scott-Elliot, 1891. *Kcd.*— Dr F. W. Grierson, 1882. *Wgt.*—G. C. Druce, 1883.
20. *c.* Floribundus, *Dfs.*—Dr Singer, 1843. *Kcd.*—F. R. Coles, 1882. *Wgt.*—J. M'Andrew, 1883.
21. *d.* Penicillatus, *Dfs.*—Miss Hannay, 1891. *Kcd.*—F. R. Coles, 1883 *Wgt.*—G. C. Druce, 1883.
Fissifolius, *Dfs.*—P. Gray, 1850. *Kcd.*—F. R. Coles, 1883.

SUBSPECIES—Diversifolius. *Schrank.*
18. *b.* Godronii, *Kcd.*—F. R. Coles, 1883.

---
* Arranged after W. P. Hiern's Monograph. The numbers are those of the *London Catalogue*, Ninth Edition, except R. elongatus, which is No. 18 of Eighth Edition.

SUBSPECIES—Capillaceus Thuill.
  17. Trichophyllus, *Dfs.*—Dr Singer, 1843. *Kcd.*—F. R. Coles, 1883.
  18. a. Drouettii, *Dfs.*—Dr Singer, 1843. *Kcd.*—F. R. Coles, 1883.
  16. Pseudo-fluitans, *Dfs.*—P. Gray, 1850.
      Salsuginosus, *Dfs.*—Miss M. J. Hamilton, 1891.
SUBSPECIES—Fluitans. *L.*
  16. Fluitans, *Dfs.*—J. Sadler, 1858.

LOCALITIES: *Nithsdale*—Kingholm Merse, 19, *b.*, M. J. H.; Nith, at Dumfries, 21 (?), J. Fn.; 20, *b. c. d.*, Dr Sn., P. Gr., Hn., S. E.; Maxwelltown Loch, 20, *b.*, F. W. G.; Moniaive, 14, J. Cr. *Annandale*—Lochmaben, 20, P. Gr., W. P. H.; Moffat, 20, *b*, S. E.; 13, J. Sd. *Eskdale*—Castle O'er, J. Wl., and R. Bl.

Appears May 3 to end of season, J. T. J. Of these forms, 20 (vars. *b.* and *c.*) are common in mud of shallow streams, 1-6 inches deep; 20 (var. *d.*) in eddies or nearly quiet backwaters of rivers, 9-18 inches deep; 16 in deep water and strong currents. Elevation 600 feet, as at Capplegill, J. T. J.

VISITORS: Parhydra aquila, Notiphila riparia, Hydrotea dentipes, and small Diptera ab., S. E.

Ranunculus lingua. *Linn.* (Great Spearwort).

RECORDS: *Dfs.*—Dr Balfour, 1839. *Kcd.*—J. Fraser, 1843. *Wgt.*—?
LOCALITIES: *Nithsdale*—Newabbey, Dr Gl.; Mabie Moss, P. Gr, G. C.; Lochar Moss, P. Gr., Hg., G. C. *Annandale*—Lochmaben, Exc. (?)

Ranunculus flammula. *Linn.*

(Small Spearwort, Steep. *Fide* J. SHAW).

RECORDS: 27. Flammula, *Dfs.*—G. N. Lloyd, 1837; *Kcd.*—P. Gray, 1843; *Wgt.*—G. C. Druce, 1883.
  Var. *b.*, pseudo-reptans, *Dfs.*—W. S. Hogg, 1882.
  28. Reptans, *Dfs.*—P. Gray, 1850 ?

LOCALITIES: Very common in all the valleys and ponds to at least 700 feet; var. *b.* from Garrick, Thornhill, J. Fn., Dr Dv.; Lochmaben, P. Gr., W. S. Hg., S. E. (28 is also given by P. Gray from Lochmaben, but is probably this variety).

Appears May and June, J. T. J. Moist marshy soil, or in water (to six inches deep) on peat mosses, as well as other mud; usually in sun, but in part wind-sheltered.

VISITORS: Platychirius clypeatus, Hyelemyia strigosa, Anthomyia radicum, Sulciventris, Hydrotea dentipes, Hyetodesia sp., Chortophila sp., etc., S. E.

Ranunculus ficaria. Linn. (Pilewort).

RECORDS: *Dfs.* and *Kcd.*—P. Gray, 1850. *Wgt.*—G. C. Druce, 1883. *b.* Incumbens, *Dfs.*—P. Gray, 1850.

LOCALITIES: Very common to lower level of peat mosses; var. *b.*, Dumfries, P. Gr.

Appears March 22, J. T. J., G. Bl. On moist or fairly dry leaf mould, holms, road sides, boulder clay, whinstone soils, etc.; usually half-shaded in woods (appearing before the leaves), and prefers bare soil in slightly wind-sheltered places.

VISITORS: Apis, Platychirius albimanus, Melanostoma mellinum, Empis punctata, bilineata, Hyetodesia basalis, Chilosia sp., Anthomyia sp., Meligethes brassicæ, Mantua napi, S. E.

Ranunculus sceleratus. Linn.

RECORDS: *Dfs.*—J. Cruikshank, 1839. *Kcd.*—Mrs Gilchrist-Clark, 1867. *Wgt.*—J. M'Andrew, 1889.

LOCALITIES: *Along the shore*—Port Yerrick, Creetown, Dowalton, J. M'A.; Ross, G. C.; Mullock Bay, G. M'C.; Southerness, Arn.; Kingholm Merse, J. Cru., P. Gr., Hn.; Caerlaverock, Dr Gl.; Lochar Mouth, Priestside, S. E.; Powfoot, J. Fn.

In shallow water or moist estuarine or alluvial mud.

VISITORS: Small Anthomyidæ, Empidæ, Dolichopodidæ, S. E.

Ranunculus auricomus. Linn. (Goldielocks).

RECORDS: *Dfs.*—J. Sadler, 1858. *Kcd.*—P. Gray, 1843. *Wgt.*—J. M'Andrew, 1889.

LOCALITIES: *Nithsdale*—Glen, F. W. G.; Th., Gullyhill, R. H. M; Lincluden, S.E.; Glencairn, Pl. J. Cr.; Cowhill, Ad. and S. D. J.; Sanquhar, Dr Dv. *Annandale*—Milk, S. E.; Dryfe, G. Bl.; Evan Water, J. T. J.; Beld Craig, J. Sd. *Eskdale*—Byreburn, Tarras; Esk to Langholm, S. E., to some 400 feet.

Appears April 19, G. Bl., and beginning of May, J. T. J. Moist leaf mould, alluvium, roadsides, whinstone soils chiefly by rivers; half or fully shaded and usually wind-sheltered.

VISITORS: Scatophaga stercoraria, Anthomia sulciventris Apion pavidum, S. E.

Ranunculus acris. *Linn.* (Buttercup).

RECORDS: *Dfs.* and *Kcd.*—P. Gray, 1850. *Wgt.*—G. C. Druce, 1883.
  b. Tomiophyllus (?).
  c. Vulgatus, *Dfs.*—Miss Hannay, 1893.
  d. Boraeanus, *Dfs.*—Miss Hannay, 1893.

LOCALITIES: Common in all the valleys to lower level of peat mosses.

Appears May 7, J. T. J. Dry (rarely moist) holms, roadsides, boulder clay, shingles, sandy soils, &c.; in sun and exposed to wind, though part sheltered by long grass.

VISITORS: Allantus nothi, Kmetes sidaris, Rhamphomyia albosegmentata, Lonchoptera lutea, Hydrotea dentipes, Hyetodesia incana, Anthomyia, Meligethes Æneus, S.E. (on R. acris and R. repens).

Ranunculus repens. *Linn.*

RECORDS: *Dfs.* and *Kcd.*—P. Gray, 1850. *Wgt.*—G. C. Druce, 1883.

LOCALITIES: Common along shore and in all valleys to lower level of peat haggs.

Appears April 27, G. Bl.; May 7, J. T. J. Very dry or dry sandstone soils, gravel, shingles, gardens; in sun and exposed or wind-sheltered (prefers short turf or bare ground).

VISITOR Syrphus balteatus, S. E.

Raununculus bulbosus. *Linn.*

RECORDS: *Dfs.*—P. Gray, 1850. *Kcd.*—J. M'Andrew, 1882. *Wgt.*—J. M'Andrew, 1890.

LOCALITIES: *Nithsdale*—Common to 600 feet; Glencairn, J. Cr.; Trigony, R. A.; Spango (800 feet), Dr Dv. *Annandale*—To 400 feet, J. T. J. *Eskdale*—To 700 feet, R. Bl.

Appears May 2, G. Bl.; May 27, J. T. J. Dry sandstone or gravel, roadsides, whinstone; exposed to sun and wind.

VISITORS: Apis, Bombus lucorum, Allantus Nothi, Hyetodesia sp., Anthomyidæ ab. S. E.

Ranunculus hirsutus. *Curt.*

RECORDS: *Dfs.*—J. Fingland, 1883. *Kcd.*—Dr Gilchrist, 1860. *Wgt.*—Dr Macnab (?), 1836. Var. *b.* Parvulus, *Dfs.*—G. F. Scott-Elliot, 1891.

LOCALITIES: *Nithsdale*—Kirkconnel, Dr Gl. *Annandale*—Ruthwell, J. Fn.; Auchencas, Moffat, var. *b.*, S. E. *Eskdale*—Kirkburn, Bentpath, S. E.

Ranunculus arvensis. Linn.

RECORDS: (Escape) *Dfs.*—J. Sadler, 1857. *Kcd.*—Gen. Irving, 1882.
LOCALITY: Moffat, J. Sd.

Caltha palustris. Linn. (Marsh Marigold).

RECORDS: *Dfs.* and *Kcd.*—P. Gray. 1850. *Wgt.*—G. C. Druce, 1883.
   b. Guerangerii, *Dfs.*—Dr F. W. Grierson, 1882.
   c. Minor, *Dfs.*—J. T. Johnstone, 1891.
LOCALITIES: Very common to 2000 feet.
Appears April 5, G. Bl.; April 9, J. T. J. In water or moist, muddy, or marshy ground.
VISITORS: Apis, Bombus lucorum, Brachycentrum sabulosum, Brachycoma sp., Cheilosia, Anthomyia sp., Micropteryx calthella, S.E.

Trollius Europæus. Linn. (Luckangowns, Butterblobs, Globe Flowers, Ballflowers.)

RECORDS: *Dfs.*—Dr Burgess, 1789. *Kcd.*—P. Gray, 1850. *Wgt.*—Dr Macnab, 1835.
LOCALITIES: *Nithsdale*—Lochanhead, Hg., M. W., Th.; Glen, P. Gr., Th.; Cluden Bridge, Hg.; Speddoch, G. C.; Routen Brig, P. Gr., S. E.; Glencairn, Pl.; Scaur, Exc.; Thornhill, R. A.; Cowhill, Ad. and S. D. J.; Lag, G. F.; Nith and Euchan, Sanquhar, Dr Dv. *Annandale*—Halldykes, Quhitewollen, G. Bl.; Kirkpatrick-Juxta, F. A. H.; Middlegill, Stidrig, Archbank, W. Br.; Beattock, Grey Mare's Tail, Black's Hope, Loch Skene, J. T. J. *Eskdale*—Lyneholme, R. Bl.
VISITORS: Allantus nothi, Eristalis asnea, Platychirius albimana, Rhingia rostrata, Anthomyia, Apion pavidum, S. E.
Appears May 18 to July 4, J. T. J.

Aquilegia vulgaris. Linn. (Columbine).

RECORDS: (Escape) *Dfs.*—Dr Singer, 1843. *Kcd.* and *Wgt.*—Arnott, 1844.
LOCALITIES: *Along the shore*—Glasserton, Balmae, Arn.; Rockcliffe, Th. *Nithsdale*—Glencaple, C. E. M.; Tinwald, J. Sn.; Routen Brig, E. M. C.; Jarbruck (400 feet), J. Cr.; Scaur, Hn.; Thornhill, R. A.; Knockenjig (400 feet), Dr Dv. *Annandale*—Garpol, Dr Sn., Kd., J. T. J.; Auchencat (950 feet), J. T. J.

Helleborus viridis. *Linn.*

RECORDS : (Escape) *Dfs.*—R. Bell, 1892. *Wgt.*—Dr Macnab, 1836.
LOCALITY : Billholm, R. Bl.

Aconitum napellus. *Linn.*

RECORD : *Dfs.* and *Kcd.*—J. M'Andrew, 1882.
LOCALITIES : *Nithsdale*—Cluden Mills, Hn. ; Nithbank, Thornhill, R. A.
VISITOR : Bombus sp., R. A.

Berberis vulgaris. *Linn.* (Barberry).

RECORD : (Planted) *Dfs.* and *Kcd.*—P. Gray, 1850. *Wgt.*—(?).
LOCALITIES : *Nithsdale*—Kirkconnel, M. W. ; Dockfoot, M. J. H. ; Lincluden, P. Gr. : Jarbruck, J. Cr. *Annandale*—Moffat, J. T. J. *Eskdale*—common, S.E.

VISITORS : Bombus lucorum, pratorum, Andrena albicans, Syrphus Ribesii, Platychirius peltatus, S. E.

Nymphea alba, *Linn.* (White Water Lily).

RECORDS : *Dfs.* and *Kcd.*—P. Gray, 1850. *Wgt.*—G. C. Druce, 1883.
LOCALITIES : *Nithsdale*—Loch Kindar, M. W., E. M. C. ; Maxwelltown Loch, P. Gr. ; Loch Urr (700 feet), J. Cr. ; Blackwood, Friars' Carse, S. E. ; Stroanshalloch (1250 feet), J. Cr. ; Auchenknight, R. A. *Annandale*—Broomhill, Lochmaben, P. Gr., S. E. *Eskdale*—Penton, C. Y.

In ponds 2 to 10 feet deep.

VISITORS : Bombus (lucorum ?) ab., Apis ab., Scatophaga litorea, Anthomyidæ. S. E.

Nuphar luteum. *Linn.* (Yellow Water Lily).

RECORDS : *Dfs.* and *Kcd.*—P. Gray, 1850. *Wgt.*—G. C. Druce, 1883.

b. Intermedium *(Ledebour)*, *Dfs.*—Dr Davidson, 1886.

LOCALITIES : *Nithsdale*—Loch Kindar, M. W. ; Loch Urr (700 feet), Trostan Lane, J. Cr. ; Stroanshalloch (1250 feet), also var. *b.*, Dr Dv., J. Cr. *Annandale*—Halleaths, Shillahill, G. Bl. : Castle Loch, Lochmaben, Dr Sn. ; Craiglands, J. T. J.

Ponds (var. *b.* with peaty bottom).

Platystemon Californicum. *Benth.*

RECORD : (Escape) *Dfs.*—J. Corrie, 1887.
LOCALITY : Moniaive, J. Cr.

Papaver rhoeas. *Linn.*

RECORDS : (Escape) *Dfs.*—J. Wilson, 1883. *Kcd.*—J. M'Andrew, 1882. *Wgt.*—J. M'Andrew, 1891.
LOCALITIES : *Annandale*—Hoddam Castle, J. Wl. ; Beattock, J. T. J. On waste ground.

Papaver dubium. *Linn.* (Poppy).

RECORDS : *a.* Lamothei, *Dfs.*—P. Gray, 1850. *Kcd.*—J. M'Andrew, 1882. *Wgt.*—G. C. Druce, 1883.
LOCALITIES: *Nithsdale*—G. & S.W. Railway line, Cummertrees, Ak. ; Dumfries, S. E. ; Maxwelltown Station, Th. ; Mildamhead, P. Gr. ; Brownhall, M. J. H. ; Closeburn, Auldgirth, R. A. ; Sanquhar, Kirkconnel, Dr. Dv. *Annandale*—Annan Station, C. E. M. ; Beattock, Adam's Holm, J. T. J. *Eskdale ?*—(Riddings, S.E.)
Dry cinders or waste ground free of other plants and fully, exposed.
VISITORS : Platychirius albimana, Sepsis cynipsea, Hydrotea dentipes, Anthomyia sulciventris, Meligethes æneus, S. E.

Papaver argemone. *Linn.* (Long-headed Poppy).

RECORDS : (Escape) *Dfs.*—Miss E. D. Adams, Miss S. D. Johnstone, 1890. *Wgt.*—G. C. Druce, 1883.
LOCALITY : Holywood Station, Ad. and S. D. J.

Meconopsis Cambrica. *Vig.* (Welsh Poppy).

RECORDS : (Escape) *Dfs.*—Dr Balfour, 1838. *Kcd.*—Dr Gilchrist ? 1867. *Wgt.*—J. M'Andrew, 1893.
LOCALITIES : *Nithsdale*—Dumfries, M. J. H. ; Nith above Friars' Carse, S. E. *Annandale*—Wamphray Bridge, S. E. ; above Moffat, S. W. Ca. ; Adam's Holm, J. T. J.
Appears May 14, J. T. J. Garden Weed, moss-covered sandstones, shingles.

Chelidonium majus. *Linn.* (Great Celandine).

RECORDS : (Escape) *Dfs.*—A. Sibbald, 1838. *Kcd.*—W. Brand, 1838. *Wgt.*—C. C. Bailey, 1883.

LOCALITIES: *Nithsdale*—Newabbey, W. Br., M. W.; Clarencefield, Hg.; Cluden Craigs and Bridge, P. Gr.; Cowhill, Ad., S. D. J.; Keir, A. Sb.; Stepends Kirk, Waterside, R. A. *Annandale*—Dornock, J. M'A.; Halleaths, Castle Loch, Lochmaben, J. M'A., J. T. J.

On dry waste ground, exposed to wind, or half-sheltered by hedges.

Glaucium luteum. *Scop.*

RECORDS: *Dfs.*—Dr Burgess, 1789. *Kcd.*—P. Gray, 1850. *Wgt.*— Dr M'Nab, 1836.

LOCALITIES: *Along the shore*—Mull, G. M'N.; Drummore, Arn.; Port William, J. M'A.; Creetown to Gatehouse, Dr Bl., J. M'A., Hn., S. E.; Ross, G. C.; Mullock Bay, J. M'A., F. R. C.; Balcary, G. M'C.; Saltflats, J. Fr.; Port o' Warren, P. Gr.; Rascarrel, J. C. W., J. Fr.; Southerness, J. M'A.; Arbigland, J. M'A., Th.; Carsethorn, C. E. M., Th.; Newbie, Dr Br., J. Fr., S. E.

On shingle, sand, or shell bank, stony boulder clay, just above high-water mark, and fully exposed.

VISITORS: Platychirius peltatus, Anthomyia radicum, Hydrotea dentipes, Meligethes æneus, S. E.

Fumaria officinalis. *Linn.* (Fumitory).

RECORDS:

SUBSPECIES CAPREOLATA. *Linn.*—
69. Pallidiflora, *Dfs.*—(Herbarium). *Wgt.*—G. C. Druce, 1883.
70. Boræi, *Dfs.*—G. F. Scott-Elliot, 1891. *Kcd.* and *Wgt.*— G. C. Druce, 1883.
71. Confusa, *Dfs.*—Dr F. W. Grierson, 1882. *Kcd.*—J. M'Andrew, 1882.

SUBSPECIES OFFICINALIS. *Linn.*

73. Densiflora, *Dfs.*—J. T. Johnstone, 1890. *Wgt.*—G. C. Druce, 1883.
74. Officinalis, *Dfs.*—G. F. Scott-Elliot, 1891. *Wgt.*—G. C. Druce, 1883.

LOCALITIES: *Nithsdale*—Kirckonnel, M. W.; Glencaple, 71, M. J. H.; Dumfries, P. Gr., 71; F. W. G., C. E. M., Th.; Cluden Mills, 71, S. E.; Glencairn, 71, J. Cr.; Holywood, 74, S. E.; Sanquhar, Dr Dv. *Annandale*—Annan, C. E. M.; Ecclefechan, 70, S. E.; Lochmaben, P. Gr.; Beattock, 71, S. E.; Hydropathic, Edgemoor Selcoth (to 500 feet), J. T. J.

Dry, corn, or turnip fields; cinders of railways, light, sandy soils, waste ground; usually exposed to sun and wind, and on bare ground.
VISITOR : Pieris brassicæ, S. E.

Corydalis claviculata. *D. C.*

RECORDS : *Dfs.*—G. Gordon, 1837. *Kcd.*—P. Gray, 1850. *Wgt.*—G. C. Druce, 1883.

LOCALITIES : *Nithsdale*—The Brow, G. Go., P. Gr., J. Fn.; Kirkconnel Moss, G. C.; Waterloo Hill, R. H. M.; Glen, P. Gr.; Jarbruck, Craigdarroch (450 feet), J. Cr.; Cleuch House Linn, Dr Dv.; King's Quarry, Newton Quarry, Auldgirth Hill, R. A.; Tynron, J. Sh.; Drumlanrig, J. Cru. *Annandale*—Craiglands, Gallowhill (800 feet), J. T. J. *Eskdale*—Langholm, C. Y.

Appears June 10, J. T. J. Moist or nearly dry humus of woods; usually half-shaded and wind-sheltered, climbing over dead branches or brambles, or on cottage roofs.

VISITORS : Apis, Bombus muscorum, S. E., J. C. W.; Derhamellus, S. E.; Terrestris, J. C. W.; Apathus quadricolor, Andrena Trimmeriana, Anthomyia, S. E.

Corydalis lutea. *D. C.*

RECORD : (Escape) *Dfs.*—J. T. Johnstone, Miss Ethel Taylor, 1891.

LOCALITIES : *Nithsdale*—Scaur Brig, R. A. *Annandale*—Bearholm, J. T. J. *Eskdale*—Old Wall, Todhillwood, E. Ty.

Appears May 17, J. T. J.

VISITOR : Bombus hortorum S. E.

Corydalis solida. *Hook.*

RECORD : (Escape) *Dfs.*—Miss Ethel Taylor, 1891.

LOCALITIES : *Nithsdale*—Newlands, R. A. *Eskdale*—Woodhouselees, E. Ty.

Cheiranthus cheiri. *Linn.* (Wallflower).

RECORDS : (Escape) *Dfs.*—Miss Harvey, 1850. *Kcd.*—Robert Burns. *Wgt.*—J. M'Andrew, 1882.
LOCALITIES : Dundrennan, Lincluden, Caerlaverock.

Barbarea vulgaris. *R. Br.* (Winter Cress, Yellow Rocket).

RECORDS : 84. Vulgaris, *Dfs.*—Dr Singer, 1843. *Kcd.*—P. Gray, 1838. *Wgt.*—G. C. Druce, 1883.
87. Intermedia, *Dfs.*—? *Kcd.*—Miss Hannay, 1891.
88. Praecox, *Kcd.*—Dr F. W. Grierson, 1882.

LOCALITIES: *Nithsdale*—Whinnyhill, 84, F. W. G.; Cargen, 84; Newton, 87, Hn., S. E.; Glen, 84, S. E., P. Gr.; Tinwald, Dr Sn.; Glencairn, J. Cr., Pl.; Cample, R. A.; Sanquhar, Dr Dv. *Annandale*— Annan, M. J. H.; Wamphray to Moffat, J. T. J., S. E.; Lochmaben, J. Wl.

Appears April 28 to June, J. T. J. Along rivers and streams, on holms, roadsides, whinstone soils; in sun or half-shaded and wind-sheltered.

VISITORS: Apis, Halictus mori, Andrena coitana, Platychirius albimana, Syritta pipiens, Siphona cristata, Anthomyia radicum, Micropteryx calthella, S. E.

Nasturtium officinale. *R. Br.* (Watercress).

RECORDS: *Dfs.*—J. Cruickshank, 1839. *Kcd.*—P. Gray, 1850. *Wgt.*—G. C. Druce, 1883.

LOCALITIES: Very common in all valleys, reaching 2000 feet; Loch Skene.

Appears May 20, J. T. J. Usually in slow streams under a foot deep, or moist ground; in sun or half-shaded and usually wind-sheltered.

VISITORS: Volucella bombylans, Eristalis arbustorum, sepulchralis Helophilus pendulus, Syritta pipiens, Platychirius clypeatus peltatus, Pyrophona rosarum, Scatophaga stercoraria, Empis livida, Pipizella sp., Anthomyia sp., Chrysogaster, Pollenia rudis, Lucilia Cæsar, Dolichopus, Hyelemyia strigosa, Caricea tigrina, Telephorus rusticus, bicolor, Thyonis morcida, Byturus rosæ, S. E.

Nasturtium sylvestre. *R. Br.*

RECORDS: *Dfs.*—Dr Burgess, 1789. *Kcd.*—J. M'Andrew, 1882.

LOCALITIES: Carlingwark, J. M'A.; along Æ below Kirkmichael House, Dr Br.

Nasturtium palustre. *D. C.*

RECORDS: *Dfs.*—G. N. Lloyd, 1837. *Kcd.*—J. Cruickshank, 1839. *Wgt.*—G. C. Druce, 1883.

LOCALITIES: *Nithsdale*—Kingholm Quay, P. Gr., S. E.; Lochar Moss, G. N. Ll.; Cargenbanks, Hn.; Glencairn, Pl.; several places by Nith, Dumfries, Hn.; S. E.; Carnsalloch, C. E. M.; Broomrigg, Hn.; Friars' Carse, Dr Dv., J. Fn., S. E.; Newlands, R. A. *Annandale*— Lochmaben, Exc.; Broom, St. Mungo, G. Bl.

Along rivers or lakes on alluvial stony mud, shingles, or in dock walls.

VISITORS: Anthomyia radicum, Hyelemyia strigosa, Meligethes æneus, S. E.

Nasturtium amphibium. *R. Br.*

RECORD: *Dfs.*—Dr Burgess, 1789.

LOCALITY: In the burn that runs into the Lochar opposite the Manse of Mouswald.

Arabis hirsuta. *R. Br.* (*A. Sagittata.* D. C.)
(Rock Cress).

RECORDS: *Dfs.*—J. Sadler, 1857. *Kcd.* and *Wgt.*—F. R. Coles, 1882.

LOCALITIES: *Nithsdale*—Dalveen Pass (900-1200 feet), J. Fn., S. E.; Kello (600 feet), Dr Dv. *Annandale*—Carterton, Lamonbie Mill, G. Bl.; Beeftub, J. Sd.; Newton Burn (800 feet), J. T. J., S. E.; Spoonburn (900 feet), S. E.; Black's Hope (2200 feet), J. T. J.; Grey Mare's Tail (900-1500 feet), J. Sd., J. T. J.; Craigmichen (800 feet), J. T. J.; also Moffat sandbeds (300 feet), J. T. J.

Appears May 14, J. T. J. Usually moist whinstone or sandstone rocks, in shade or half-shaded and well wind-sheltered by narrowness of ravines; sometimes exposed on shingles and sandbeds.

VISITORS: Platychirius albimana, Syritta pipiens, Empis bilineata, Scatophaga sp., Anthomyia sp., S. E.

Arabis thaliana. *Linn.* (Thale Cress).

RECORDS: *Dfs.*—J. Shaw, 1882. *Kcd.*—F. R. Coles, 1882. *Wgt.* —J. M'Andrew, 1891.

LOCALITIES: *Nithsdale*—Holywood, Ad. and S. D. J.; Portrack, S. E.; Auldgirth Linn, Exc.; Tynron, J. Sh.; Cample, R. A.; Sanquhar, Dr Dv. *Annandale*—Shieldhill, S. E.; Lochmaben, J. Wl.; Moffat (to 1000 feet), J. T. J.

Appears March 23, J. T. J.; March 26, G. Bl. Dry cinders or sandy soil, old walls, roadsides, etc.; fully exposed to wind and sun.

VISITOR: Platychirius clypeatus, S. E.

Cardamine amara. *Linn.* (Bitter Cress).

RECORDS: *Dfs.*—Dr Burgess, 1789. *Kcd.*—J. Fraser, 1843? *Wgt.* —Dr Macnab, 1839.

LOCALITIES: *Nithsdale*—Common by Cluden and Cairn to Moniaive, Hn., P. Gr., S. E., J. Cr.; Glencaple to Troqueer and Dumfries, Hn., C. E. M.; Milnhead, J. Cru.; Auldgirth, S. E., R. A.; Tibbers Castle, J. Wl.; Sanquhar, Dr Dv. *Annandale*—Annan, S. E.; Jardine Hall, Th.; Tundergarth, G. Bl.; Milke and Æ, Dr Br., S. E.; Beld Craig, Hg.;

Moffat, J. T. J. *Eskdale*—Glenzier, E. Ty.; Esk, Tarras, Ewes (to 400 feet), S. E.

Appears April 27, J. T. J.; April 30, G. Bl. In wet mud by streams, holms, shingles, etc.; in sunny or shady, but not windy places.

VISITORS : Dacnusa sp., Chilosia peltata, Rhingia rostrata, Scæna Ribesii, Anthomyidæ ab., Micropteryx calthella, Mantua napi, Antholobium triviale, Phyllobius calcaratus, S. E.

Cardamine pratensis. *Linn.*

(Cuckoo Flower, Ladies' Smock, Karson, *fide* J. Shaw).

RECORDS : *Dfs.*—J. Sadler, 1857. *Kcd.*—P. Gray, 1839. *Wgt.*—G. C. Druce, 1883.

LOCALITIES : Very common in all the valleys to about 900 feet.

Appears April 15, J. T. J.; April 14, G. Bl. Flowers April 13, R. A. Usually moist holms, roadsides, boulder clay, whinstone soils, carboniferous and other sandstones; in sun, half-shaded or quite shaded, and windy or sheltered places.

VISITORS : Chilosia peltata, Bibio sp., Helophilus pendulus, Scæna ribesii, Anthomyidæ, Mantua napi, Zostenophorus tæniatus, S. E.

Cardamine impatiens. *Linn.*

RECORD : *Dfs.*—J. Sadler, 1857.

LOCALITIES : *Nithsdale*—Churchyard, Irongray, Hn. *Annandale*—Middlegill; roadside, Kirkpatrick-Juxta, J. T. J.; Garpol, Beld Craig, J. Sd.

Appears June 13, J. T. J.

Cardamine hirsuta. *Linn.*

RECORDS : 98. Hirsuta, *Dfs.* and *Kcd.*—P. Gray, 1850. *Wgt.*—Dr Balfour, 1843.
99. Flexuosa, *Dfs.*—Dr Davidson, 1886. *Kcd.*—Dr Grierson, 1882. *Wgt.*—G. C. Druce, 1883.

LOCALITIES : 98 very common in all the valleys; appears March 23, J. T. J. 99 appears May 4, J. T. J. *Nithsdale*—Whinnyhill, F. W. G.; Dumfries, P. Gr.; Auldgirth, S. E.; Euchan, Dr Dv. *Annandale*—Auchenbraith Linn, Æ water, Milke, Wellburn, Brackenside, etc., common at Moffat, S. E., J. T. J. *Eskdale*—Todhillwood, E. Ty.; common Esk to Langholm, S. E.; and Castle O'er, J. Wn., R. Bl.

98 on dry sandy and gravelly soils, roadsides, shingles, garden refuse, old walls, whinstone; in sun or half-shaded and usually windy places; prefers bare ground.

99 on moist leaf mould, whinstone or alluvial ground, usually mud; in shade or half-shaded and always wind-sheltered, often in long grass.

VISITORS: Melanostoma sp., Anthomyia radicum, sulciventris, Apion germari, S. E.

Sisymbrium officinale. *Scop.* (Hedge Mustard).

RECORDS: *Dfs.* and *Kcd.*—P. Gray, 1850. *Wgt.*—G. C. Druce, 1883.

LOCALITIES: *Nithsdale*—Kirkconnel, M. W.; Cargen, Dumfries, P. Gr., Hn., S. E.; Glencairn, Pl.; Sanquhar, Dr Dv. *Annandale*—Very rare east end of Castle Loch, Brocklerigg, G. Bl.; Wamphray, J. M'A., J. T. J.; waste ground, Moffat, J. T. J. *Eskdale*—near Liddel, S. E.; Kirkandrews, E. Ty.

Appears July 2, J. T. J. On pretty dry roadsides, railways, river banks; in sun and part wind-sheltered by hedges, etc.

VISITORS: Syritta pipiens, Anthomyia radicum, S. E.

Alliaria officinalis. *Andrz.*
(Jack by the Hedge, Sauce alone).

RECORDS: *Dfs.*—P. Gray, 1843. *Kcd.*—J. Cruickshank, 1839. *Wgt.*—J. M'Andrew, 1891.

LOCALITIES: *Nithsdale*—Whinnyhill, Exc.; Troqueer, Hn., S. E.; Grovehill, Virgin Hall, J. Fn., R. A.; Glencairn, Pl. *Annandale*—rare Lockerbie, G. Bl.

VISITORS: Syrphus umbellatarum, Platychirius albimana, Empis livida, Hyelemyia strigosa, Anthomyia radicum, Haltica oleracea, S. E.

Erysimum cheiranthoides. *Linn.*

RECORD: (Escape) *Dfs.*—Dr Davidson, 1891.
LOCALITY: Kirkconnel Station, Dr. Dv.

Brassica monensis. *Huds.* (Isle of Man Cabbage).

RECORD: *Dfs.*—Rev. T. Bell, 1882. *Kcd.*—J. Fraser, 1843. *Wgt.*—Dr Balfour, 1843.

LOCALITIES: *Along the shore*—rare Powfoot, J. Fn., S. E.; Annan shore, T. Bl., S. E.

On shingles and sand against small banks, fully exposed. (Railway as casual, Dr Dv.)

VISITORS: Halictus mori, Formica cunicularis, Syritta pipiens, Anthomyia radicum, Chortophila sp., Haltica oleracea, Meligethes aeneus, S. E.

Brassica tenuifolia. *Boiss.*

RECORD: (Escape) *Dfs.*—G. F. Scott-Elliot, 1891.
LOCALITY: Glentarras, S. E.

Brassica campestris. *Linn.*

RECORDS: (All Escapes) *Dfs.*—P. Gray, 1839. *Kcd.*—J. M'Andrew, 1882.
SUBSPECIES—132. Napus, *Kcd.*—G. F. Scott-Elliot, 1891. *Dfs.* —Dr Davidson, 1886.
   133. Rutabagga, *Kcd.*—G. F. Scott-Elliot, 1891.
   134. Rapa—*a.* Sativa, *Wgt.*—G. C. Druce, 1883.
     c. Briggsii, *Dfs.*—Miss M. J. Hamilton 1891.

LOCALITIES: *Nithsdale*—Lochar, P. Gr.; Cargen Glen, Th., C. E. M.; Newton, 132 and 133, S. E.; Glencairn, Pl.; Sanquhar Castle, Dr Dv. *Annandale*—Annan, 133, S. E. *Eskdale*—Bentpath, S. E.; Castle O'er, J. Wn., R. Bl.

Usually in arable fields, roadsides, etc.

Brassica sinapis. *Vis.* (Charlock).

RECORDS: *Dfs.* and *Kcd.*—P. Gray, 1850. *Wgt.*—J. M'Andrew, 1882.
LOCALITIES: Very common in all the valleys to 800 feet.

Dry sandy or gravelly soils, whinstones, boulder clay, holms, granite; in sun or half-shaded, and usually sheltered by corn, etc.

VISITORS: Bombus pratorum, Eristalis tenax, pertinax, tumidata, Syrphus bifasciatus, Empis punctata, Syritta pipiens, Chlorosia formosa, Platychirus albimana, Anthomyia radicum, Mantua napi, Dicrorampha, S. E.

Brassica alba. *Boiss.*

RECORDS: (Escape) *Kcd.*—P. Gray, 1843. *Wgt.*—G. C. Druce, 1883.
LOCALITY: Lochrutton, P. Gr.

Brassica nigra. *Koch.*

RECORD: (Escape) *Dfs.*—Dr Burgess, 1789.
LOCALITY: Duncow, Dr Br.

Cochlearia officinalis. *Linn.* (Scurvy Grass).

RECORDS:
SUBSPECIES—112. Officinalis, *Dfs.*—P. Gray, 1843. *Kcd.*—Dr Macnab, 1835. *Wgt.*—G. C. Druce, 1883.
    b. Montana, *Dfs.*—G. F. Scott-Elliot, 1891.
115. Danica, *Kcd.*—Rev. J. Fraser, 1882.
117. Anglica, *Dfs.*—Miss Ethel Taylor, 1891. *Kcd.*—Dr Macnab, 1835. *Wgt*—J. M'Andrew, 1893.

LOCALITIES: *Nithsdale*—Kirkconnel, M. W.; Glencaple, and along Nith to Dumfries, 112 and 117, Hn., S. E.; Martown, Dibbin Hill (1600 feet), J. Cr.; Shinnelhead, Lamgarroch (1000-1700 feet), J. Sh.; Glenmaddie, Euchan, Dr Dv.; Cample Cleugh, J. Fn. *Annandale*—common by shore, S. E.; Balgray, G. Bl.; Æ water head, J. Fn.; Queensberry, Penbreck, S. E.; Wamphray to Beeftub, by Annan, S. E.; Hartfell, ab., S. E.; Black's Hope, Corriefron, Grey Mare's Tail, Whitecoombe, ab., to 2500 feet, J. Sd., Hg., S. E., J. T. J. *Esk-dale*—Kirkandrews, 117, E. Ty.; common by Esk to Langholm, S. E.; Langholm Hill, Archie Hill, Ewes, S. E.

Appears April 9, G. Bl.; May 5, J. T. J. Always moist places, in mud, etc., and fully exposed.

VISITORS: Anthomyia radicum, Meligethes æneus, S. E.

Camelina saliva. *Crantz.* (Gold of Pleasure).

RECORD: (Escape) *Dfs.*—Herbarium Greville, 1830.
LOCALITIES: Jardine Hall, Grev.; Tinwald, Dr Sn.

Hesperis matronalis. *Linn.* (Dame's Violet).

RECORDS: (Escape) *Dfs.*—J. Corrie, 1891. *Kcd.* or *Wgt.?*—Dr Balfour, 1836.
LOCALITIES: By Cairn (300 feet), J. Cr.; Scroggs, G. Bl.

Draba verna. *Linn.* (*Erophila vulgaris.* D. C.)
(Whitlow Grass).

RECORDS: *Dfs.* and *Kal.*—P. Gray, 1850. *Wgt.*—J. M'Andrew, 1890.

LOCALITIES: *Nithsdale*—Newton, P. Gr., S.-E.; Mile House, Lincluden, Hn.; Peelton, Woodlea, Blackstone (300-340 feet), J. Cr.; Tynron, J. Sh.; Cample Brig, R. A.; Glenmaddie (1350 feet), Dr Dv.; Durrisdeer, Exc.; Sanquhar Castle, Dr Dv. *Annandale*—Lochmaben, Capelgillfoot, J. T. J.; Beeftub, Kd.; Hydropathic, Moffat, S.-E. *Eskdale*—Glenzier burn, E. Ty., S.-E.; Tarras foot, Meggatdale (500 feet), S.-E.

Appears March 20. I have also seen it flowering in December, J. T. J. Dry turf-covered dykes, stony roads; usually in sun, and in part wind-sheltered.

VISITORS: Anthomyia pluvialis, Dolichopodæ, Latridius porcatus, E. Ty., S.-E.

Subularia aquatica. *Linn.* (Awlwort).

RECORDS: *Dfs.*—Greville Herbarium, 1830. *Kal.*—F. R. Coles, 1882.

LOCALITIES: Threave Castle, on Dee, F. R. C.; Loch Skene, Grev. Herb., J. T. J.

Appears August 7, J. T. J. On bare mud or shallow water amongst Equisetum stems.

VISITOR: A small fly, Limosina sp.? s.

Thlaspi arvense. *Linn.* (Penny Wort).

RECORDS: (Escape) *Dfs.*—Dr Davidson, 1891. *Kal.*—Arnott, 1844. *Wgt.*—J. M'Andrew, 1889.

LOCALITIES: Chapelhill Carco, Dr Dv.

Teesdalia nudicaulis. *R. Br.*

RECORDS: *Dfs.*—J. Wilson, 1882. *Kal.*—J. M'Andrew, 1882. *Wgt.*—Dr Macnab, 1837.

LOCALITIES: *Nithsdale*—Ferney Cleugh, Locharbriggs, J. Wl.; Evan Water, Beattock, J. T. J., S.-E.

Appears May 6, J. T. J.

VISITORS: Small Diptera, S.-E.

Hutchinsia petræa. *R. Br.*

RECORD : *Dfs.*—Rev. Dr Little, 1850?
LOCALITY : Kirkpatrick-Juxta.?

Iberis amara. *Linn.* (Candytuft).

RECORDS : (Escape) *Dfs.*—G. Bell, 1893. *Kcd.*—Misses Hannay and R. H. Masterman, 1891.
LOCALITIES : Shillahill, G. Bl.; Kirkconnell Lodge, Hn., R. H. M.

Capsella bursapastoris. *R. Br.* (Shepherd's Purse).

RECORDS : *Dfs.*—P. Gray, 1838. *Kcd.* and *Wgt.*—J. M'Andrew, 1882.
LOCALITIES : Very common in all the valleys but not on peat.

Appears March 30, J. T. J.; April 7, G. Bl. Usually dry roadsides, holms, gardens, sandy soil, etc.; not on peat, rarely in shade or wind; prefers bare ground.

VISITORS : Apis, Andrena bicolor, Syritta pipiens, Syrphus corollæ, ribesii, Siphona geniculata, Opomyza germinationis, Anthomyia radicum, Hydrotea dentipes, S.-E.

Lepidium ruderale. *Linn.*

RECORD : (Escape) *Wgt.*—J. M'Andrew, 1887.

Lepidium campestre. *R. Br.*

RECORDS : *Dfs.*—Dr Balfour, 1843. *Kcd.*—P. Gray, 1843. *Wgt.*— Rev. G. Wilson, 1893.
LOCALITIES : By River Nith, Dr Bl.; Dumfries, P. Gr.

Lepidium smithii. *Hook.*

RECORDS : *Dfs.*—J. Cruickshank, 1839. *Kcd.*—P. Gray, 1841. *Wgt.*—Dr Macnab, 1835.

LOCALITIES : *Nithsdale*—Kirkconnell, M. W.; Cargen, Hn.; Dumfries, P. Gr., Hn., S.E.; Glencairn, Pl., J. Cr.; Sanquhar, Dr Dv. *Annandale*—Adam's Holm, Moffat, J. T. J. *Eskdale*—Rare, Kirkandrews, E. Ty.; Gretna Green, S.-E.

Appears April 28, J. T. J.

VISITORS : Apis ab. and suff., Melanostoma, Anthomyidæ, Chortophila sp., S.-E.

Senebiera coronopus. *Poir.*

RECORDS: *Kcd.*—F. R. Coles, 1882. *Wgt.*—J. M'Andrew, 1882.

LOCALITIES: *Along the shore*—Portpatrick, Drummore, Port Logan, J. M'A.; Deemouth, F. R. C.

In crannies of stones in harbours; on mud flats.

Isatis tinctoria. *Linn.* (Woad).

RECORD: (Escape) *Wgt.*—Dr Balfour, 1835.

LOCALITY: Drummore.

Cakile maritima. *Scop.* (Sea Rocket).

RECORDS: *Dfs.*—Dr Davidson or J. Fingland, 1886. *Kcd.* and *Wgt.*—J. Fraser, 1843.

LOCALITIES: *Along the shore*—rare Portpatrick, Arn., J. M'A; Portwilliam, G. C., S. E.; Creetown, S. E.; Borgue shore, J. M'A.; Southerness, J. Fn.; Powfoot to Newbie, Dr Dv., J. Fn.

Coarse gravel or stony shingles; exposed.

VISITORS: Anthomyia ab.; Meligethes æneus, S. E.

Crambe maritima. *Linn.* (Seakale).

RECORDS: *Kcd.*—G. N. Lloyd, 1837. *Wgt.*—A. Sibbald, 1823.

LOCALITIES: *Along the shore*—Mull, Sb.; West Tarbert, Dr. Bl., Arn., Dr M'N.; Garliestown, Hn.; Creetown, Borgue, J.M'A.; Ross, G. N. Ll., G. C., J. M'A.; Balmae, G. N. Ll.; Balcarry, Auchencairn; G. M'C., J. M'A.

In shingle or rocks.

VISITORS: Lucilia Cæsar, cornicina, Meligethes æneus, S.-E.

Raphanus raphanistrum. *Linn.* (Radish).

RECORDS:

SUBSPECIES—162. Raphanistrum, *Dfs.* and *Kcd.*—P. Gray, 1838 and 1847. *Wgt.*—Dr Balfour, 1835.
163. Maritimus, *Kcd.*—P. Gray, 1847. *Wgt.*—Dr Balfour, 1835.

LOCALITIES: *Along the shore*—163 Mull, Dr Bl., Dr M'N., J. M'K.; Portpatrick, J. M'A.; Drummore, Arn.; Portwilliam, S.-E.; Port o' Warren, J. Fr., P. Gr., Fq., S.-E. *Nithsdale*—162 Harleybank Ford, P. Gr.; Newton, S.-E.; Cowhill, Ad. and S. D. J.; Sanquhar, Dr Dv. *Annandale*—Moffat, S.-E. *Eskdale*—Burnfoot, S.-E.

On shingles of shore or on rocks and in waste ground or arable land; inland.
VISITORS: Apis, Bombus sp., Anthomyidæ, Meligethes œneus, S.-E.

Reseda luteola. Linn. (Dyer's Weed, Mignonette).

RECORDS: *Dfs.*—By the Nith ab., Dr Burgess, 1789. *Kcd.* and *Wgt.*—J. M'Andrew, 1682.
Also found at Ecclefechan. (Exc. 82).

Helianthemum vulgare. Gaert. (Rockrose).

RECORDS: *Dfs.*—J. Wilson, 1882. *Kcd.*—Rev. J. Fraser, 1863. *Wgt.*—Arnott, 1848.
LOCALITIES: Common along the Kirkcudbright coast, Glenlair, Wd. Herb.; Rascarrel, J. C. W. *Nithsdale*—Lochanhead, F. W. G.; Littleknowe, Glen, Th., C. E. M.; Grove Hills, S.-E.; Craigneston (500 feet), Barndennoch (650 feet), J. Cr.; Dalveen, Exc.; Scaur, Tynron Doon, R. A.; Spango Bridge, Dr Dv. *Annandale*—Wellburn, Carr; Beeftub, Spout Craig, Corehead, burn between Crofthead and Selcoth, J. T. J. (A peculiar and anomalous distribution, possibly due to presence of limestone fragments.)
Appears June 14, J. T. J. On granite, whinstone, or boulder clay soils with a dry sunny southern exposure; in windy or part sheltered places.
VISITORS: Melanostoma mellina, Cynomyia mortuorum, Platychirius clypeatus, Lucilia Cæsar, Rhamphomyia, Dolichopods, Chortophila, and five unnamed species, S.-E.

Viola palustris. Linn. (Marsh Violet).

RECORDS: *Dfs.*—J. Cruikshank, 1836. *Kcd.*—P. Gray, 1844. *Wgt.*—J. M'Andrew, 1882.
LOCALITIES: *Nithsdale*—Lincluden, P. Gr.; Newton, Locharmoss, S.-E.; Cowhill, Ad. and S. D. J.; Twomerkland (a white variety), J. Cr.; Sanquhar, Dv. *Annandale*—Elsieshiels, S.-E.; Millhill, Eskrigg, G. Bl.; Echo Tower, Drumcrieff, Moffat, J. T. J. *Eskdale*—Solway Moss, E. Ty.
Appears April 19, G. Bl.; May 4 to 12, J. T. J. Wet or moist places; peaty, alluvial, or humoid loam; exposed to sun; partly sheltered by grass or ditch sides, etc.
VISITOR: Siphona geniculata, S.-E.

Viola cornuta. Linn.

RECORD: (Escape) *Dfs.*—Railway Station, Miss Hannay, 1892.

Viola odorata. *Linn.* (Garden Violet).

RECORDS: *Dfs.* and *Kcd.*—P. Gray, 1850; *Hgt.*—J. M'Andrew, 1890.

LOCALITIES: *Nithsdale*—Netherwood, Nunbank, Glencaple Road, Gillfoot, R. A., C. E. M.; New Quay, P. Gr.; Lincluden, Nunholm, Hn.; Clarencefield, Th.

An escape establishing itself in moist, shady and sheltered places on leaf mould.

Viola hirta. *Linn.*

RECORD: *Kcd.*—Arnott, 1848.
LOCALITY: Confirmed on Criffel, A. B. Hall, 1891.

Viola canina. *Linn.* (Dog Violet).

RECORDS: Sylvatica, Fr.
   Var. *a.*, *Dfs.*--J. Cruickshank, 1836. *Kcd.* --P. Gray, 1850. *Hgt.*—G. C. Druce, 1883.
   Var. *b.*, Flavicornis, Smith; *Dfs.*—Miss Milligan, 1892; Canina, *Kcd.*—F. R. Coles, 1885.

LOCALITIES: *Nithsdale* —Common near Dumfries, P. Gr., F. W. G.; Park Road, var. *b.*, C. E. M.; Troqueer, Hn.; Moniaive, J. Cr.; Sanquhar, Dv. *Annandale*—Water of Æ, var. *b.*, S.-E.; common to 300 feet Milke, Kirtle, G. Bl.; to 1300 feet Beef-tub, and 2300 feet Midlawburn and Whitecoombe, var. *a.* and *b.*, J. T. J., S.-E. *Eskdale*—Very common to 500 feet, S.-E.; Castle O'er, J. Wl., and R. Bl.

Prefers dry rather than moist leaf mould, roadsides, and other soils; sunny slopes; usually in part wind-sheltered. Becomes white in shady moist places.

Appears April 5, G. Bl.; April 14 to 28, J. T. J. Var. *b.* prefers ground free of other plants, short grass, steep banks, etc.

VISITOR: Bombus muscorum, lucorum Hn.; Platychirius clypeatus, Empis bilineata, S.-E.

Viola tricolor. *Linn.* (Pansy).

RECORDS: 181. Eutricolor, *Dfs.* and *Kcd.*—P. Gray, 1850. *Hgt.* —G. C. Druce, 1883.
   182. Arvensis, *Dfs.*—Dr F. W. Grierson, 1882. *Hgt.* —G. C. Druce, 1883.
   183. Curtisii, *Dfs.*—G. F. Scott-Elliot, 1892.
   184. Lutea, var. *a.*, *Dfs.*—W. Stevens, 1848. *Kcd.*— P. Gray, 1850. *Hgt.*—G. C. Druce, 1883.
       Var. *b.*, Amoena, *Dfs.*—J. Sadler, 1854. *Kcd.*— F. R. Coles, 1882.

LOCALITIES: *Nithsdale*—Very common near Dumfries (181, 182, 184), P. Gr., C. E. M., F. W. G.; Racks, 182, S.-E.; Cowhill, 184, Ad. and S. D. J.; (reaching 1200 feet) Moniaive, J. Cr.; Sanquhar, 181, 182, Dr Dv.; Wanlockhead, 184, W. St., Dr Gl., F. W. G. *Annandale*—Very common Newbie, S.-E.; Æ Water, 183, S.-E.; Dryfe, Milke, G. Bl.; Beattock, 183, J. T. J.; Auchencas, 182, Penbreck, 184, S.-E.; Garpel, Beef-tub, and Grey Mare's Tail, 184 (to 2400 feet), Black's Hope, var. *b.*, J. Sd., J. T. J.; Loch Skene, (2000 feet), S.-E. *Eskdale*—Common Canobie, E. Ty.; Wauchope, 184, var. *b.*, S.-E.; Castle O'er, 181, 184, J. Wl., and R. Bl.

Appears April 5 till September, J. T. J. The *preferences* of these forms seem to be:—181 for pretty dry leaf mould, roadsides, etc., partly shaded and wind-sheltered, and usually below 500 feet; 182 distinctly dry ground on railway cinders, waste roadsides with full exposure and below 500 feet; 183 slightly moist *sandy* soil, bare of other plants, and fully exposed; 184 rather dry slopes on boulder clay or whinstone in full sun exposure, and only partly wind-sheltered by shortish grass, commonest above 500 feet; 184 var. *b.* similar to last, but the least sheltered and sunniest spots, specially at highest altitudes.

Polygala vulgaris. *Linn.* (Milkwort).

RECORDS: 185. Euvulgaris, *Dfs.*—P. Gray, 1863. *Ked.* and *Wgt.*—J. M'Andrew, 1882 and 1892.
186. Oxyptera, *Ked.*—J. M'Andrew, 1890 (not typical, but not vulgaris, A. Bennet). *Wgt.*—J. M'Andrew, 1893.
187. Serpyllacea, *Dfs.*—P. Gray, 1845. *Ked.*—F. R. Coles, 1883. *Wgt.*—G. C. Druce.

LOCALITIES: *Nithsdale*—Common Kirkconnel, M. W.; Carnsalloch, Hn.; Moniaive, J. Cr.; Sanquhar, Dr Dv. *Annandale*—Very common Queensberry, 187, P. Gr.; Hindhill, S. W. Ca.; Hartfell, J. T. J.; Whitecoombe, 187, S.-E. *Eskdale*—Solway Moss and very common, E. Ty., S.-E.; Blochwell, S.-E.; Castle O'er, J. Wl. and R. Bl.

Appears April 27, G. Bl.; May 15 to June 7, J. T. J. Prefers dry but often marshy ground, chiefly on whinstone, boulder clay or peat; in sun, rarely shaded; usually windy places, or partly sheltered by short grass or broken ground. 187 on specially dry, sunny, windy, and bare places.

Dianthus armeria. *Linn.* (Deptford Pink).

RECORDS: (Escape) *Dfs.*—Auldgirth, Messrs Fingland and Davidson, 1882. *Ked.*—F. R. Coles, 1883. *Wgt.*—J. M'Andrew, 1893.

Silene inflata. *Sm.* (Bladder Campion).

RECORDS: 201. Cucubalus, *Dfs.* and *Kcd.*—P. Gray, 1850. *Wgt.*
—J. M'Andrew, 1882.
var. *b.*, puberula, *Dfs.*—J. Sadler, 1857. *Kcd.*—J. M'Andrew, 1882.
202. Maritima, *Dfs.*—P. Gray, 1844. *Kcd.* and *Wgt.*—J. M'Andrew, 1882.

LOCALITIES: 202, common on shore, Carlaverock, Powfoot, M. J. H.; Seafield to Redkirk, Old Gretna, etc., S.-E.; also, inland at Shillahill, G. Bl.; Grey Mare's Tail, Whitecoombe, P. Gr., J. T. J., S.-E. *Nithsdale*—201, not common; Kirkconnel, M. W.; Glencaple, C. E. M.; Terregles, P. Gr., F. W. G.; Nunwood, var. *b.*, S.-E.; Glencairn, J. Cr.; Tynron, J. Sh.; Cowhill, Ad. and S. D. J.; ab. on G. & S.-W. Ry., S.-E.; Thornhill, Cample, var. *b.*, R. A.; Elliock Brig, Dr Dv. *Annandale*—Ecclefechan, Lochmaben, Exc.; Balstack of Milke, Lockerbie, and var. *b.*, G. Bl.; common on Caledonian Railway, Wamphray, Beattock, Hunterheck, Craigmichen, var. *b.*, J. T. J. *Eskdale*—Dickstree, E. Ty; Burnfoot, S.-E.; Castle O'er, J. Wl. and R. Bl.

202 on shingle, rocks, or sand, often dead seaweed drift line; fully exposed to wind and sun. 201, on moist roadsides, sandy alluvium, cindery embankments; almost always wind-sheltered in full sun exposure.

VISITORS: 201, Platychirius manicatus, ab., Cordylura, sp. ab., Anthomyids, five or six unnamed kinds; 202, Eristalis pertinax, Empis vitripennis, Cordylura, ab., Chortophila, ab , seven unnamed kinds.

Silene noctiflora. *Linn.*

RECORD: (Escape) *Dfs.*—Miss Hannay, 1893.
LOCALITY: Cluden Mills.

Lychnis vespertina. *Sib.* (White or Evening Campion).

RECORDS: *Dfs.*—J. Cruickshank, 1836. *Kcd.*—J. M'Andrew, 1882. *Wgt.*—G. C. Druce, 1883.

LOCALITIES: *Nithsdale*—Kirkconnel, M. W.; Cargen, Hn.; Maxwelltown, Dalawoodie, Hn., F. W. G.; Tinwald Road, S.-E.; Nithside, M. J. H.; Newton, S.-E., C. E. M.; Dunscore Road, Th.; Woodlea (400 feet), Penpont, J. Cr.; Tynron, J. Sh.; Holywood, S.-E.; Cowhill, Ad. and S. D. J.; Auldgirth Saw Mill, Exc.; Thornhill, R. A., Dr Dv. *Annandale*—Priestside, Browhouses, below cliff at Torduff, S.-E.; by Railway, Moffat, J. T. J. *Eskdale*—Old Gretna, S.-E.

Appears June 11 to 14, J. T. J. Usually moist or dry places; roadsides, cornfields, on boulder clay, etc.; in full sun or half in shade; partly wind-sheltered by hedges, cliffs, etc.

VISITORS: Bombus pratorum, Platychirius manicatus, peltatus, Empis livida, Chordelura, sp., Anthomyidæ, Tipulæ, Moths, S.-E.

Lychnis diurna. *Sibth.* (Red or Day Campion).

RECORDS: *Dfs.*—P. Gray, 1850. *Kcd.*—J. Cruickshank, 1836. *Wgt.*—J. M'Andrew, 1882.

LOCALITIES: *Nithsdale*—Kirkconnel, M. W.; Mavisgrove, J. Cru.; Cluden and Routen Brig, C. E. M., Dr Dv., Glencairn, J. Cr.; Nith and Sanquhar, Hn., Dr Dv. *Annandale*—To Whitstonehill of Milke, Dryfe, very common, Kirtle, G. Bl.; Grey Mare's Tail (1750 feet), Black's Hope, J. T. J. *Eskdale*—Kirkandrews, E. Ty.; common Langholm, S.-E.; Castle O'er, J. Wl. and R. Bl.

Appears April 28, G. Bl.; May 4 to 23, J. T. J. Damp or dry places; leaf mould, holms, roadsides, occasionally whinstone or other rocks; usually half-shaded, more white in deep shade, or deeper colour in sun; almost always wind-sheltered.

VISITORS: Bombus lucorum, muscorum, S.-E.; terrestris, J. C. W.; Platychirius albimanus, J. C. W.; manicatus, S.-E.

Lychnis floscuculli. *Linn.* (Ragged Robin).

RECORDS: *Dfs.* and *Kcd.*—P. Gray, 1850. *Wgt.*—J. M'Andrew, 1882.

LOCALITIES: Common in all the valleys as far as Sanquhar, Dr Dv., and Glencairn, J. Cr.; seven miles up the Dryfe, G. Bl.; Beef-tub, Grey Mare's Tail, S.-E.; and Castle O'er, J. Wl. and R. Bl.

A white variety has been found at Palnackie, J. T. J.; Wamphray, G. Bl., and in the Wauchope, S.-E.

Appears June 11, J. T. J. Usually damp or wet ground; most abundant on boulder clay or whinstone detritus; always in sun and partly wind-sheltered by rush or grass, otherwise exposed.

VISITORS: Bombus lucorum, ab., muscorum, Rhingia rostrata, Hydrotea, sp., Platychirius, sp., Homalomyia, S.-E.

Lychnis githago. *Lam.* (Corncockle).

RECORDS: (Escape) *Dfs.*—Tynron, J. Shaw, 1882. *Kcd.*—Mrs Gilchrist-Clark, at the Ross (1867?). *Wgt.*—G. C. Druce, 1883.

LOCALITIES: Also found at Annan, Fn.; Moniaive, J. Cr.; Sanquhar Station, Dv.; Dumfries, P. Gr.; Gilnockie, S.-E.

An escape introduced with seed corn.

Lychnis viscaria. *Linn.*

RECORDS: (Escape) *Kcd.* and *Dfs.*—Rev. J. Singer, 1843.

LOCALITIES: Port o' Warren, Lot's Wife, P. Gr., Fq., J. M'A., (Moffat Hills, J. Sn.?).

Sagina nodosa. *Fenzl.*

RECORDS: *Dfs.* and *Kcd.*—P. Gray, 1850. *Wgt.*—G. C. Druce, 1883.

LOCALITIES: Glencaple, F. W. G, J. Fn., J. Wl.; Kenneth Bank, C. E. M.; Jardine Hall, Th.; Dalveen, Dv., J. Wl.; Torduff Point, S.-E.; Lochmaben, Exc.; Carterton, G. Bl.; Craigmichen Scaur, J. T. J.

Appears July 25, 26, J. T. J. Rather wet ground; stony or sandy soil or concrete; exposed.

Sagina subulata.* *Presl.*

RECORDS: *Dfs.*—J. T. Johnstone, 1890. *Wgt.*—C. C. Bailey, 1883.

LOCALITIES: Sandhills, Torrs, Warren, C.C.B., on the Beef-tub Road, between 1100 and 1300 feet, J. T. J., S.-E.; Castlehill, Glencairn, J. Cr.

Appears August 4, stony roads, J. T. J.; also on damp roadsides where it has a different growth, J. T. J.

Sagina procumbens. *Linn.* (Pearlwort).

RECORDS: 239. Maritima, var. *a.*, *Dfs.*—G. F. Scott-Elliot, 1892. *Kcd.*—J. M'Andrew, 1883. *Wgt.*—Arnott, 1843.
Var. *c.*, Densa, *Wgt.*—J. M'Andrew, 1890.
240. Apetala, *Kcd.* and *Wgt.*—J. M'Andrew, 1883 and 1890.
242. *Dfs.* and *Kcd.*—P. Gray, 1850. *Wgt.*—G. C. Druce, 1883.

LOCALITIES: 239, *a.* Annan mouth, S.-E.; Kirkcolm, Arn.; 239, common West Tarbert, J. M'A.; 240, South Dromore, J. M'A.; 242, very common in all the valleys, particularly below 700 or 800 feet.

Appears May 14, J. T. J. 242 dry or very dry places on walls, hard roads, cinders of railways, waste ground, shingle; shade or in sun; usually bare ground and exposed to wind; 240 apparently prefers turf and shelter of herbage, etc.; 239, var. *a.*, only seen once in wall crevices; 239, var. *c.*, apparently bare, much exposed places.

VISITORS: Ants ab.

---

* I cannot help considering this the right name for Mr Johnstone's plant, after careful comparison with the specimens in the British Museum. Rev. E. F. Linton supposes it to be Sagina procumbens, var. spinosa.

Arenaria verna. *Linn.*

RECORDS : *Dfs.*—Drumlanrig, J. Cruickshank, 1836 ; *Kcd.*—J. A. Brown, 1836.

LOCALITY :—Craig near Piper's Cove, Torrheugh Cliffs, Colvend ; has been confirmed by Rev. J. Fraser, Dr Grierson, and Mr M'Andrew.

Arenaria peploides. *Linn.* (Sea Purslane).

RECORDS : *Dfs.*—J. Cruickshank, 1839. *Kcd.*—Rev. J. Fraser, 1866. *Wgt.*—G. C. Druce, 1883.

LOCALITIES : Common on shore—Rerwick, J. Fr. ; Colvend, Hn., C. E. M., Th.; Caerlaverock, J. Cru.; common on shore, Seafield, S.-E. ; Powfoot, M. J. H.; Gretna, S.-E., etc. Usually in sand or gravelly shingle and fully exposed.

VISITORS : Anthomyia radicans ab., Sapromyza rorida ab., S.-E.

Arenaria serpyllifolia. *Linn.* (Sandwort).

RECORDS : *Dfs.*—J. Cruickshank, 1831. *Kcd.*—P. Gray, 1850. *Wgt.*—G. C. Druce, 1883, var *c.*, leptoclados. *Dfs.*—Messrs Johnstone and M'Andrew, 1889.

LOCALITIES : *Nithsdale*—Racks Station and G. & S.W. line occasionally, S.-E.; Dalbeattie Road, Hn.; Newton, Th., C. E. M., S.-E.; Auldgirth Station, Exc. ; Sanquhar, Dr Dv. *Annandale*—Newbie, Powfoot, M. J. H. ; Hollis Linn Brig, Beattock. J. T. J. *Eskdale*— Kirkandrews, E. Ty. ; N.B. Railway line, Liddel Brig to Langholm, ab. Glentarras Distillery, S.-E. ; var. *c.*, New Edinburgh Road, Moffat, J. T. J., J. M'A.

Appears June 5. Cinders and stones of railways, sandstone or whinstone walls, gravel, hard macadam, short turf; always dry or very dry spots ; fully exposed.

VISITORS : Thrips, Syritta pipiens, and two unnamed kinds, S.-E.

Arenaria trinervis. *Linn.* (Sandwort).

RECORDS : *Dfs.*—J. Cruickshank, 1839. *Kcd.*—P. Gray, 1850. *Wgt.*—J. M'Andrew, 1885.

LOCALITIES : *Nithsdale*—Cargenbrig, C.E.M. ; Brownhall, J. Cru. ; Craigs, M.J.H. ; White Bridge, Hn. ; Cluden Mill, Th., S.-E.; Blackwood Linn, Exc. ; Sanquhar, Dr Dv. *Annandale*—Dryfe Road, Lockerbie, J.T.J ; Tundergarth, S.-E. ; Lamonbie, G. Bl. ; Craiglands, Beattock, Drumcrief, Old Edinburgh Road, Old Well Road, J. T. J. *Eskdale*—Two miles from Langholm, near Irvine, S.-E.

Appears April 27, G. Bl., May 22, J.T.J. Damp or wet places on humus, roadside mud, sandy or rocky soil, shaded or half-shaded, sheltered.

VISITORS: Sphegina clunipes ab., Platychirius scutatus, Empis ignota, Ichneumon, sp., Meseoleius, sp., S.-E.

Cerastium vulgatum. *Linn.* (Mouse-ear Chickweed).

RECORDS: 221. Tetrandum, *Kcd.* and *Wgt.*—J. M'Andrew, 1892 and 1890.
223. Semidecandrum, *Dfs.*—Dr Davidson, 1886? *Kcd.* P. Gray, 1850. *Wgt.*—J. M'Andrew, 1883.
224. Glomeratum, *Dfs.* and *Kcd.*—P. Gray, 1850. *Wgt.* —G. C. Druce, 1883.
225. Triviale, var. *a.*, *Dfs.* and *Kcd.*—P. Gray, 1850. *Wgt.*—G. C. Druce, 1883,
Var. *c.*, pentandrum, *Kcd.*—F. R. Coles, 1883.
Var. *d.*, alpestre, *Dfs.*—G. F. Scott-Elliot, 1892.

LOCALITIES: 221 only from Dromore, J. M'A., Southerness, Exc.; 223, Port Logan (capsules very long, gland hairs very few, A. Bennett), J. M'A.; Sanquhar, Dr Dv.? (not seen). *Nithsdale*—224 and 225 very common to Moniaive (J. Cr.) and Sanquhar (Dr Dv.) *Annandale* —224 and 225 very common to 2000 feet, Black's Hope, J.T.J. 225, Whitecoombe, Auchencat Burn, Craigboar, S.-E. *Eskdale*—224 and 225 very common to Castle O'er, J. Wl. and R. Bl., and Pikethow 1500 ft., S.-E.

Appears May 4 to 25, J. T. J. 224 on dry railway lines, hard road-sides, waste ground, always exposed. 225 pretty dry or damp places, roadsides, turf on walls, railways, waste ground, shingles, and all soils except peat; sometimes half-sheltered by grasses and half-shaded. 225 dry whinstone rocks above 1400 feet (cf. C. alpinum).

VISITORS: Andrena albicans, R. Sc.; Syrphus arcuatus, Syrphus sp., Platychirius manicatus, Hydrella griseola, and three other species, S.-E.

Cerastium arvense. *Linn.*

RECORDS: *Dfs.*—J. Cruickshank, 1836. *Kcd.*—P. Gray, 1844. *Wgt.*—G. C. Druce, 1883.

LOCALITIES: *Nithsdale*—Kirkconnel, M. W.; Racks, Hn., S.E.; Portland Place, P. Gr.; Castle-Douglas Road, Maxwelltown, Newton, S.-E.; New Bridge, P. Gr.; Portrack (?) S.-E. *Annandale*— Annan, Hn.

Dry, gravelly soil mixed with cinders, turf on walls, leaf mould; exposed or rarely shaded; windy spots.

VISITORS : Opomyza germinationis, Meromyza sp., and three other kinds (caught rather late in season), S.-E.

Cerastium alpinum. *Linn.*

RECORD: *a.* lanatum, *Dfs.*—Rev. J. Singer, 1843.

LOCALITIES : *Annandale.*—1450 to 2200 feet on Black's Hope and Whitecoombe, J. T. J., S.-E.

Appears June 22 to July 1, J. T. J. On dry bare whinstone rocks, fully exposed.

VISITOR : Chortophila, several kinds, ab., S.-E.

Stellaria nemorum. *Linn.* (Stitchwort).

RECORDS : *Dfs.*—Dr Burgess, 1789. *Kcd.*—J. Cruickshank, 1839. *Wgt.*—J. M'Andrew, 1891.

LOCALITIES : *Nithsdale*—Gillichill, R. H. M.; Castle-Douglas Road, C. E. M.; Lincluden Abbey, Cluden Brig and Mills, P. Gr., Hn., S.-E.; Dalawoodie, Hn.; Woodlands, Irongray, S.-E.; Bennan, Tynron, J. Sh., T. Br., J. Cr.; Enoch Castle, J. Wl.; Brewery, Nithbank, R. A. ; Auldgirth, Drumlanrig, Dr Dv. *Annandale*—Kirtle, Carr., Springkell, Hoddam Castle, Broomholm, Glen Æ Wood, Br.; Milke, Thundergarth, Whitestonehill, G. Bl. ; Marchbankwood to Craiglands, Garpel and Beld Craig, Sd.; Barnhill, Wellburn, J. T. J. *Eskdale*—Glenzier, E. Ty.; Prior's Linn, Penton, Byreburn, Tarras, very abundant from Canobie to Langholm, Bexburn, S.-E. (500 feet).

Appears April 23, G. Bl. May 2 to 29, J. T. T. Damp or wet, rarely dry leaf mould, roadside or other soils mixed with leaf mould ; usually in shade and always wind-sheltered.

VISITORS : Melanostoma sp., Empis 2 or 3 sp., Dolichopids, Chlorops sp., Anthomyidæ, Tipulæ, very ab. (about 20 unnamed species). Meligethes æneus very ab., S.-E.

Stellaria media. *Vill.* (Chickweed).

RECORDS: *Dfs.* and *Kcd.*—P. Gray, 1850. *Wgt.*—G. C. Druce, 1883.

Var. neglecta, *Dfs.*—Dr F. W. Grierson, 1882.

LOCALITIES : Very common in all the valleys reaching Moniaive, J. Cr., and Sanquhar, Dv., to a height of 1400 feet ; Annan and Moffat, J.T.J., and 1500 feet ; Esk, S.-E.

Appears March 23, J. T. J. Damp or wet springheads, more rarely dry places ; waste ground, roadsides, shingles of rivers and shore and other soils (except peat); usually exposed to sun, sometimes in shade; often wind-sheltered.

VISITORS: Chalcid, 3 sp., Proctotypes, Phora, Dolichopids, 2 sp., Chlorops (numerous small flies), Meligethes æneus, ab., S.-E.

Stellaria uliginosa. *Murr.*

RECORDS: *Dfs.* and *Kcd.*—P. Gray, 1850. *Wgt.*—G. C. Druce, 1883.

LOCALITIES: *Nithsdale*—Arbigland, Th.; Maxwelltown Station, Tinwald, Hn.; Tynron, Sh.; Auldgirth Brig, S.-E.; Sanquhar, Dv., common upper part of valley. *Annandale*—Troutbeck of Milke, Gillenbie Mains of Corrie, G. Bl.; common Putts and Annan shingles, Moffat, Beeftub (1000 feet), J. T. J.; Hartfell Burns (2000 feet), Loch Skene, and very ab. springs, etc., J. T. J., S.-E. *Eskdale*—Canobie, Tarras, Langholm, ab. (1400 feet), Whitehope, S.-E.

Appears April 27, G. Bl.; May 4, J. T. J. Wet places, mud of spring heads, ditches; on roadsides, whinstone, boulder clay, etc., usually sunny places; partly wind-sheltered by other plants or ditch sides.

VISITORS: Chlorops, Anthomyids, Tipulæ, Dolichopids (many unnamed), S.-E.

Stellaria graminea. *Linn.*

RECORDS: *Dfs.* and *Kcd.*—P. Gray, 1850. *Wgt.*—G. C. Druce, 1883.

LOCALITIES: *Nithsdale*—Very common Kirkconnel, all roads about Dumfries, Moniaive, J. Cr.; Sanquhar, Dv. *Annandale*—Very common Gretna Green, Annan, S.-E.; Dryfe, Milke, G. Bl.; Beld Craig, Sd.; common Moffat, J. T. J. *Eskdale*—Common Glenzier, Canobie, E. Ty.; common Langholm, S.-E.; Castle O'er, J. Wl., R. Bl. (500 feet).

Appears May 1, G. Bl.; June 7 to 28, J. T. J. Dry or damp places; roadsides and all other soils; sunny, occasionally shaded; always windsheltered, in long grass or by hedges, etc.

VISITORS: Empis livida, Syritta pipens, Platychirius peltatus, Dolichopids (4 unnamed).

Stellaria glauca. *With.*

RECORDS: *Dfs.*—J. Cruickshank, 1839. *Kcd.*—P. Gray, 1844. *Wgt.*—J. M'Andrew, 1891.

LOCALITIES: *Nithsdale*—Maxwelltown Loch, P. Gr.; Tinwald, J. Sn.

(and J. Cru?). *Annandale*—Lochmaben, F. M. W., Dv.; Dumcrieff, Moffat, S. W. Ca.

Wet, half peaty ditches; exposed to wind and sun.

Stellaria holostea. *Linn.*

RECORDS: *Dfs.* and *Kcd.*—P. Gray, 1850. *Wgt.*—G. C. Druce, 1883.

LOCALITIES: *Nithsdale*—Common roads about Dumfries, M. J. H., C. E. M., Th., Hn.; Dunscore, P. Gr.; Moniaive, J. Cr.; Tynron, J. Sh.; Sanquhar, Dv. *Annandale*—Annan, Water of Æ, ab., S.-E.; Lockerbie House to Troutbeck of Milke, G. Bl.; Beld Craig, J. Sd.; common Moffat, Correifron, etc., J. T. J. *Eskdale*—Very ab. Woods lee, E. Ty.; Tarras, S.-E. (to 1500 feet).

Appears April 5 to May 15, J. T. J. Rather dry roadside banks, on old turf of walls, cinders of railways, shingles, gravelly soil, more rarely holms and leaf mould; usually in sun, sometimes shaded; often quite unsheltered by other plants, and rarely in long grass.

VISITORS: Empis pennata, Siphona cristata, Platychirius manicatus, and two or three other flies. Meligethes æneus, Telephorus bicolor, S.-E.

Buda rubra. *Dum.* (Sandspurrey).

Records: 260. Rubra, *Dfs.*—J. Sadler, 1857. *Kcd.*—P. Gray, 1844. *Wgt.*—Arnott, 1848.
261. Marina a., Genuina, *Dfs.*—G. F. Scott-Elliot, 1892. *Kcd.*—Rev. J. Fraser, 1866. *Wgt.*—G. C. Druce, 1883. *c.* Neglecta, *Wgt.*—G. Horn, 1873.
262. Marginatum, *Dfs.*—G. F. Scott-Elliot, 1892. *Kcd.* —Rev. J. Fraser, 1866. *Wgt.*—G. C. Druce, 1883.
263. Rupestre, *Kcd.*—Dr Craig Christie, 1868. *Wgt.*— G. Horn, 1873.

LOCALITIES: Seacoast pretty common. 261 *a.* from Annan shore, Rerrick, Orchardton, and Port Mary. 261 *c.* from Port Logan, G. Horn and J. M'Andrew. 262 from Kirtle, S.-E.; Rerrick, J. Fr.; Portwilliam and Glasserton, Hn. 263 from Port Logan, J. M'A.; Portpatrick, R. R.; Mull of Galloway, G. Ho.; Kirkandrew, C.C. Inland less common. 260 road from Brow to Stank, Bankhead, S.-E.; Glencaple, Hn.; Laught, R.A.; Bruce's Castle, Lochmaben, F. M. W.; Junction, Selkirk, and Craigbeck Road, Broomhill, New Edinburgh Road, J. T. J.; Beattock, S.-E.

Appears May 5, J. T. J. Wet places, rarely fairly dry stones; bare

FLORA OF DUMFRIESSHIRE. 31

estuarine mud of seashore with Armeria, stony shingles of seashore, sandy places or rock crevices (263), hard roadsides (260).

VISITORS : Apis very ab., Nemotelus notatris, Lucilia cæsar, Scatophaga litorea, stercoraria, Nemopoda stercoraria, Chlorops, and three other kinds of flies, S.-E.

Spergula arvensis. *Linn.* (Corn Spurrey)

RECORDS : *Dfs.* and *Kcd.*—P. Gray, 1850. *Wgt.*—F. R. Coles, 1872 (both varieties, *a.* sativa, and *b.* vulgaris).

LOCALITIES : Common in arable land in all valleys to Moniaive (J. Cr.) and Sanquhar (Dv.) ; at least 700 feet Moffat (S.-E.), Castle O'er, Eskdale (J. Wl. and R. Bl.).

Appears May 24 to June 2, J. T. J. Fairly dry or moist places; waste ground in fields, rocks by shore, boulder clay, sandy or even half peaty soil ; almost always exposed to sun and wind.

VISITORS : Syritta pipiens, Platychirius manicatus, scutatus, albimanus, rhingia rostrata, Empis vitripennis, Anthomyia radicum, Scatophaga stercoraria, Chortophila, etc., S.-E.

Montia fontana. *Linn.* (Blinks).

RECORDS : *a.* minor, *Dfs.* and *Kcd.*—P. Gray, 1850. *Wgt.*—G. C. Druce, 1883. *b.* rivularis, *Dfs.*—J. Saddler, 1857. *Kcd.*—F. R. Coles, 1883. *Wgt.*—G. C. Druce, 1883.

LOCALITIES : *Nithsdale*—Cargen, F. W. G.; Glencaple, C. E. M.; Broomrigg, Dr Dv.; Barndennoch, Tynron, J. Sh.; Glencairn, J. Cr.; Sanquhar, Dr Dv. *Annandale*—Jardine Hall, Th.; Dryfe Cemetery, G. Bl.; Andrew's Well, var. *b.*, S. W. Ca. ; Grey Mare's Tail, Sd. ; common Hartfell and Loch Skene (reaching 2300 feet), S.-E. *Eskdale*— common at Langholm, etc., S.E.

Appears April 27, G. Bl.; May 17, J. T. J. Var. *a.*, on wet or often dry mud ; *b.*, in the water or much protected by other plants ; both at origin of springs, shingles of rivers, on peaty or other mud ; always in sun, and usually preferring mud bare of other plants.

VISITORS : Chlorops sp., Anthomyia,? S.-E.

Hypericum androsæmum. *Linn.* (Tutsan).

RECORDS : *Kcd.*—J. M'Andrew, 1882. *Wgt.*—Professor Balfour, 1843.

LOCALITIES : Kirkcolm, J. H. Bl.; Cruggleton, Garliestown, Cuicton Lake, Hn.; Flarick Glen, Ravenshall, J. M'A.

Grassy cliffs or roadside bank facing sea ; sheltered from wind.

Hypericum perforatum. *Linn.* (St. John's Wort).

RECORDS: Var. *a.*, *Dfs.*—Dr Little, 1834. *Kcd.*—Miss Harvey, 1830? *Wgt.*—J. M'Andrew, 1882.
Var. *b.*, angustifolium, *Dfs.*—J. T. Johnstone, 1891.
Var. *c.*, lineolatum, *Dfs.*—J. T. Johnstone, 1891.

LOCALITIES: *Nithsdale*—Kirkconnel, M.W.; Glencaple road, M.J.H.; Troqueer, C. E. M.; Nith, by Dumfries, Hn.; Portrack, S.-E.; Cluden, Th., Hn.; Moniaive, J. Cr.; Sanquhar, Dv. *Annandale*—Johnstone, Dr Lt.; Seafield, Kirtle mouth, S.-E.; Ecclefechan, Exc.; Lochmaben, S.-E.; Boreland, Whitstonehill, and very common Milke, Dryfe, G Bl. Var. *b.*, Wamphray, Barnhill road, var. *c.*, Hydropathic, J. T. J.; Beld Crag, J. Sd. *Eskdale*—Common, E. Ty.; Eskdalemuir Kirk, S.-E.

Appears from July 3 to August 15, J. T. J. Dry or moist banks on railway cuttings, leaf mould, sandy gravel, or sandstone; usually sunny or rarely shaded places, partly wind-sheltered by other plants, hedges, etc.

VISITORS: Apis, Bombus lucorum, pratorum, Derhamellus, S.-E.; ab. muscorum, J. C. W.; Platychirius albimanus, peltatus, J. C. W.; Empis livida, ab.; Rhingia rostrata, Syrphus balteatus (and J. C. W.), topiarius, Eristalis pertinax, Syritta pipiens (and J.C. W.), S -E.; Hydrotea sp., Cheilosia sp., Anthomyia radicum, Chortophila, Chlorops, and eight other forms, J. C. W., see p. 246.

Hypericum quadrangulum. *Linn.*

RECORDS: Var. *a.*, dubium, *Dfs.*—J. T. Johnstone, 1891. *Kcd.*— F. R. Coles, 1882. *Wgt.*—J. H. Balfour, 1845.
Var. *b.*, Maculatum, *Dfs.*—Miss Ethel Taylor, 1892. *Kcd.*—J. M'Andrew, 1882.

LOCALITIES: *Nithsdale*—Troqueer road, Th., C. E. M.; Auldgirth Station, Exc.; Shaw wood, R. A. *Annandale*—From Scroggs upwards on Milke, G. Bl.; Johnstone Parish, J. T. J. *Eskdale*—Between Liddell Brig and Canobie Station, var. *a.*, S.-E.; Canobie, var. *b.*, E. Ty.; Castle O'er, J. Wl., R. Bl. (to 800 feet).

Damp embankments on boulder-clay, roadsides; shaded or open to sun; more exposed to wind than perforatum?

VISITORS: Apis, Bombus lucorum, muscorum, pratorum, Leptis tringaria, Sepsis cynipsea, Anthomyia radicum, Chortophila, etc., S.-E.

Hypericum quadratum. *Stokes.*

RECORDS: *Dfs.*—Dr F. W. Grierson, 1882. *Kcd.*—F. R. Coles, 1882. *Wgt.*—J. M'Andrew, 1885.

LOCALITIES: *Nithsdale*—Glencaple, C. E. M.; Clarencefield, F. W.

G.; Townhead, R. A.; Sanquhar, Dv. *Annandale*—Milke Water, Catch Hall Loaning, G. Bl.; Archbank Bridge, Alton, Craigieburn, Breconside Moor (up to 600 or 700 feet), J. T. J. *Eskdale*—Woodslee Orchard, Liddel Bridge, Burnfoot, S.-E.

Appears July 5, J. T. J. Wet or pretty dry slopes; roadsides, railway banks, shaded or partly shaded; wind-sheltered in long grass, etc.

VISITOR: Andrena albicans, Hn.

Hypericum humifusum. *Linn.*

RECORDS: *Dfs.*—Rev. Dr Little, 1834. *Kcd.*—P. Gray, 1848. *Wgt.*—Stranraer (J. A. Brown?), 1836, Edin. Herb.

LOCALITIES: *Nithsdale*—Kenneth Bank, Glencaple, C. E. M.; Kirkconnel, M. W.; Cargenhillside, Hn.; Dalskairth Hills, P. Gr.; Barndennoch (600 feet), J. Cr.; Auldgirth Station, Exc.; G. & S.-W. railway, Gateside, Euchan Cottage, Dr Dv. *Annandale*—Kirtle mouth, S.-E.; Limestone Rigg, Shillahill, Scroggs, Middlebie, G. Bl., Halleaths, S.-E.; Lochmaben, Hn.; Raehills, Exc.; Beld Craig, Sd.; common Moffat, J. T. J. *Eskdale*--Glenzier Mill, E. Ty.; Langholm, C. Y.; Wauchope 2 miles from Langholm (to 700 feet), S.-E.

Appears from June 23, J. T. J. Dry bare places on roadsides, sandstone, rather peaty loam; usually in sun and partly wind-sheltered.

VISITORS: Anthomyidæ (four visitors), S.-E.

Hypericum pulchrum. *Linn.*

RECORDS: *Dfs.*—Rev. Dr Little, 1834. *Kcd.*—P. Gray, 1850. *Wgt.*—G. C. Druce, 1883.

LOCALITIES: *Nithsdale*—Common Kenneth Bank, C. E. M.; Cargen, F. W. G.; Lochabbey Wood, C. E. M.; Nunholm, Hn.; Cluden, Th.; common Moniaive, J. Cr.; common Sanquhar, Dv. *Annandale*—Annan, Springkeld, S.-E.; Ecclefechan, Raehills, Exc.; Dryfesdale, G. Bl.; Garpel, S.-E.; Beeftub, Sd.; Breconside (reaching 2000 feet), Loch Skene, S.-E. *Eskdale*—Common Kirkandrews, E. Ty.; Langholm Hill, C. Y.; Castle O'er, J. Wl. and R. Bl.; Stennies and Meggat (to 2200 feet), S.-E.

Appears June 30 to July 6, J. T. J. Dry or moist whinstone rocks, roadsides, more rarely on humus or boulder clay; in sun or part or wholly shaded; usually at least partly wind-sheltered.

VISITORS: Apis, Sericomyia borealis, Siphona cristata, Drymeia hamata, Hyetodesia basalis, Dolichopidæ, Chortophila (4 unnamed), S.-E.

Hypericum hirsutum. *Linn.*

RECORDS: *Dfs.*—Dr Davidson, 1882. *Kcd.*—P. Gray, 1844.

LOCALITIES: *Nithsdale*—Castle-Douglas Road and Glen, J. Wl., Th., C. E. M.; Redpaths, Waterside, R. A.; Bankhead, Newark, Ardoch, Sanquhar, Dr Dv. *Annandale*—Scroggs, G. Bl.; Wamphray Glen, Moffat, J. T. J. *Eskdale*—Liddel Bridge, Bilholm, S.-E.

Appears August 12. Moist or dry sloping banks and meadows; carboniferous sandstone, boulder clay, leaf mould, whinstone soils; usually in shade, and always in well sheltered spots.

VISITORS: Bombus pratorum, lucorum, Syrphus balteatus, Empis vitripennis, Scatophaga inquinita, stercoraria, Chortophila, S.-E.

Hypericum elodes. *Linn.*

RECORDS: *Kcd.*—Rev. J. Fraser, 1843. *Wgt.*—Dr Balfour, 1843.

LOCALITIES: Portpatrick, Arn.; Port Logan, J. H. Bl.; Ditch, Newton-Stewart to Glenluce, Mau.; Loch Cree, G. M'N.; New-Galloway, J. M'A.; Minnigaff, Sn.; Auchencairn Bay, J. C. W.; Barscraigh, J. Fr.; Barnhourie, R. R., P. Gr.; Kirkbean, Sn.

Wet peaty soil in ditches; fully exposed.

VISITOR: Hydrotea dentipes, Hn.

Linum perenne. *Linn.*

RECORDS: (Escape) *Dfs.*—Dr Davidson, 1886. *Kcd.*—Rev. J. Fraser, 1843.

LOCALITIES: Kirkcudbright, J. Fr.; Brighouse Bay, F. R. C.; Ross, Rockville, G. C.; Sanquhar (?), Dv.

On hillsides near high water mark.

Linum angustifolium. *Huds.*

RECORDS: (Escape) *Kcd.*—Colvend, Miss C. E. Milligan, 1892.

Linum usitatissimum. *Linn.*

RECORDS: (Escape) *Dfs.*—P. Gray, 1850. *Kcd.* and *Wgt.*—J. M'Andrew, 1892.

LOCALITIES: Near Dumfries, P. Gr.; Milton's Mill, W. Br.; Drummond's Yard, Moffat, 1891 and 1892, J. T. J. (Escape).

Linum catharticum. *Linn.* (Purging Flax).

RECORDS: *Dfs.* and *Kcd.*—P. Gray, 1850. *Wgt.*—G. C. Druce, 1883.

LOCALITIES: *Nithsdale*—Very common near Dumfries and Cluden, Th., C. E. M., F. W. G.; Moniaive, common, J. Cr.; Sanquhar, common, Dr Dv. *Annandale*—Very common Dryfe, Milke, Corrie, G. Bl.; Caledonian Line, common, S.-E.; Beeftub (to 2300 feet), Sd.; Loch Skene, S.-E. *Eskdale*—Common Glenzier, E. Ty.; common Esk, Wauchope, S.-E.; Castle O'er (to 1400 feet,) J. Wl. and R. Bl.; Ewes Water, S.-E.

Appears June 12, 25, J. T. J. Dry or moist places on gravelly or stony soil, sandstone soil, railway cinders, common old moraines. Almost always in full sun; in short turf or bare spots exposed to wind, though in valleys.

VISITORS: Platychirus manicatus, Empis vitripennis, Chlorops, sp., and two doubtful forms, S.-E.

Radiola millegrana. *Sm.* (Allseed).

RECORDS: *Dfs.*—Dr Burgess, 1789. *Kcd.*—J M'Andrew, 1882. *Wgt.*—Arnott, 1848.

LOCALITIES: Port Logan, Portpatrick, Arn.; Loch Ken, C. E. M., J. M'A., Ravenston, New-Galloway, J. M'A., F. R. C.; Auchencairn, J. C. W.; Lochmaben, Little Dormont, on north side of road from Brow to Stank, Ruthwell, Br.

Sandy paths on moor, Br.; damp gravelly roadsides, J. M'A.; shingle of banks, C. E. M.

Lavatera arborea. *Linn.*

RECORD: (Escape) *Wgt.*—J. M'Andrew, 1890.

Malva moschata. *Linn.* (Musk Mallow).

RECORDS: *Dfs.*—J. Shaw, 1882. *Kcd.*—P. Gray, 1844. *Wgt.*—G. Graham, 1836.

LOCALITIES: *Along the shore*—Portwilliam, J. M'A.; Senwick Bay, J. Sh.; Balmac, G. N. Ll.; Colvend, Th.; Tongland, F. R. C.; Mullock Bay, Almorness, J. M'A. *Nithsdale*—Near Dumfries, P. Gr.; Locharbriggs, Hn.; Craigencoon, Tynron, J. Sh.; (Sanquhar, Dv.?); Cowhill, Ad. and S. D. J. *Annandale*—Douglas Bridge, Lockerbie, G. Bl.; Torduff shore, S.-E.; Milke, above Scroggs, G. Bl.; Raehills, Exc.; Gardenholm, Moffat, J. T. J. *Eskdale*—Langholm, C. Y.

Shingles of shore, roadsides, river banks, full exposure or partly wind-sheltered.

VISITORS: Apis, Bombus pratorum, Derhamellus (all Miss Hannay).

Malva rotundifolia. *Linn.*

RECORDS: (Escape) *Dfs.*—Mrs Carthew-Yorstoun (1880?) *Wgt.*—J. M'Andrew, 1892.

LOCALITIES: Cairnryan, J. M'A.; Thornhill, R.A.; roadside, Canobie Manse, Langholm, C. Y.; Gilnockie siding, S.-E.

Malva sylvestris. *Linn.*

RECORDS: *Dfs.*—Miss Adams and Miss Johnstone, 1889. *Ked.*—P. Gray, 1865. *Wgt.*—J. M'Andrew, 1890.

LOCALITIES: *Nithsdale*—Carsethorn, F. W. G., Hn.; Glencaple Road, M. J. H.; Glen, P. Gr.; Whitebridge and Cluden, Hn., S.E., Th.; Glencairn, J. Cr.; Cowhill, Ad. and S. D. J.; Holywood Kirk, S.-E.; Doocot House, R.A. *Annandale* -Browhouses, S.-E.; Powfoot, M. J. H.; Annan wall, Hn.; Jardine Hall, Th.; Lochmaben, M. J. H.; Beattock, J. T. J. *Eskdale*—Cummertrees, Gretna, S.-E.; Henry's Town, E. Ty.

Dry, stony, or gravelly waste soil; in sun or shade; usually partly wind-sheltered.

VISITORS: Bombus derhamellus, Hn., lucorum, S.-E.; Platychirius manicatus, Syritta pipiens, Hn.

Malva borealis. *Wallm.*

RECORD: (Escape) *Dfs.*—Mrs Thomson, 1893.
LOCALITY: Cluden Mill.

Althea officinalis. *Linn.* (Marsh Mallow).

RECORDS: (Escape) *Dfs.*—J. T. Johnstone, 1889. *Ked.*—Dr Burgess, 1789.

LOCALITIES: Moffat, J. T. J. (escape); Arbigland (in cultivation now), Dr Br., M. W.

Tilia Europea. *Linn.* (Lime Tree).

RECORDS: *Dfs., Ked., Wgt.*—J. M'Andrew, 1882, commonly planted at Drumlanrig (250 to 600 feet), J. H. D.; Langholm (500 to 600 feet), R. Do.; Moffat, common, J. T. J.

Dry (J. H. D.) or damp (R. Do.) places on holms or good loam; requires plenty of light and air, but shelter from strong winds.

Geranium sanguineum. Linn. (Bloody Crane's Bill).

RECORDS: *Kcd.*—P. Gray, 1841. *Wgt.*—Edin. Herb., 1843.

LOCALITIES: Very common along the shore from the Mull to Southerness and Carsethorn, Hn., F. W. G.; Glen, abundant, S.-E.

Full sun and wind exposure on grassy cliffs or broken undercliffs. Sheltered or shaded inland.

VISITORS: Apis, Bombus muscorum, Hn.; Halictus albipes, Prosopis hyalinata, Licus ferrugineus, Rhingia rostrata, Drymeia hamata, Hydrotea, Anthomyia radicum, Chortophila, S.-E.

Geranium pheum. Linn. (Dusky Crane's Bill).

RECORDS: *Dfs.*—Castlemilk, J. H. Balfour, 1839. *Kcd.*—J. M'Andrew, 1882. *Wgt.*—Sir H. Maxwell, 1889.

LOCALITIES: Near Moniaive, J. Cr.; Kirkburn, Lockerbie, Roberthill, G. Bl.; Dumcrieff, S. W. Ca.; Kirkandrews, E. Ty.; Bilholm, J. Wl. and R. Bl.

Geranium sylvaticum. Linn. (Wood Crane's Bill).

RECORDS: *Dfs.*—Drumlanrig, W. Stevens, 1848. *Kcd.*—P. Gray, 1850. *Wgt.*—J. M'Andrew, 1891.

LOCALITIES: *Nithsdale*—By Nith and Cluden, P. Gr.; Langlands, Hn.; Cluden, M. J. H.; common Moniaive (300 to 600 ft.), J. Cr.; Tynron, J. Sh.; Penpont, P. Gr.; Thornhill, T. Br.; Carron Glen (a pale pink variety), W. St., Th., C. E. M.; Glenquhargen, Exc.; Sanquhar, Dv. *Annandale*—Jardinehall, Th.; Springkeld, S.-E.; common to Whitestonehill of Milke, Boreland of Dryfe, G. Bl.; very common Dinwoodie to Wamphray, Caledonian Railway, S.-E.; Garpol, Beld Craig, J. Sd.; reaching waterfall at head Black's Hope, and nearly 2000 feet, Loch Skene, J. T. J., S.-E. *Eskdale*—Abundant Woodslee, Canobie, E. Ty.; extraordinarily abundant from Canobie to Langholm, Byreburn, Tarras, Wauchope, Meggat, S.-E.; Castle O'er, J. Wl. and R. Bl.; Mosspaul, Eskdalemuir Kirk, S.-E.

Appears May 25 to June 12, J. T. J. Usually moist leaf-mould, carboniferous sandstone or whinstone soils, preferring a slope ; almost always shaded or part shaded and wind-sheltered.

VISITORS: Apis abundant, Bombus pratorum abundant, muscorum, Halictus cylindricus, Nomanda lateralis, Empis tessellata, pennata, vitripennis, sp.; Platychirius peltatus, manicatus, five Anthomyids, etc., S.-E.

Geranium pratense. *Linn.* (Field Crane's Bill).

RECORDS: *Dfs.* and *Kcd.*—P. Gray, 1850. *Wgt.*—J. M'Andrew, 1890.

LOCALITIES: *Nithsdale*—Carnsalloch, Hn.; Cowhill, Ad. and S. D. J.; Cluden, F. W. G., Th., C. E. M.; Glencairn, J. Cr.; Thornhill, Sanquhar, Dr Dv. *Annandale* —Scroggs, Boreland, Sibbaldbie, Cowburn of Corrie, G. Bl.; Lochanburn, Kd.; common Moffat, J. T. J. *Eskdale* —Canobie, E. Ty.; Liddel Bridge, Burnfoot, S.-E.; Castle O'er, J. Wl. and R. Bl.; Meggat (to 500 feet), S.-E.

Appears from July 8 to 29, J. T. J. Moist or dry holms, or sandy alluvial, leaf mould; usually full sun; sheltered from high winds.

VISITORS: Hive bee (stealing honey from back), Bombus pratorum, muscorum, Andrena albicans, Platychirius manicatus, Anthomyia radicum, and three other kinds, S.-E.

Geranium robertianum. *Linn.* (Robert).

RECORDS: *Dfs.* and *Kcd.*—P. Gray, 1850. *Wgt.*—G. C. Druce, 1883.

LOCALITIES: Very common in all valleys, but not, as a rule, beyond the limit of wooded glens, that is, about 700 or 800 feet.

Appears May 18 to June 11, J. T. J. Moist or dry leaf mould of woods, roadsides, old mossy walls, usually shade or half-shade; windsheltered by woods or banks, prefers ground bare of other plants.

VISITORS: Pieris napi, J. C. W.; Bombus muscorum, pratorum, Empis pennata very abundant, tessellata, Anthomyids, Meligethes æneus, S.-E.

Geranium lucidum. *Linn.*

RECORDS: *Dfs.* and *Kcd.*—P. Gray, 1850 and 1844. *Wgt.*—J. M'Andrew, 1882.

LOCALITIES: *Nithsdale*—Craigs, P. Gr.; Glen, Th., C. E. M.; Penpont, J. Sh.; Blackwood, Exc.; Clauchries, Craighope Linn, Cample, J. Fn., Dv.; Glenquhargen, Exc.; Black Linn, R.A. *Annandale*— Craigieburn, Beeftub (1400 feet), J. T. J.

Appears May 18, J. T. J. Moist atmosphere on whinstone, or porphyrite R. A.; rocks or walls; in full shade and shelter.

VISITORS: (Archbank Garden) Syrphus cinctellus abundant, Platychirius manicatus abundant, Melanostoma mellina, three other Syrphids, S.-E.

Geranium pyrenaicum. *Linn.*

RECORDS : *Dfs.*—(Sown?) J. Shaw, at Tynron, 1882.

Geranium molle. *Linn.* (Dootae).

RECORDS : *Dfs.* and *Kcd.*—P. Gray, 1850. *Wgt.*—G. C. Druce, 1883.

LOCALITIES : Very common in all the valleys, but not often beyond the limit of roads and arable land; Capelgill (600 feet), J. T. J.

Appears May 1, G. Bl. May 5 to June, J. T. J. Dry roadsides, waste ground, field corners; usually in sun; exposed to wind in short turf or sheltered by hedges and banks.

VISITORS : Andrena parvula, Platychirius albimanus, Hydrellia, and other Anthomyids, S.-E.

Geranium dissectum. *Linn.*

RECORDS : *Dfs.* and *Kcd.*—P. Gray, 1850. *Wgt.*—G. C. Druce, 1883.

LOCALITIES : *Nithsdale*—Common roads about Dumfries, Hn., C. E. M., Th.; Lochanhead, F. W. G.; Sanquhar, Dv. *Annandale*—Common, Annan, S.-E.; common, Lockerbie, G. Bl.; common, Moffat, J. T. J. *Eskdale*—Common, Scotch Dyke, Canobie, E. Ty.; Langholm and Tarras, S.-E. (not so frequent as molle, and apparently rarer above 500 feet).

Appears May 21 to June 21, J. T. J. Dry or rather wet roadsides, field margins, gravelly and cindery soils; usually in sun or half shaded, almost always sheltered by grass or herbage.

VISITORS : Platychirius manicatus and two doubtful Anthomyids, S.-E.

Geranium columbinum. *Linn.*

RECORDS : (Escape) *Kcd.*—Mrs Gilchrist-Clark (1867?).
LOCALITY : Ross, G. C.; Southerness, C. E. M.

Erodium cicutiarum. *L'her.*

RECORDS : Var. *a.*, Vulgatum, *Dfs.*—J. Fingland, 1887. *Kcd.*—P. Gray, 1848. *Wgt.*—J. M'Andrew, 1887.
Var. *b.*, Chærophyllum, *Dfs.*—G. F. Scott-Elliot, 1892.

LOCALITIES : *Along the shore*—Monreith, Frighouse, J. M'A.; Sandyhills Bay, P. Gr.; Douglas Hall, J. M'A., Th.; Southerness, M. J. H.;

Carsethorn, Th., C. E. M.; Arbigland, Th.; Powfoot, M. J. H., S.-E.; Annan Waterfoot, J. Fn. *Inland*—Railway between Langholm and Gilnockie, S.-E.

Dry shingles or turf; full sun and wind exposure.

VISITORS: Anthomyia radicum, Hydrellia griseola, Chortophila, S.-E.

Erodium moschatum. *L'her.*

RECORD: *Kcd.*—Field Club Excursion, 1893.

LOCALITY: Southerness?

Erodium maritimum. *L'her.*

RECORD: *Wgt.*—Herbarium Greville, at Edinburgh, 1836.

LOCALITIES: Glenluce, Portwilliam, J. H. Bl., T. B. Bl., J. M'A.; Monreith Bay, J. M'A.; Garliestown, Hn.

Dry, sunny shore, in grass.

Oxalis acetosella. *Linn.* (Wood Anemone).

RECORDS: *Dfs.*—J. Cruickshank, 1839. *Kcd.*—P. Gray, 1850. *Wgt.*—G. C. Druce, 1883.

LOCALITIES: Very common in all valleys reaching Moniaive, J. Cr., and Sanquhar, Dv.; to 2500 feet, Loch Skene; and 1700 feet, Eweslees-knowe, J. Rae.

Appears April 1, G. Bl.; 13 to 28, J. T. J. Wet or rarely dry humus of woods or roadsides; in full shade and wind-sheltered almost always.

VISITORS: Anthomyia radicum, abundant, S.-E.

Oxalis corniculata. *Linn.*

RECORDS: (Escape) *Dfs.*—Established many years, Ivy House, Garden, Moffat, on gravel walks, J. T. Johnstone, 1892. *Wgt.*—Rev. G. Wilson, 1893.

Impatiens noli-me-tangere. (Balsam).

RECORDS: (Escape) *Dfs.*—Canobie, J. Fingland, 1885. *Kcd.*—J. Matthewson, Dalbeattie, 1882.

LOCALITIES: Canobie, J. Fn.; Dalbeattie, J. Mt.; Almorness, Th.

Impatiens parviflora. *D. C.*

RECORD: (Escape) *Kcd.*—Fully established Auchencairn Bay, G. F. Scott-Elliot, 1891.

Acer campestre. *Linn.* (Maple).

RECORDS: *Dfs.* and *Kcd.*—J. M'Andrew, 1882. *Wgt.*—G. C. Druce, 1883.

LOCALITIES: (Planted 250 to 340 feet) Drumlanrig, J. H. D.; Elliock, Dv.; bowling green, Moffat, etc., J. T. J.; Scroggs, G. Bl.; Langholm woods, not common, R. Do.

In old woods, on free deep loam; sheltered rather than exposed, J. H. Dx.

Acer platanoides. *Linn.*

RECORD: (Escape) *Kcd.*—F. R. Coles, 1883 (in Herb., Dumfries).

Acer pseudo-platanus. *Linn.* (Sycamore).

RECORDS: *Dfs.*—J. T. Johnstone, 1889. *Kcd.*—J. M'Andrew, 1882. *Wgt.*—G. C. Druce, 1883.

LOCALITIES: (Planted 200 to 800 feet) Drumlanrig, J. H. Dx.; Elliock, Dv.; common, Lockerbie, G. Bl.; common, Moffat, J. T. J.; common, Canobie, E. Ty.; common, Langholm, R. Do.

Prefers dry, strong loam free from stagnant water, but in all soils; prefers sheltered though growing in exposed places, J. H. Dx., R. Do.

Ilex aquifolium. *Linn.* (Holly).

RECORDS: *Dfs.* and *Kcd.*—P. Gray, 1850. *Wgt.*—J. M'Andrew, 1886.

LOCALITIES: (Planted frequently from 150 to 800 feet) Drumlanrig, J. H. Dx.; Langholm, common, R. Do., etc.

Prefers dry light loam and shelter, J. H. Dx.

Euonymus Europæus. *Linn.* (Spindle Tree).

RECORDS: (Planted) *Kcd.*—J. M'Andrew, 1882. *Dfs.*—J. H. Dixon, 1892.

LOCALITIES: Senwick, J. M'A.; Kirkconnel, M. W.; Broomfields, Glencairn, J. Cr.; Drumlanrig, J. H., Dx.; Gilnockie, E. Ty.

Rhamnus catharticus. *Linn.* (Hag or Hackberry).

RECORDS: (Planted) *Dfs.*—Rev. J. Singer, 1843. *Kcd.*—J. M'Andrew, 1882.

LOCALITIES: Dalbeattie, J. Mt., J. M'A.; Keir, Tinwald, J. Sn.; Langholm, R. Do.

Grows freely in moist ground, R. Do.

Rhamnus frangula. *Linn.* (Buckthorn).

RECORDS: (Planted) *Dfs.*—Tinwald, Keir, Dr Burgess, 1789; common, Langholm, R. Doughty. *Kcd.*—Dec at Slogarie, J. M'Andrew, 1882.

Ulex Europæus. *Linn.* (Whin, Gorse, Furze).

RECORDS: *Dfs.* and *Kcd.*—P. Gray, 1850. *Wgt.*—G. C. Drnce, 1883.

LOCALITIES: Very common in all the valleys to 1000 feet. Flowers more or less all the year, but chief bloom from March 17 to April 10. Dry broken banks on alluvium, sandy soil, whinstone and trap rocks, etc.; fully exposed.

VISITORS: Apis, Hn.; Bombus lucorum, May 3, S.-E.; Platychirius clypeatus, Ascia podagrica, Rhyphus fenestralis, S.-E.

Ulex nanus. *Forst.*

RECORDS: 337. Gallii, *Dfs.*—Rev. T. Bell, 1882. *Kcd.*—F. R. Coles, 1884. *Wgt.*—R. M. Stark, 1885.
338. Nanus, *Dfs.*—J. F. Fingland (?). *Kcd.*—G. N. Lloyd, 1831.

LOCALITIES: *Nithsdale*—Newabbey, Dumfries, J. M'A. *Annandale* Eskrig Moor, G. Bl.; Dornock, T. Bl.; Commonside, Crook's Pool, Moffat, J. T. J.

Appears September 1 to October 29, J. T. J. On dry, sandy holms, shingles, etc.; fully exposed (337 apparently by sea).

Genista tinctoria. *Linn.* (Greenweed, Dyer's Weed).

RECORDS: *Dfs.*—Mr Patrick, 1789. *Kcd.*—G. Graham, 1836. *Wgt.* —Dr Balfour, 1836.

LOCALITIES: *Nithsdale*—Loch Kindar, G. C.; Dalskairth, P. Gr.; Glen, C. E. M., Hn.; Grove Hills, Th. and S.-E.; Scaur, Hn.; Trigony, R. A. *Annandale*—Blacket House, Middlebie, Pat.; Scroggs roadside, Tundergarth, G. Bl.; Beld Craig, Kd.; Evan Water, W. Bn.; Peter's Moss, Old Carlisle Road, Langside, Craigsland Burn, Craigbeck, Beattock Hill (up to 400 feet), J. T. J. *Eskdale*—Glenzier, Kirkandrews, E. Ty.; railway Langholm to Canobie, S.-E.

Appears July 4 to 8, J. T. J. On dry sloping banks in short grass over whinstone, boulder clay, granite, roadsides, etc.; fully exposed.

VISITORS: Bombus muscorum, lucorum; June 5, S.-E.

Genista anglica. *Linn.* (Petty Whin).

RECORDS : *Dfs.*—J. Sadler, 1857. *Kcd.*—P. Gray, 1846. *Wgt.*—C. C. Bailey (?) 1883.

LOCALITIES : *Nithsdale*—Old Quay, J. M'A.; Terregles, P. Gr.; Trigony, R. A.; Wanlockhead, Dr Dv. *Annandale*—Dornock, T. Bl.; Cemetery Wood, Eskrig Wood, Lamb Fair Hill, G. Bl.; Wamphray, Pag.; Garpol, J. Sd.; Beattock Hill, J. Sd., J. T. J. *Eskdale*—Langholm, C. Y.

Appears April 27, G. Bl. In shade on pretty dry peaty alluvial.

Cytisus scoparius. *Link.*

RECORD : *Dfs.* and *Kcd.*—P. Gray, 1850. *Wgt.*—G. C. Druce, 1883 ; var. *b.*, prostrata, no record.

LOCALITIES : Common in all the valleys to 900 feet.

Appears April 25, G. Bl.; May 4, 14, J. T. J. Prefers pretty dry slopes on sandy or gravelly soil, cinders, whinstone, etc.; fully exposed.

VISITORS : Bombus muscorum, lucorum ; May 5, S.-E.*

Ononis arvensis. *Linn.* (Restharrow).

RECORDS : 340. Repens, *Dfs.*—Rev. W. Little, 1834. *Kcd.*—J. M'Andrew, 1882. *Wgt.*—G. C. Druce, 1883.
 a. Inermis, *Dfs.*—Dr Davidson, 1891.
 341. Spinosa, *Dfs.*—Dr F. W. Grierson, 1882. *Kcd.*—J. M'Andrew, 1882. *Wgt.*—J. M'Andrew, 1893.

LOCALITIES: *Nithsdale*—Kingholm, Caerlaverock, C. E. M., F. W. G., S.-E., Hn.; Carnsalloch, Hn.; Lincluden, Hn.; Cowhill, Ad. and S. D. J.; Auldgirth Bridge, F. W. G., S.-E.; Ardoch, Sanquhar (340 *a.*), Dr Dv.; *Annandale*—Very common by shore from Annan to Nethertown and Old Gretna, F. W. G., S.-E.; Johnstone, W. St.; Adam's Holm, Three Water Foot, J. T. J. *Eskdale*—Liddel below railway, Canobie Bridge, S.-E.

Appears July 28, J. T. J. On dry sandy or gravelly soils, alluvium of a sandy character, shingles of rivers and shore ; fully exposed to wind and sun.

VISITORS : Apis, J. C. W.; Bombus lucorum, ab. and suff., (Seaford), June 19, S.-E.

---

* **Cytisus laburnum.** VISITORS: Apis, Hn.; Bombus lucorus, Syrphus Ribesii, S. bifasciatus, Hydrotea dentipes ; May 13, S.-E.

Ononis reclinata. *Linn.*

RECORD : *Wgt.*—Dr Graham, 1836.

LOCALITY : Mull and Galloway, G. Gr., Arn. Supposed to be extinct, J. M'A.

Medicago lupulina. *Linn.* (Nonsuch).

RECORDS : *Dfs.*—J. T. Johnstone and T. M'Andrew, 1891. *Kcd.*—J. M'Andrew, 1882. *Wgt.*—J. M'Andrew, 1887.

LOCALITIES : Common along the Shore from Creetown to Orchardton, J. M'A.; Southerness to Kirkbean, Exc. *Nithsdale*—Racks, S.-E.; Friars' Carse, Auldgirth, R. A. *Annandale*—Very abundant railway, Solway Bridge, S.-E.; Lockerbie, S.-E.; Beattock Station, J. M'A.

Appears July 11, J. T. J. On shingle and sand by sea, dry railway banks on cinders inland; usually exposed to wind and sun, or part wind-sheltered.

VISITORS : Polyommatus icarus (suck.), very abundant ; Cænonympha pamphilus (suck.), very abundant ; Platychirius clypeatus, Hyetodesia jucana, S.-E.

Medicago denticulata. *Willd.*

RECORD : (Garden escape) *Dfs.*—J. M'Andrew, 1882.

LOCALITIES : *Nithsdale*—Dumfries, J. M'A. *Annandale*—Moffat, in garden, J. T. J. *Eskdale*—Gilnockie, S-E.

Medicago maculata. *Willd.*

RECORDS : *Dfs.*—G. F. Scott-Elliot, 1892. *Wgt.* (Garliestown)—J. M'Andrew, 1889.

LOCALITY : *Eskdale*—On cinders of railway between Glentarras and Gilnockie, S.-E.

Medicago sativa. *Linn.*

RECORD : *Wgt.*—Edinburgh Herbarium (confirmed), Miss Hannay, 1893.

Melilotus arvensis. *Willd.*

RECORDS : (Escape) *Dfs.*—G. F. Scott-Elliot, 1892. *Kcd.*—Miss Hannay, 1893.

LOCALITY : Cinders near Glentarras, S.-E.; casual plant in garden, Moffat, J. T. J.; Dumfries, Hn.

Appears August 27.

Trigonella ornithopodioides. *D. C.*

REPORTED: *Dfs.*—P. Gray, 1876.
LOCALITY: Downs below Caerlaverock Castle?

Trifolium arvense. *Linn.* (Hare's Foot Clover).

RECORDS: *Dfs.*—P. Gray, 1850. *Kcd.*—1841-4. *Wgt.* -J. M'Andrew, 1886.
LOCALITIES: *Nithsdale*—Holywood Station, S.-E. *Annandale*—Barnhill sandpit, J. M'A.
Appears August 5. On dry, sandy, gravelly mounds fully exposed.
VISITORS: Bombus lucorum, abundant and suff., July 7; Platychirius clypeatus, Hydrotea dentipes, June 30. R. Sc., S.-E.

Trifolium pratense. *Linn.* (Red Clover).

RECORDS: *Dfs.*—Dr Davidson, 1886. *Kcd.*—J. M'Andrew, 1882. *Wgt.*—G. C. Druce, 1883.
LOCALITY: Very common to level of arable land (90 feet).
Appears May 26 to June 16, J. T. J. Grows on all soils, but best in somewhat sheltered strong soil amongst long grass, etc.
VISITORS: Bombus hortorum, lucorum, S.-E.; muscorum, Hn., S.-E.; lapidarius, Hn.

Trifolium medium. *Linn.* (Zigzag Clover).

RECORDS: *Dfs.*—Dr Burgess, 1789. *Kcd.*—P. Gray, 1850. *Wgt.*—G. C. Druce, 1883.
LOCALITIES: *Nithsdale*—Common to Holywood, S.-E.; Sanquhar, Dr Dv. *Annandale*—Kirtle Bridge, Annan, S.E.; Glenkill Burn, Dr Br.; Wamphray-Dinwiddie, S.-E.; Moffat, common, J. T. J. *Eskdale*—Canobie-Langholm, Lynholm, S.-E.
On dry, cindery soil of railways, hay fields, etc., exposed.
VISITORS: Apis, S.-E.; Bombus lucorum, pratorum, July 7, S.-E.

Trifolium hybridum. *Linn.* (Alsike Clover).

RECORD: *Dfs.*—Dr Davidson, 1890. *Kcd.* and *Wgt.*—J. M'Andrew, 1893.
LOCALITIES: An escape as at Cowhill, Ad. and S. D. J.; Wamphray, Gilnockie, S.-E.; Sanquhar, Dr Dv.
Roadsides in neighbourhood of Moffat; abundant, J. T. J.

Trifolium repens. *Linn.*

RECORDS: *Dfs.*—Dr Davidson, 1886. *Kcd.*—J. M'Andrew, 1882. *Wgt.*—G. C. Druce, 1883.

LOCALITIES: Common in all the valleys to at least 1400 feet.

Appears May 29 to June 18, J. T. J. On dry roadsides, whinstone soils, shingles, etc.; in bare spots or short grass (cf. Pratense), exposed to sun.

VISITORS: Apis ab. and suff., S.-E.; Eristalis pertinax, Hn.

Trifolium ochroleucum.

RECORD: (Escape) *Dfs.*—G. F. Scott-Elliot, 1892.

LOCALITY: *Eskdale*—Burnfoot, S.-E.

Trifolium procumbens. *Linn.* (Hop Clover).

RECORDS: *Dfs.*—P. Gray, 1850. *Kcd.*—J. M'Andrew, 1882. *Wgt.*—G. C. Druce, 1883.

LOCALITIES: *Nithsdale*—Along shore, M. J. H.; Maxwelltown Station, C. E. M.; Holywood, S.-E.; Sanquhar, Dr Dv. *Annandale*—Kirtlebridge, abundant, S.-E.; Lockerbie, G. Bl.; Moffat, J. T. J. *Eskdale*—Kirkandrews, E. Ty.; Castle O'er, J. Wn. and R. Bl.

Appears July 5, J. T. J. On dry sandy soils, cinders of railways, limestones, roadsides in sun, but partly wind-sheltered.

VISITORS: Platychirius clypeatus, Melanostoma mellina, June 7, S.-E.; Syrphus corollæ, July 7, S.-E.

Trifolium minus. *Sm.* (Small Clover).

RECORDS: *Dfs.*—Dr Davidson, 1886. *Kcd.*—J. M'Andrew, 1882. *Wgt.*—G. C. Druce, 1883.

LOCALITIES: Very common in all the valleys to about 1000 feet.

Appears May 17 to 31, J. T. J. On dry, sandy soils, roadsides, holms, shingles of sea and rivers; in sun and wind or sheltered by long grass, etc.

Trifolium striatum.

RECORD: *Wgt.*—J. M'Andrew, 1890.

Lotus corniculatus. *Linn.*

RECORDS :
SUBSPECIES—379. *a.*, *Dfs.* and *Kcd.*—P. Gray, 1850. *Wgt.*—C. C. Bailey, 1883.
  b., Crassifolius, *Kcd.*—Mrs Gilchrist-Clark, 1867. *Wgt.* C. C. Bailey, 1883.
381. Uliginosus, *Dfs.*—W. Stevens, 1848. *Kcd.*—P. Gray, 1850. *Wgt.*—G. C. Druce, 1883.

LOCALITIES : *Nithsdale*—379, very common from seashore to Sanquhar, etc. 381, very common Whinnyhill, Exc.; Cargen Glen, F. W. G.; Dalskairth, P. Gr.; Cumnock--Sanquhar, W. St., Dr Dv. *Annandale*—Very common Seafield S.-E.; Moffat (to 2300 feet), S.-E.: Loch Skene, S.-E., J. T. J. 380, common Lockerbie, S.-E. *Eskdale*—Canobie, S.-E.; Langholm. 380, Glentarras, S.-E.

Appears—379, April 25, G. Bl.; May 14-27, J. T. J. 380, middle of June, S.-E. 379 *b.*, shingles and short turf by the sea. 379 *a.*, on dry, stony soils, railways, whinstone rocks and soils, shingles of sea and rivers, usually in sun and exposed to wind in short turf or somewhat sheltered by hedges, banks, etc. 380, on moist or wet roadsides, holmlands, gravelly soil, etc., often part shaded and usually wind-sheltered by long grass and position.

VISITORS: Bombus muscorum, S.-E., J. C. W.; lucorum, Apis, S.-E., 1892 ; also on 380, Bombus lucorum, Andrena bicolor, Syrphus corollae, Telephorus fulvus, Hn.

Anthyllis vulneraria. *Linn.*

(Kidney Vetch ; Ladies' Fingers)

RECORDS : *Dfs.*—Miss F. A. Hope, 1881. *Kcd.*—P. Gray, 1850. *Wgt.*—G. C. Druce, 1883.

LOCALITIES : *Nithsdale*—Racks Station, Grove Hills, Cluden Mills, C. E. M., S.-E., Hn.; Cowhill, Ad. and S. D. J.; Glen, Hn.; Isle, Auldgirth bridge, S. E.; Thornhill, R. A.; shooting range, Sanquhar, Dr Dv. *Annandale*—Shillahill bridge, G. Bl.; Shieldhill, S.-E.; Caledonian Railway from Lockerbie, Beattock summit very abundant, G. Bl., S.-E.; Annan Wamphray, J. Wg., and very common to Moffat, F. A. H., J. T. J., S.-E. *Eskdale*—Torduff Point, S.-E. ; Langholm, C. Y. ; second milestone, Wauchope, Burnfoot, S.-E.; Castle O'er, J. Wn. and R. Bl.

Appears May 26 to June 12, J. T. J. On dry, sandy, or stony ground, holmlands, river shingles, trap rocks, exposed to wind and sun.

VISITOR : Bombus muscorum, Hn.

Astragalus hypoglottis. *Linn.*

RECORD: *Kcd.*—J. M'Andrew, 1882. *Wgt.*—J. M'Andrew, 1882.

LOCALITIES: *Along the shore*—Mullpoint, Burrowhead, J. M'A.; Brighouse Bay, F. R. C.; Ross, G. C.

On dry hillsides.

Astragalus glycyphyllos. *Linn.* (Milk vetch).

RECORDS: *Dfs.* (?), *Kcd.*—P. Gray, 1848. *Wgt.*—J. M'Andrew, 1887.

LOCALITIES: *Along the shore*—Portwilliam, G. C., J. M'A.; Blackneuk, Millstone Quarry, Colvend, P. Gr., J. Fr., C. E. M.; Orroland, Barcheskie, G. M'C.; Port o' Warren, S.-E.; Southwick, St.

On shingle or granite rock by shore.

Oxytropis uralensis. *D. C.*

RECORD: *Wgt.*—Arnott, 1848.

LOCALITY: Mull of Galloway, Arn.; West Tarbet, J. M'A.

Ornithopus perpusillus. *Linn.* (Bird's Foot).

RECORDS: *Dfs.*—P. Gray, 1850. *Kcd.*—1841-4. *Wgt.*—Dr Balfour, (1836?).

LOCALITIES: *Nithsdale*—Southerness, Kenneth Bank, C. E. M.; Racks Station, S.-E.; Locharbriggs, F. W. G.; Dalawoodie, J. Fn., S.-E.; Holywood, Ad. and S. D. J.: Cample Bridge, Kirkland, Roschill, R. A. *Annandale*—Margin loch, Halleaths, G. Bl.; Beattock-Moffat road, J. T. J.; Lochhouse Tower, J. T. J. *Eskdale?*

Appears June 19 to July 1. On dry sandy or gravelly soil bare of other plants, half-shaded by beech trees or in sun exposed to wind.

VISITOR: Platychirius albimanus, July 7, S.-E.

Coronilla varia. *Linn.*

RECORD: (Escape) *Dfs.*—Mrs Thompson, 1893.

LOCALITY: On damp soil in shade, Fourmerkland, Jardine Hall, Th.

Vicia hirsuta. *Koch.* (Tares).

RECORDS: *Dfs.* and *Kcd.*—P. Gray, 1850. *Wgt.*—G. C. Druce, 1883.

LOCALITIES: *Nithsdale*—Along the Nith, S.-E.; Cargen, F. W. G.,

FLORA OF DUMFRIESSHIRE. 49

Hn.; Cluden Mills, Hn.; Maxwelltown Station, C. E. M.; Holywood Station, Hn., S.-E.; Sanquhar Castle, Mennoch, Dr Dv. *Annandale*—Seafield, S.-E.; Kirtle mouth, on shore, S.-E.; along Caledonian line and Moffat (common), J. T. J. and S.-E.

Appears May 17 to June 21, J. T. J. On dry sand, gravel, shingles of rivers and shore, cinders of railways, boulder clay (usually bare of other plants); exposed to sun and wind.

VISITORS: Scatophaga stercoraria, Hn.; Dolichopodidæ, abundant, S.-E.

Vicia tetrasperma. *Mœnch.**

RECORD: 394. Gemella (seeds, 4), *Kcd.*—G. Macnab, 1837.
395. Gracilis (seeds, 5-6), *Kcd.*—(see J. M'Andrew, page 18), 1841-4.

LOCALITY: Torrs Point, Kirkcudbright. 394, G. M'N., J. M'A.; Southerness, 395 (authority?)

Vicia cracca. *Linn.*

RECORD: *Dfs.*—P. Gray, 1850. *Kcd.*—J. M'Andrew, 1882. *Wgt* —G. C. Druce, 1883.

LCOALITIES: *Nithsdale*—Very common on hedges about Dumfries, Holywood Station, F. W. G., S.-E.; Sanquhar, Dr Dv. *Annandale*—Very common on whole Caledonian line and Moffat, J. T. J., S.-E. *Eskdale*—Along Esk and Liddel, Langholm, S.-E.; Castle O'er (to 900 feet), J. Wn. and R. Bl.

Appears June 21 to July 3. On dry, sandy gravel, cinders, whinstone rocks, over hedges (requires bare ground or space free of other plants); in sun and usually windy places.

VISITORS: Bombus muscorum, Hn., S.-E.; lucorum, Apis, S.-E., Hn.; Platychirius clypeatus, Hn.; Empis livida, Hn.

Vicia sylvatica. *Linn.*

RECORDS: *Dfs.*—J. Saddler, 1857. *Kcd.*—J. Cruickshank, 1839. *Wgt.*—G. Macnab, 1836.

LOCALITIES: *Nithsdale*—Cargen Glen, F. W. G.; Crawick Woods, Knockenhair, Kello, Dr Dv. *Annandale*—Above and below Scroggs, Dryfe Bridge, G. Bl.; Grey Mare's Tail, J. Sd., J. T. J.; Saddleyoke (1700 feet), J. T. J. *Eskdale*—Byreburn, and between Canobie Bridge and Langholm, S.-E.; Becksburn, S.-E.; seashore, Portwilliam, G. C. Dr., Hn., S.-E.; Drumore, J. M'A.

* P. Gray—"within three miles of Dumfries, both sides" is probably a mistake.

Appears June 22 to August 6, J. T. J. On pretty moist humus mixed with sandstone or whinstone, either shaded and sheltered inland, or fully exposed on seashore.*

VISITORS: Bombus muscorum, Hn.; hortorum, S.-E.

Vicia orobus. D. C.

RECORDS: *Dfs.*—Mr Winch, 1789. *Kcd.*—J. M'Andrew, 1882. *Wgt.*—C. C. Bailey, 1873.

LOCALITIES: *Nithsdale*—Glenquhargen, Hn.; Euchan and Kellowoods, Sanquhar, Wi., Dr Dv., R. A. *Annandale*—Kirtle woods, Carr.; Beeftub, Corehead (800 feet), J. T. J.; Grey Mare's Tail, J. Sd.; along river Ken and river Dee, J. M'A.

Appears June 21, J. T J. On alluvial near rivers; half shaded and exposed to wind.

VISITORS: Probably B. Muscorum, but not actually seen sucking, S.-E.

Vicia sepium. Linn.

RECORDS: *Dfs.* and *Kcd.*—P. Gray, 1850. *Wgt.*—G. C. Druce, 1883.

LOCALITIES: *Nithsdale*—Very common near Dumfries, P. Gr.; common Sanquhar, Dr Dv. *Annandale*—Common Annan, S.-E.; common Moffat, J. T. J. *Eskdale*—Very common Liddel, Esk, Wauchope, S.-E.; Castle O'er, J. Wn. and R. Bl.

Appears April 19, G. Bl.; May 17 to 28, J. T. J. Usually moist roadsides, humus, turf of old walls, holmlands, cinders, boulder clay; usually half-shaded or exposed or full shade; slightly wind-sheltered by long grass or position.

VISITORS: Bombus muscorum, Hn., S.-E.; Lucorum, S.-E.; Hortorum, Hn.

Vicia sativa. Linn. (Vetch).

RECORDS: 403. Sativa, *Dfs.* and *Kcd.*—P. Gray, 1850. *Wgt.*—G. C. Druce, 1883.
    404. Angustifolia, *Dfs.*—Dr Davidson, 1886. *Kcd.*—J. M'Andrew, 1882. *Wgt.*—G. C. Druce, 1883.
    a. Segetalis, *Dfs.*—Miss Hannay, 1893.
    b. Bobartii, *Dfs.*—Dr Davidson, 1890. *Kcd.*—F. R. Coles, 1883.

---

* A rather stunted form found on shingle at Portwilliam, Wigtownshire, is var. condensata, G. C. Druce.

LOCALITIES : Very common in all the valleys to at least 900 feet.
Appears April 27, G. Bl.; May 17 to June 21, J. T. J. On pretty dry roadsides, waste ground, sandy soils, cinders, etc.; exposed to sun, but usually in part wind-sheltered.
VISITORS : Bombus muscorum, hortorum, S.-E.

Vicia lutea. *Linn.*

RECORDS: (Escape) *Dfs.*—Miss Hannay, 1893. *Kcd.*—J. M'Andrew, 1890. *Wgt.*—J. M'Andrew, 1890.
LOCALITIES : Cluden Mills, Hn.; New England Bay, North of Drummore, Wigtown, J. M'A.; Rerrick Shore, J. M'A.

Vicia lathyroides. *Linn.*

RECORDS : (Escape) *Kcd.*—Rev. J. Fraser, 1882. *Wgt.*—(South of Drummore), J. M'Andrew, 1891.

Vicia bithynica. *Linn.*

RECORD : (Escape) *Dfs.*—Miss Hannay, 1893.
LOCALITY : Cluden Mills, Hn.

Lathyrus aphaca. *Linn.* (Yellow Vetchling).

RECORD : (Escape) *Dfs.*—Miss Milligan, 1892.
LOCALITY : On exposed shingles below Cluden Mills, Hn., C. E. M.

Lathyrus pratensis. *Linn.*

RECORDS: *Dfs.* and *Kcd.*—P. Gray, 1850. *Wgt.*—G. C. Druce, 1883.
LOCALITIES : *Nithsdale*—Very common near Dumfries, S.-E., C. E. M.; very common Sanquhar, Dr Dv. *Annandale*—Annan, very common, S.-E.; Lockerbie common, S.-E.; Moffat common, J. T. J. *Eskdale*—Very common Woodslee and Canobie, E. Ty., S.-E ; Langholm, S.-E.; Castle O'er (to 600 feet), J. Wn. and R. Bl.
Appears June 22, J. T. J. On dry or moist roadside banks, cinders of railways, gravel, sandy holms, whinstone rocks, boulder clay, in sun ; usually sheltered by long grass, hedges, etc.
VISITORS : Apis, Bombus lucorum, muscorum (S.-E., 1892), lapidarius, June 8, 1893, S.-E.; Allantus nothi, Hn.

Lathyrus sylvestris. *Linn.* (Everlasting Pea).

RECORDS : (Escape) *Dfs.*—Miss Hannay, 1893. *Kcd.*—Flora Scotica, 1789. *Wgt.*—J. M'Andrew, 1893.

LOCALITIES: *Along the shore*—Creetown, Gatehouse, J. M'A., Hn.; Barcheskie, Burnfoot, Orroland, J. H., Bl., J. M'A. *Nithsdale*—Shingle below Cluden Mills, Hn.

Lathyrus macrorrhizus. *Wimm.*

RECORDS : *Dfs.*—P. Gray, 1850. *Kcd.*—J. M'Andrew, 1882. *Wgt.* —G. C. Druce, 1883.

LOCALITIES : *Nithsdale*—Common Newton, Nith above Dumfries, S.-E.; Glen, Hn.; Routen Brig, Hn., S.-E.; Sanquhar, Dr Dv. *Annandale*—Milke, S.-E.; Beld Craig, Wellburn, Black's Hope, Whitecoombe (to 2300 feet), Midlaw Burn, J. T. J., S.-E. *Eskdale*—Canobie, Langholm, Blackknowe Burn, Stennies, S.-E.; Castle O'er, J. Wn. and R. Bl.

Appears April 15, G. Bl.; May 4 to 18, J. T. J. On dry, moist or wet whinstone, humus, roadsides, sandy soils, usually in shade and shelter, at low altitudes and partly sheltered in valleys at high altitudes or fully exposed.

VISITORS: Bombus lucorum, S.-E.; muscorum, Hn., S.-E.; May 13.

Lathyrus palustris. *Linn.*

"Galloway." Hooker in *British Flora*.

Lathyrus maritimus. *Bigel.*

RECORD: *Kcd.*—J. Wilson?

Prunus communis. *Huds.* (Sloe).

RECORDS: 419. Spinosa, *Dfs.*—P. Gray, 1850. *Kcd.*—P. Gray, 1848. *Wgt.*—G. C. Druce, 1883.
420. Insititia, *Dfs.*—Dr Burgess, 1789. *Kcd.*—J. M'Andrew, 1882. *Wgt.*—Dr Balfour, 1836.
421. Domestica (Escape), *Dfs.*—J. T. Johnstone, 1890. *Kcd.*—J. M'Andrew, 1882.

LOCALITIES: 419, common in all the valleys. *Nithsdale*—Tinwald, 420, W. Sn.; Fourmerkland, Steilston, Holywood, Dr Br.; Jarbruck; Tynron, 420, J. Sh.; Sandrum Cample, 420, R. A. *Annandale*— Lockerbie, 420, G. Bl.; Gardenholm, Moffat, 421, J. T. J.

Appears April 6, J. Sh.; to May 12, J. T. J. On dryish or moist (W. Do.) ground, boulder clay, whinstone fragments, etc.; fully exposed, A. M., S.-E. (often planted); 420 and 421 seem to be always planted or escapes in woods, by roads, etc.; usually in moist soil and half-sheltered.

Prunus cerasus. *Linn.* (Cherry Gean).

RECORDS · (Escape) 422, Avium, *Dfs.*—Dr Davidson, 1890. *Kcd.*—
P. Gray, 1846. *Wgt.*—G. C. Druce, 1883.

LOCALITIES : Common in woods in lower parts of all the valleys.

Appears April 18. On dry sandstone soil, sandy gravel, humus, etc.; usually in sun and part wind-sheltered, W. Do., S.-E., A. M.

VISITORS : Apis, Apathus vestalis, S.-E.

Prunus padus. *Linn.* (Bird Cherry).

RECORDS : *Dfs.*—Dr Burgess, 1789. *Kcd.*—P. Gray, 1850. *Wgt.*
—Rev. G. Wilson, 1893.

LOCALITIES : *Nithsdale*—Cargen Bridge, Th., R. H. M.; Scaur, Hn.; Manse-wood, Nith Bank, R. A.; Sanquhar, Dr Dv. *Annandale*—To Boreland, Dryfe, to Whitstone Hill, Corrie, G. Bl. ; Lochwood, Middlegill, W. Bn.; Moffat, J. T. J.; Corehead, S.-E. *Eskdale*—Penton Linn, Broomholm, Dr Br., S.-E.; Langholm, S.-E.; Castle O'er, J. Wn. and R. Bl.

Appears April 23, G. Bl.; May 4, J. T. J. Apparently always planted ; prefers wet or dry sunny loam and sheltered spots, W. Do., A. M.

Spirea salicifolia. *Linn.*

RECORDS : (Escape) *Dfs.*—Mr Keddie, 1854. *Kcd.*—J. M'Andrew, 1882. *Wgt.*—Miss Hannay, 1893.

LOCALITIES : *Nithsdale*—Old Cottage Glen, C. E. M., Th.; Jarbruck, J. Cr.; Elliock, Dr Dv.; Rashbriggs, R. A. *Annandale*—Kirkpatrick-Juxta, Kd.; Johnstone, roadside near Skemrigg, J. Wi., J. T. J.; Moffat, J. T. J. *Eskdalemuir*—Near Church, S.-E.

Appears June 24 to August 20, J. T. J. An escape; usually in hedges, etc.

Spirea ulmaria. *Linn.* (Queen of the Meadow).

RECORDS : *Dfs.* and *Kcd.*—P. Gray, 1850. *Wgt.*—G. C. Druce, 1883.

LOCALITIES : *Nithsdale*—Glen, Th.; Ruthwell, Dr Gl.; Sanquhar, Dr Dv. *Annandale*—Very common, reaching 1000 feet at the Beeftub and 1700 feet Corrcifron, S.-E. *Eskdale*—Very common to Castle O'er, S.-E., J.Wn., and R. Bl.

Appears July 9. On wet or moist holms, boulder clay, whinstone soils, etc.; in sun or half-shaded, in part wind-sheltered.

VISITORS: Apis, abundant and sufficient; Bombus lucorum, abundant and sufficient, S.-E., E. Ty., Hn.; Allantus nothi, E. Ty.; Eristalis æneus, Tenax, Horticola, J. C. W.; Notiphila cinerea, Morellia hortorum, Platychirius sp., Dolichopus sp., E. Ty.; Melanostoma mellina, Hn.; Meligethes æneus, viridescens, J. C. W.

Spirea filipendula. *Linn.* (Dropwort).

RECORD: *Dfs.*—Rev. W. Bennet, 1893.

LOCALITY: Kirkpatrick-Juxta, W. Bn. (requires confirmation)

Geum urbanum. *Linn.* (Avens).

RECORDS: *Dfs.* and *Kcd.*—P. Gray, 1850. *Wgt.*—G. C. Druce, 1883.

LOCALITIES: Common in all the valleys to 600 feet.

Appears May 18 to June 14, J. T. J. On dry or moist roadsides, old sod walls, humus; in shade or more rarely sun; usually wind-sheltered by hedges, etc.

VISITORS: Siphona geniculata, Anthomyia radicum, S.-E., May 21, Hydrotea dentipes, Hyelemyia strigosa, June 5, S.-E.

Geum rivale. *Linn.* (Water Avens).

RECORDS: *Dfs.* and *Kcd.*—P. Gray, 1850. *Wgt.*—G. C. Druce, 1883.

LOCALITIES: *Nithsdale*—Very common by Cluden, Hn.; Glen, C. E. M.; Nithside, Hn., C. E. M., S.-E., R. H. M.; Scaur, Hn.; Sanquhar, Dr Dv. *Annandale*—Very common Annan, Milke, Kirtle (reaching 2200 feet), Black's Hope, S.-E. *Eskdale*—Very common Tarras, Esk to Langholm, S.-E.; Castle O'er, J. Wn. and R. Bl.

Appears April 5 to May 14, J. T. J. On wet or rarely dry humus, holms, roadsides, boulder clay, shingles, etc.; usually shaded or half-shaded and wind-sheltered.

VISITORS: Bombus muscorum, Hn., S.-E.; hortorum, Hn., S.-E.; Rhingia rostrata, abundant reg., S.-E.; May 16 and 22.

Geum intermedium. *Ehrh.*

RECORDS: *Dfs.*—E. F. Linton, 1890. *Wgt.*—G. C. Druce, 1883.

LOCALITIES: *Nithsdale*—Woodlands, S.-E.; Glen, C. E. M., Th.; Caitloch (429 feet), Dalmakerran (450 feet), J. Cr.; Nithbank Wood, R. A.; Ryehill, Elliock, Burnsands, Dr Dv. *Annandale*—Annan road,

Th.; Annan to Ecclefechan, S.-E.; Railway near Sandbed Mill, G. Bl.; Dykefarm, New Mills, J. T. J.; Spoonburn, E. F. L. *Eskdale*—Wauchope, S.-E.

Appears April 27, G. Bl., to June 7 (where urbanum and rivale grow together, usually in moist shade).

Rubus idæus. *Linn.*

(Raspberry, Hindberry, *fide* J. Shaw).

RECORDS: *Dfs.*—Miss Harvey, 18—? *Kcd.*—P. Gray, 1850. *Wgt.*—G. C. Druce, 1883.

    b. Leesii, *Dfs.*—A. Craig Christie, 1887.

LOCALITIES: *Nithsdale*—Very common Craighope Linn, Harv.; Sanquhar, Dr Dv. *Annandale*—Very common Beld Craig (var. *b.*, C.C.), J. T. J. *Eskdale*—Very common Castle O'er, J. Wn. and R. Bl.

Appears May 30 to June 16, J. T. J. On moist or pretty dry humus, whinstone soils, roadside banks, boulder clay, etc.; usually shaded or half-shaded and in part wind-sheltered.

VISITORS: Apis, abundant; Bombus muscorum, abundant and suff., pratorum, Hn., S.-E.; Apathus vestalis, S.-E.; Syrphus cinctellus, Ribesii, S.-E. (In cultivation, chiefly B. lucorum, S.-E.)

Rubus fruticosus. *Linn.*

SUBSPECIES SUBERECTUS :—

    429. Fissus, *Kcd.*—J. M'Andrew, 1837.
    430. Suberectus, *Dfs.*—G. N. Lloyd, 1837. *Kcd.*—P. Gray, 1848. *Wgt.*—G. C. Druce, 1883.
    432. Plicatus, *Dfs.*—Dr Davidson, 1890. *Wgt.*—J. M'Andrew, 1890.

SUBSPECIES RHAMNIFOLIUS :—

    441. Imbricatus (No. 428 of 8th edition, *London Catalogue?*) *Dfs.*—Dr Davidson, 1890.
    442. Carpinifolius, *Dfs.*—J. Sadler, 1858.
    444. Lindleianus, *Dfs.*—E. F. Linton, 1890.
    447. Rhamnifolius, *Dfs.*—G. F. Scott-Elliot, 1893. *Kcd.*—F. R. Coles, 1883. *Wgt.*—G. C. Druce, 1883.

SUBSPECIES SUBSILVATICUS :—

    449. Pulcherrimus, *Dfs.*—G. F. Scott-Elliot, 1893.

SUBSPECIES SILVATICUS :—

    463. Macrophyllus, *Dfs.*—G. N. Lloyd, 1837. *Kcd.*—J. M'Andrew, 1882.

SUBSPECIES VESTITUS :—
    468. Sprengelii, *Kcd.* and *Wgt.*—C. C. Bailey, 1889 and 1890.

SUBSPECIES RADULA :—
    483. Radula, *Kcd.* and *Wgt.*—G. C. Druce, 1883.

SUBSPECIES KOEHLERIANUS :—
    506. Koehleri, *Dfs.*—J. Fingland, 1887. *Kcd.*—J. M'Andrew, 1887. *Wgt.*—G. C. Druce, 1883.
    Var. *b.*, Pallidus, *Dfs.*—E. F. Linton, 1890.

SUBSPECIES BELLARDIANUS :—
    517. Hirtus, *Kcd.*—J. M'Andrew, 1887.

LOCALITIES : *Nithsdale*—Lochar Moss, 430, G. N. Ll.; Newton, 506, *b.*, S.-E.; Dumfries, 463, G. N. Ll.; Auldgirth, 430, Exc.; Crawick, 506, 441, Dr Dv.; Sanquhar, 432, 442, Dr Dv. *Annandale*—Applegarth, 463, G. N. Ll.; Lochmaben, 432, 449, S.-E.; Moffat, 506 *b.*, E. F. L.; Craigbeck Bridge, 444, E. F. L.; Beattock Garpol, 442, J. Sd.; Craigieburn, 442, J. T. J.

Appears June 16 to 25, J. T. J.

VISITORS : Apis, Bombus muscorum, Hn., S.-E.; terrestris, S.-E.; pratorum, Hn.; hortorum, J. C. W.; Derhamellus, Hn.; Eristalis pertinax, Hn.; Sericomyia borealis, S.-E.; Pieris napi, S.-E., J. C. W.; Platychirius albimanus, Syrphus balteatus, topiarius, Anthomyia radicum, J. C. W.

Rubus cæsius. *Linn.* (Dewberry).

SUBSPECIES :—
    524. Balfourianus, *Kcd.*—Field Club, 1893.
    523. Corylifolius, *Kcd.*—J. M'Andrew, 1882. *Wgt.*—G. C. Druce, 1883.
    525. Cæsius, *Kcd.*—P. Gray, 1868. *Wgt.*—G. C. Druce, 1883.

LOCALITIES : *Along the shore*—Ross, G. C.; Port Ling, J. Fr.; P. Gr., J. M'A.; Arbigland, 524, Exc.

Rubus saxatilis. *Linn.*

RECORDS : *Dfs.*—J. Sadler, 1858. *Kcd.*—J. Cruickshank, 1836. *Wgt.*—G. C. Druce, 1883.

LOCALITIES : *Nithsdale*—Dalskairth, P. Gr.; Minnygrile (600 feet), Glencrosh (700 feet), J. Cr.; Nithlinns, Thornhill, R. A.; Sanquhar, Dr Dv.; Merkland Glen, Exc. *Annandale*—Bold Craig, S.-E.; Garpol, J.

Sd.; Grey Mare's Tail, J. Sd.; to 1000 feet, J. T. J. *Eskdale*—Wauchope, S.-E.

Appears May 17 to June 23. On wet or dry whinstone mixed with humus; in shade and wind-sheltered or exposed.

Rubus chamæmorus. *Linn.* (Cloudberry).

RECORD: *Dfs.*—Rev. W. Singer, 1843.

LOCALITIES: Almost invariably found about 1450 feet, *e.g.*, Queensberry, J. Sn.; Wanlockhead, Garland, Dr Dv.; Dalveen Pass Head, Th.; Bellybucht, R. A.; Hartfell, Black's Hope, J. T. J., S.-E.; Dobb's Linn, J. T. J.; Loch Skene, J. Sd.; Loch Craig, J. T. J.; Loch Fell, Archie Hill, Moodlaw Loch, Pikethow, Causey Grain (1600 feet), White Hope, S.-E.

Appears May 18 to June 17. On fairly dry peat; fully exposed (1400-2400 feet).

VISITORS: Empis, spp. (apparently new to Britain), very abundant; Anthomyia radicum, abundant; Siphona cristalis, Hydrotea dentipes, S.-E.; Grey Mare's Tail, May 17.

Fragaria vesca. *Linn.* (Strawberry).

RECORDS: *Dfs.* and *Kcd.*—P. Gray, 1850. *Hgt.*—G. C. Druce, 1883.

LOCALITIES: Very common in all the valleys to 1800 feet.

Appears April 4, G. Bl.; May 4 to 15, J. T. J. On dry, mossy walls, whinstone rocks, cinders of railways, humus, roadsides, etc.; usually half-shaded and partly wind-sheltered.

VISITORS: Ascia podagrica, Anthomyia sulciventris, abundant, radicum, S.-E.

Potentilla fragariastrum. *Ehrh.* (Barren Strawberry).

RECORDS: *Dfs.* and *Kcd.*—P. Gray, 1850. *Hgt.*—C. C. Bailey, 1883.

LOCALITIES: *Nithsdale*—Common Dumfries, P. Gr., Hn.; White Bridge, C. E. M.; Cowhill, Ad. and S. D. J.; Thornhill, R. A.; Sanquhar, Dr Dv. *Annandale*—Common Dryfe, Douglas Hill, Milke, G. Bl.; Lockerbie, Æ, S.-E.; Moffat (elevation up to 2000 feet), Correifron, etc., J. T. J. *Eskdale*—Ewes water, Th.; Esk, Tarras, S.-E.; Castle O'er, J. Wn. and R. Bl.

Appears March 23 to 31. On dry stony banks, sandstones, whinstones, old walls, sandy holms; usually half-shaded and slightly wind-sheltered.

Potentilla reptans. *Linn.*

RECORDS : *Dfs.*—Dr Davidson, 1886. *Kcd.*-- G. Gordon, 1836. *Wgt.*—G. C. Druce, 1883.

LOCALITIES : *Nithsdale*—Common by shore from Southerness to Kirkbean, G. Go., Exc., C. E. M.; Cowhill, Ad. and S. D. J.; Trigony, R. A.; Sanquhar Castle, Dr Dv., R.A.   *Annandale*—Ruthwell, Dr Gl.; Dumfries road bridge, Moffat, J. T. J.; Scroggs Mill, Dryfeholm, G. Bl. *Eskdale*—Liddel Railway Bridge, S.-E.

Appears July 5 to 13, J. T. J. On dry stony soil, shingles of shore and rivers, railway banks (Hn.), sandy hillocks, etc., bare of other plants ; exposed to sun and wind.

VISITORS : Dolichopodidæ, Syrphus ribesii, Hydrotea dentipes, Ascia podagrica, S.-E.; June 19.

Potentilla tormentilla. *Sibth.* (Tormentil).

RECORDS : *Dfs.* and *Kcd.*—P. Gray, 1850.
538. Silvestris, Neck. *Wgt.*—G. C. Druce, 1883.
339. Procumbens, Sibth. *Dfs.*--*Scottish Naturalist*, Vol. II. *Kcd.*--G. N. Lloyd, 1837. *Wgt.*— G. C. Druce, 1883.

LOCALITIES : *Nithsdale*—538, very common from the lower limit of permanent pastures to the highest summits ; 539, Sanquhar, Dr Dv. *Annandale*—538, very common ; 539, new Edinburgh road, Gardenholm wood, J. T. J. ; Hartfell, S.-E.   *Eskdale*—538, common ; 539, Liddel, near railway, Langholm Hill, S.-E.

538 appears April 24, G. Bl.; May 15-23, J. T. J.; 539, July 27, J. T. J. 538, on dry peat, whinstone soils, cinders, alluvium, boulder clay ; exposed to wind and sun, or sometimes in shade. 559, chiefly by roadsides ; part sheltered and in shade.

VISITORS : Bombus lucorum, Prosopis hyalinata, Homalomyia, Drymeia hamata, Anthomyia radicum, Helophilus frutetorum, S.-E.; Leucogonia lucorum, Hn.; Sphærophoria scripta, Syritta pipiens, J.C.W.

Potentilla maculata. *Power.*

RECORD : *Dfs.*—J. T. Johnstone, 1890.

LOCALITIES : *Annandale*—Black's Hope, Whitecoombe, Midlaw Burn (about 1500-1750 feet), Hartfell, Lochanburnside, J. T. J.

Appears June 22 to July 19, J. T. J. On dry steep whinstone rocks ; soon shaded and partly wind-sheltered by position.

VISITORS : Empis sp. (very rare), Hydrotea dentipes, Anthomyia radicum, S.-E.

Potentilla anserina. *Linn.* (Silverweed).

RECORDS : *Dfs.* and *Kcd.*—P. Gray, 1850. *Wgt.*—G. C. Druce, 1883.

LOCALITIES : Very common in all the valleys to 1700 feet.

Appears May 18 to June 4, J. T. J. On moist places bare of other plants, roadsides, mud flats by sea, shingles of shore and rivers, gardens, etc.; exposed to sun and wind.

VISITORS : Platychirius albimanus, Hyetodesia incana, Hydrotea dentipes, Anthomyia radicum, abundant ; Melanostoma sp., S.-E. ; May 19.

Potentilla comarum. *Nestl.*

RECORDS : *Dfs.* and *Kcd.*—P. Gray, 1850. *Wgt.*—J. H. Balfour, 1837.

LOCALITIES : *Nithsdale*—Very common Lochar Moss, G. C.; Troqueer, Dr Gl.; Cowhill, Ad. and S. D. J.; Friars' Carse, S.-E.; Dabton, R. A.; Sanquhar, Dr Dv. *Annandale*—Common Moffat, J. T. J. *Eskdale*—The Flow, Canobie, S.-E. (usually under 500 feet).

Appears June 25 to July 4, J. T. J. On damp peat mosses exposed to sun, and half-sheltered by reeds or exposed to wind.

VISITORS : Apis, abundant and suff. ; Bombus muscorum, lucorum, Hyetodesia incana, S.-E., June 7.

Potentilla suberecta. *Zimm.**

RECORD : *Kcd.*—1887.

Sibbaldia procumbens. *Linn.*

RECORDS : (Escape) *Dfs.*—Dr Burgess, 1789. *Kcd.*—Mrs Gilchirst Clark, 1867.

LOCALITIES : The Ross, G. C.; Broomholm, Langholm, Dr Br.*

Alchemilla vulgaris. *Linn.*

RECORDS : *Dfs.*—J. Cruickshank, 1836. *Kcd.*—P. Gray, 1850. *Wgt.*—G. C. Druce, 1883.
 b. Glabra, *Dfs.*—" Closeburn," 1865, Herb., Dumfries.

LOCALITIES : Very common in all the valleys, and reaching 2600 ft

---

* Requires confirmation.

Appears April 16, G. Bl.; May 4 to 12, J. T. J. On wet, moist, more rarely dry (and then clayey soil) roadsides, old walls, whinstone rocks by waterfalls, shingles, humus; usually wind-sheltered and exposed to sun or half-shaded.

VISITORS: Andrena albicans, Scatophaga stercoraria, lutaria, Anthomyia radicum, Chortophila sp., Empis trigramma, Empis sp., Athalia spinarum, Dolerus sp., Hyetodesia incana, Escophanes occupator, Telephorus discoideus, S.-E.

Alchemilla alpina. *Linn.*

RECORD: *Dfs.*—Rev. W. Singer, 1843. *

Alchemilla arvensis. *Scop.* (Parsley Piert).

RECORDS: *Dfs.* and *Kcd.*—P. Gray, 1850. *Wgt.*—G. C. Druce, 1883.

LOCALITIES: *Nithsdale*—Kenneth Bank, Glencaple, C. E. M.; White Bridge, Jardington, Routen Brig, S.-E.; Thornhill, R. A.; Sanquhar, Dr Dv. *Annandale*—Browhouses, S.-E.; Tundergarth Linn sides, G. Bl.; Lockerbie railway and roadsides, S.-E., G. Bl.; Halleaths, S.-E.; Moffat, J. T. J. *Eskdale*—Langholm Hill, Wauchope, S.-E.; Castle O'er, J. Wn. and R. Bl.

Appears June 22, J. T. J. On dry bare stony ground, roadsides, boulder clay, seashore, cinders, sandy gravel; fully exposed to wind and sun.

Sanguisorba officinalis. *Linn.* (Burnet).

RECORDS: *Dfs.*—Dr Burgess, 1789. *Kcd.*—Mr Maughan, 1796. *Wgt.*—J. M'Andrew, 1889.

LOCALITIES: *Nithsdale*—Kirkconnel, Mau.; Mabie, P. Gr.; Kelton, P. Gr.; Carnsalloch, Hn.; Tynron, J. Sh. *Annandale*—Jardine Hall, Th.; Brow Well, Dr Gl.; Seafield railway, S.-E.; Jardine Hall, Th. Muirhead Bridge, Lochmaben, Corriemains, and from half-a-mile below Scroggs to Whitstanehill, Dryfe, G. Bl.; Barnhill sandpit, J. T. J.; Roundstonefoot (600 feet), J. T. J. *Eskdale*—Glenzier, E. Ty.; Liddel Bridge, on railway, Meggat Water, abundant, S.-E.; Castle O'er, J. Wn. and R. Bl.; Eskdalemuir Kirk, S.-E.

On moist or fairly dry holms, railway banks, roadsides, in full exposure.

VISITORS: Cynomyia mortuorum, abundant; Lucilia Cesar, abundant; Onesia sepulchralis, Hyetodesia incana, Scatophaga stercoraria, S.-E.; June 19.

* Requires confirmation.

Poterium sanguisorba. *Linn.* (Salad Burnet).

RECORDS: *Dfs.*—G. N. Lloyd, 1837. *Kcd.*—J. M'Andrew, 1882.

LOCALITIES: *Nithsdale*—Caerlaverock, F. W. G.; Ruthwell, Dr Gl.; Castle Loch, Sanquhar, G. N. Ll. *Annandale* — Brow Well, Comlongon, Lin.; Dornock, Applegarth, T. Bl.; Caledonian railway near Lockerbie, G. Bl. *Eskdale* — Glenzier, E. Ty.; railway near Cogriebridge, J. T. J.

On dry cindery railway banks; in full exposure, S.-E.; sandy ground, G. N. Ll.

VISITORS: Very rare. Allantus nothi, c. p.; Platychirius, sp.; Siphona cristata, S.-E.

Agrimonia eupatoria. *Linn.*

RECORDS: *Dfs.* and *Kcd.*—P. Gray, 1850. *Wgt.*—G. C. Druce, 1883.

LOCALITIES: *Nithsdale*—Glen, F. W. G., Hn., C. E. M.; Woodlea (400 feet), roadside, Glencairn (350 feet), J. Cr.; Doocot, Carron, Morton, R. A.; Bracheads, Elliock Woods, Dr Dv. *Annandale*—Milke Water, Millbank of Dryfe, G. Bl; Garpol, Beldcraig, Moffat Water, J. T. J., S.-E. *Eskdale*—Very common along Liddel and Esk, Adam's Holm, Canobie, Penton, Wauchope, Bexburn, Glencorfe, Meggat Water, S.-E.; Castle O'er, J. Wn. and R. Bl.

Appears July 7 to 26, J. T. J. On moist (often steep) banks, usually boulder clay, alluvium, humus, whinstone soils; in shade or half-shaded and sheltered from wind in long grass or valleys.

Agrimonia odorata. *Mill.*

RECORDS: (Escape) *Kcd.* — Professor Oliver, 1887. *Wgt.* — J. M'Andrew, 1893.

Rosa spinosissima. *Linn.*

RECORDS: 556. Pimpinellifolia, *Dfs.*—Dr Burgess, 1789. *Kcd.*— P. Gray, 1848. *Wgt.*—C. C. Bailey, 1883.
557. Involuta, *Dfs*—J. Corrie, 1893.
 b. Sabini, *Wgt.*—G. C. Druce, 1883.
558. Hibernica, *Dfs.*—E. F. Linton, 1889.

LOCALITIES: *Nithsdale*—Common along shore from Southerness to Kirkbean, St., Exc.; Cluden Mills, Ad. and S. D. J.; Jarbruck (374 feet), 557, J. Cr.; Elliock Bridge, Euchan, Kello, Dr Dv., R. A. *Annandale*—Cummertrees Burn, Dr Br.; Saddleback Craigs, Correifron (1250 feet), 556, Beeftub, J. T. J.; Grey Mare's Tail, 558, E. F. L., J. T. J.

Flowers June to July. On shingle or short sandy turf by the sea, on bare rocks or whinstones inland; in full exposure to sun and wind; 558, wind-sheltered and in humid atmosphere.

VISITORS: See R. canina.

Rosa villosa. Linn.

RECORDS: 559. Mollis, *Dfs.*—J. Fingland, 1888. *Kcd.*—J. M'Andrew, 1882. *Wgt.*—G. C. Druce, 1883.
    b. Cœrulea, *Dfs.*—J. Fingland, 1888. *Kcd.* and *Wgt.*—G. C. Druce, 1883.
    c. Pseudo-rubiginosa, *Dfs.*—J. Fingland, 1888.
560. Tomentosa, *Dfs.*—J. Sadler, 1858. *Wgt.*—G. C. Druce, 1883.
    b. Subglobosa, *Dfs.*—Dr Davidson, 1886.
    d. Scabriuscula, *Dfs.*—Dr Davidson, 1886.

LOCALITIES: *Nithsdale*—Common about Dumfries, Friars' Carse, S.-E.; Thornhill, 559, 560, *d.*, J. Fn.; Nithside, Thornhill, 560, 559, *b.* and *c.*, J. Fn.; Sanquhar, 559, 560, Dr Dv.; Rigg, 560 *b.*, Dr Dv.; Mennock, 560, *d.*, Dr Dv. *Annandale*—Garpol and Beld Craig, 560, J. Sd.; common by Annan at Moffat, 559, 560, J. T. J. Adam's Holm, 560, *b.*, S.-E. *Eskdale*—Gretna, 559, S.-E.; common about Kirkandrews, S.-E.

On roadsides, by rivers, etc., usually holmlands and moderately sunny and windy places.

Rosa rubiginosa. Linn.

RECORDS: 561. Rubiginosa, *Dfs.* and *Kcd.*—P. Gray, 1850. *Wgt.*—G. C. Druce, 1883.
562. Micrantha.
    c. Hystrix, *Dfs.*—Dr Davidson, 1886 *

LOCALITIES: *Nithsdale*—Southerness to Kirkbean, Exc.; Dumfries, P. Gr.; Closeburn Woods, J. Fn.; Sanquhar, Dr Dv. *Annandale*—Railway near Dryfe Bridge, G. Bl.

Rosa canina. Linn.

RECORDS: *Dfs.* and *Kcd.*—P. Gray, 1850. *Wgt.*—G. C. Druce, 1883.
    565. Canina, *a.*, lutetiana, *Dfs.*—Dr Davidson, 1886. *Wgt.*—G. C. Druce, 1883; var. andegavensis, *Dfs.*—J. Fingland?

---

* No locality given and no specimen extant, so far as I know. *Scottish Naturalist*, Vol. 3.

c. Dumalis, *Dfs.*—Dr Davidson. *Kcd.* and *Wgt.*—G. C. Druce, 1883.
  Var. Verticillantha, *Dfs.*—Dr Davidson.
f. Biserrata, *Dfs.*—J. Fingland, 1890.
i. Urbica, *Dfs.*—Dr Davidson. *Wgt.*—G. C. Druce, 1883.
k. Arvatica, *Dfs.*—Dr Davidson.
j. Dumetorum, *Dfs.*—Dr Davidson. *Kcd.* and *Wgt.*—G. C. Druce, 1883.
l. Pruinosa, *Dfs.*—Dr Davidson.
m. Incana, *Dfs.*—Dr Davidson.
n. Tomentella, *Dfs.*—Dr Davidson.
o. Borreri, *Dfs.*—Dr Davidson.
566. Glauca, *Dfs.*—Dr Davidson.
b. Subcristata, *Dfs.*—Dr Davidson. *Wgt.*—J. M'Andrew.
e. Coriifolia, *Dfs.*—Dr Davidson.
g. Watsoni, *Dfs.*—J. Fingland, 1888.
Also, varieties Collina, Koscinciana and decipiens, *Dfs.*—Dr Davidson.

LOCALITIES: *Nithsdale*—Near Dumfries, Hn., S.-E.; Keir, J. Fn.; Trigony, J. Fn.; Thornhill, J. Fn.; Holmhill, J. Fn.; Carcoside and Holmwoods, Dr Dv.; Elliock Bridge, Dr Dv.; Sanquhar town, Dr Dv.; Crawick, Dr Dv.; Mennock, Dr Dv.; above Burnfoot, Dr Dv.; Grange, Dr Dv.; Sanquhar, Dr Dv. *Annandale*—Ecclefechan, S.-E.; Dryfe, S.-E.; Adamsholm, J. T. J., S.-E.; Frenchland Burn, J. T. J. *Eskdale* —Castle O'er, J. Wn. and R. Bl.

Appears June 8 to 21. I cannot give the exact habitats of the varieties.

VISITORS: Apis, abundant, S.-E.; Bombus pratorum, S.-E.; lucorum, muscorum, Hn., S.-E.; Allantus nothi, Andrena coitana, E. Ty.; Chrysid sp., E. Ty.; Melanostoma mellina, Hn., E. Ty.; Sericomyia Lapponum, E. Ty.; Syrphus balteatus, E. Ty.; Platychirius albimana, peltatus, sp., S.-E., E. Ty.; Eristalis pertinax, S.-E.; Hyetodesia sp., E. Ty.

Rosa arvensis. *Huds.*

RECORD: (Escape) *Dfs.*—G. F. Scott-Elliot, 1893.
LOCALITY: Birnswark to Ecclefechan, S.-E.

Rosa dicksoni. *Lindl.*

RECORD: (Escape) *Dfs.*—J. T. Johnstone, 1891.
LOCALITY: Roadside New Mills, J. T. J. Since rooted out.

Pyrus malus. *Linn.* (Crabtree).

RECORDS: *Dfs.* and *Kcd.*—P. Gray, 1850. *Wgt.*—G. C. Druce, 1883. *b.* Mitis, *Wgt.*—G. C. Druce, 1883.

LOCALITIES: Common in all the valleys to about 600 feet; on wet or dry rich loams exposed to sun and wind, W. Do., A. M.

Pyrus aucuparia. *Gaert.* (Rowan).

RECORDS: *Dfs.* and *Kcd.*—P. Gray, 1850. *Wgt.*—G. C. Druce, 1883.

LOCALITIES: Very common in all the valleys, reaching 2200 feet, Whitecoombe.

Appears May 24 to June 12. Along dry, rocky edges of wooded linns, chiefly on whinstone, fully exposed to wind and sun, (W. Do., A. M., S. E.).

VISITORS: Bombus lucorum, abundant and suff.; Empis bilineata, abundant; tessellata, abundant, sp.; Eristalis pertinax, Calliphora erythocephala, abundant; Siphona cristata, Anthomyia radicum, Dolichopus febrilis, Meligethes æneus, Epuræa æstiva, S.-E.; May 18.

Aremonia agrimonioides.

RECORD: (Escape) *Dfs.*—Mrs Gilchrist-Clark, 1890.

LOCALITY: Irongray, G. C., S.-E.

Has been established under a hedge, near a small stream, for 31 years.

Cratægus oxyacantha. *Linn.* (Hawthorn).

RECORDS: *Dfs.* and *Kcd.*—P. Gray, 1850. *Wgt.*—G. C. Druce, 1883.

LOCALITIES: Very common in all the valleys to 800 feet.

Flowers May 22 to June 12, J. T. J. Always planted, chiefly on boulder clay, sandy gravel, etc., A. M., S.-E.

VISITORS: Apis, abundant; Bombus sp., Dolichopus febrilis, Hyetodesia incana, S.-E.; May 23.

Epilobium angustifolium. *Linn.* (French Willow).

RECORDS: *Dfs.*—Dr Burgess, 1789. *Kcd.*—P. Gray, 1850. *Wgt.*—J. M'Andrew, 1893.

LOCALITIES: *Nithsdale*—Rare Lochar Moss, S.-E.; Nithside, Burnbridge, R. A.; Euchan, Gareland Cleugh, Dr Dv. *Annandale*—Eaglesfield, S.-E.; Murrayfield, Wamphray, G. Bl.; Beeftub, Rowan-tree Grain, J. T. J.; Dumcrief, Dr Br.; Middlegill, Craigieburn, Win.; Black's Hope, Correifron, J. T. J.; Whitecoombe, J. H. Bl.; Birk Hill (to 1750 feet), J. T. J. *Eskdale*—Irvine, Langholm, C. Y., E. Ty.

Appears August 4, J. T. J. On whinstone rocks, humus, roadsides, peaty ditches; usually fully exposed to wind and sun.

VISITORS : Apis, abundant and suff.; Bombus lucorum, abundant and suff.; pratorum, abundant and suff.; Vespa sylvestris, abundant and suff.; Cyrtonema stabularis, S.-E.; July 4.

Epilobium hirsutum. *Linn.* (Codlins and Cream).

RECORDS : *Dfs.*—Dr Burgess, 1789. *Kcd.*—P. Gray, 1850. *Wgt.* —G. C. Druce, 1883.

LOCALITIES : *Nithsdale*—Cargen, M. J. H., Hn.; burn, Mouswald Manse, Dr Br., S.-E.; Troqueer, Hn.; Glen Mill, Hn., S.-E.; Nithside, Th.; Kirkbog, Kirkland, R. A. *Annandale*—Ecclefechan, S.-E.; Milke Water, G. Bl. *Eskdale*—Scotch marsh, Canonbie, E. Ty.; Gilnockie, S.-E.

On wet holmlands, boulder clay, granite soils, etc.; sheltered from wind, but in sun.

VISITORS : Bombus muscorum, Hn.; Halictus albipes, Hn.

Epilobium parviflorum. *Schreb.*

RECORDS : *Dfs.*—G. F. Scott-Elliot, 1890. *Kcd.*—J. M'Andrew, 1882. *Wgt.*—G. C. Druce, 1883.

LOCALITIES : *Annandale*—Lochmaben, Exc. *Eskdale*—Kirkburn, Meggat, Stennies, S.-E.; Castle O'er, J. Wn., and R. Bl.

Epilobium montanum. *Linn.*

RECORDS : *Dfs.* and *Kcd.*—P. Gray, 1850. *Wgt.*—G. C. Druce, 1883.

Var. minus, *Dfs.*—E. F. Linton, 1889.

LOCALITIES : *Nithsdale*—Round Dumfries common, S.-E.; Sanquhar, Dr Dv. *Annandale*—Springkeld, S.-E.; Beld Craig, S.-E.; Meiklcholmside, S. W. Ca.; common on all Moffat Hills to 2000 feet, J. T. J., S.-E.; Black's Hope and Grey Mare's Tail (minus), E. F. L. *Eskdale*—Penton Linn, Wauchope, S.-E.; Castle O'er, J. Wn. and R. Bl.

Appears June 8 to 21, J. T. J. On dry or rarely wet whinstone rocks, old walls, humus, boulder clay, etc.; in shade or sun, usually in part wind-sheltered.

VISITORS: Syritta pipiens, Platychirius clypeatus, Siphona cristata, Anthomyia radicum, S.-E.

Epilobium roseum. *Schreb.*

RECORDS: *Dfs.*—Miss J. Wilson and Mr R. Bell, 1892. *Kcd.*—Dumfries Herbarium, 1865.

LOCALITIES: *Nithsdale*—Kirkbean, Dumfries Herbarium. *Annandale*—Grey Mare's Tail, S.-E. *Eskdale*—Castle O'er, J. Wn. and R. Bl.

Epilobium tetragonum. *Linn.*

RECORDS: 652. Adnatum? *Dfs.*—P. Gray, 1850. *Kcd.*—J. M'Andrew, 1882. *Wgt.*—Miss Hannay, 1890. 653. Obscurum, *Dfs.*—Dr F. W. Grierson, 1882. *Kcd.*—F. R. Coles, 1883. *Wgt.*—C. C. Bailey, 1883.

LOCALITIES: *Nithsdale*—Locharbriggs, Cluden Bridge, S.-E.; Maxwelltown Loch, Hn., S.-E.; Auldgirth, F. W. G.; Craighope Linn, Hn. *Annandale*—Kirkpatrick-Juxta, F. A. H.; Holmshaws, Beattock, Frenchland Burn (elevation 1750 feet), J. T. J.; Craigboar, S.-E.; Moffat Water, 653, E. F. L. *Eskdale*—Old Gretna, 653, S.-E.

Appears July 8, J. T. J. On wet or rarely dry whinstone rocks, sandstone walls or granite roadsides (653); in sun but partly wind-sheltered.

VISITORS: Platychirius albimana, S.-E.; clypeatus, Hn.; Hedrerigaster urticæ, Picris napi, Hn.

Epilobium palustre. *Linn.*

RECORDS: *Dfs.*—Dr Davidson, 1886. *Kcd.*—P. Gray, 1850. *Wgt.*—G. C. Druce, 1883.

LOCALITIES: *Nithsdale*—New Loch, R. A.; Sanquhar, Dr Dv. *Annandale*—Turnmoor Wood, G. Bl.; Lochmaben, Exc.; Beld Craig, Moffat, J. T. J., S.-E.; Beld Craig, S.-E. *Eskdale*—Gilnockie, Canobie Road, Langholm to Bentpath, S.-E.; Merrylaw Bog, J. Rae.

In wet marshy grass (half peat) by roadsides, on boulder clay, whinstone soils; usually exposed to sun, but partly wind-sheltered.

Epilobium alsinefolium. *Vill.*

RECORDS: *Dfs.*—*Flora Scotica*, 1789.

LOCALITIES: *Annandale*—Hartfell, *Flora Scotica*; Black's Hope,

E. F. L.; Correifron (2000 feet), J. T. J., S.-E.; Grey Mare's Tail, J. Sd. *Eskdale*—Merrylaw Bog, Ewesleesdowns, J. Rae.

Appears July 6, J. T. J. In wet mud of springs, drains, etc. ; fully exposed or half-sheltered.

Epilobium alpinum. *Linn.*

RECORD : *Dfs.*- -Rev. W. Singer, 1843.

LOCALITIES : *Annandale*—Dryfe water, near head, W. Sn.; Whitecoombe, J. H. Bl. and J. Sd.*

Epilobium parvifolium × obscurum.

RECORD : *Dfs.*—E. F. Linton, 1882.

LOCALITY : Moffat, damp roadsides, E. F. L.

Œnothera biennis. *Linn.*

RECORDS : (Escape) *Dfs.*—Dr Davidson, 1890. *Kcd.*—P. Gray, 1850.

LOCALITIES : *Nithsdale*—Dumfries, P. Gr.; Nith Mills, Sanquhar, Dr Dv.

Circæa lutetiana. *Linn.* (Enchanter's Nightshade).

RECORDS : *Dfs.*—G. Gordon, 1836. *Kcd.*—P. Gray, 1846. *Wgt.* —J. M'Andrew, 1893.

b. intermedia, *Dfs.*—Herbarium, 1865.

LOCALITIES : *Nithsdale*—Glen, F. W. G., P. Gr.; Nithside, Hn.; Caitloch (400 feet), J. Cr.; Irongray, S.-E.; Cowhill, Ad. and S. D. J.; Craighope Linn, Hn., R. A.; Sanquhar, Dr Dv. *Annandale*—Scroggs, G. Bl.; Boreland, J. Wg.; Hartfell, var *b.* Herb., Garpol, Beld Craig, etc., S. W. Ca., J. T. J. *Eskdale*—Penton Linn, Langholm, C. Y.; common by Esk to Lineholm, S.-E.; Castle O'er, J. Wn. and R. Bl.

Appears July 27, J. T. J. On damp or rarely dry leaf-mould ; in shade and wind-sheltered.

VISITORS : Platychirius albimana, S.-E.

Circæa alpina. *Linn.*

RECORD : *Dfs.*—Rev. W. Singer, 1843.

LOCALITIES : Beld Craig, Moffat, W. Sn., W. Ca.; Garpol, J. T. J.; Grey Mare's Tail, J. Sd.

On moist, steep and stony banks ; half-shaded.

* Requires confirmation.

Lythrum salicaria. *Linn.* (Loosestrife).

RECORDS: *Dfs.*—G. N. Lloyd, 1837. *Kcd.*—P. Gray, 1846. *Wgt.*—Dr Balfour, 1836.

LOCALITIES: *Nithsdale*—Lochar Moss, G. C.; Cargen, Cluden, Hn., C. E. M.; by Nith, Dumfries, Tinwald, Hn.; Maxwelltown Loch, P. Gr., Hn., S.-E.; Cowhill, Ad. and S. D. J.; Friars' Carse Loch, F. W. G., S.-E.; Auldgirth, Kirkland, R. A. *Annandale*—Lochmaben, S.-E.

In marshy peat bogs; usually in water; in sun and only sheltered by reeds, etc.

VISITORS: Vanessa Urticæ, abundant; Bombus pratorum, S.-E.; hortorum, Hn.; Apis, Hn., S.-E.; Platychirius peltatus, S.-E. June 26.

Peplis portula. *Linn.* (Water Purslane).

RECORDS: *Dfs.*—P. Gray, 1850. *Kcd.*—J. M'Andrew, 1882. *Wgt.*—Arnott, 1848.

LOCALITIES: *Nithsdale*—Dumfries, P. Gr.; Brownhall, Th. *Annandale*—Mill Loch, Lochmaben, G. Bl.; S.-E.

On moist mud of loch margins; fully exposed.

Cotyledon umbilicus. *Linn.* (Navelwort).

RECORDS: (Escape) *Kcd.*—Trans. Phil. Soc., Glasgow, 1844. *Wgt.*—Balfour, 1836.

Sedum rhodiola. *D. C.* (Roseroot).

RECORDS: *Dfs.*—Rev. W. Singer, 1843. *Kcd.*—J. M'Andrew, 1882. *Wgt.*—Arnott, 1848.

LOCALITIES: *Nithsdale*—Dumfries, Locharbriggs, Hn. *Annandale*—Craigboar, Penbreck, S.-E.; Auchencat burn-head, Hartfell Craigs, Black's Hope, Correifron, Coombe Craigs, Whitecoombe, above Loch Skene, Grey Mare's Tail (about 2200 feet), Kd., Bl., J. T. J., S.-E.

Appears May 18 to June 22. Usually bare, often dry whinstone rocks; fully exposed.

VISITORS: Cynomyia mortuorum, Empis sp. (very rare), S.-E.

Sedum telephium. *Linn.* (Orpine, Livelong).

RECORDS: *Dfs.*—C. Gordon, 1836. *Kcd.*—J. T. Syme, 1842. *Wgt.*, —G. C. Druce, 1883.

LOCALITIES: *Along the shore*—C. Go. *Nithsdale*—Nith at Dumfries, Hn., C. E. M.; from Lincluden to Cowhill, by White Bridge,

S.-E., Ad., and S. D. J.; Nunwood and Terregles roads, S.-E.; Scaur, Hn., J. Sh.; Redpaths, R. A.; Sanquhar railway, Dr Dv. *Annandale*—Railway, Lockerbie to Nethercleugh, G. Bl.; roadsides, Moffat, J. T. J. *Eskdale*—Railway, Langholm, S.-E.

Appears July, S.-E. On damp roadsides; half-shaded and part sheltered from wind.

Sedum anglicum. *Huds*.

RECORDS: *Dfs.*—Dr Burgess, 1789. *Kcd.*—P. Gray, 1846. *Wgt.*—Herb., Greville, 1836.

LOCALITIES: *Nithsdale*—Dalscone, Shambellie hills, Portland Place, Lincluden Abbey, P. Gr. *Annandale*—Annan, Powfoot, S.-E.; Cummertrees, Ak.; Kerr sandbeds, J. T. J. *Eskdale*—Glentarras, S.-E.

Appears June 22, J. T. J. On shingles of rivers and shore, old walls, rocks, roadsides, etc.; fully exposed.

VISITORS: Nemotela notatris, Oxycera sp., Anthomyia, abundant, S.-E.

Sedum villosum. *Linn*.

RECORDS: *Dfs.*—Mr Keddie, 1854. *Kcd.*—P. Gray, 1850.

LOCALITIES: *Nithsdale*—Barndennoch (630 feet), Moniaive, J. Cr.; Scaur, J. Sh.; Cogshead, Nethercog, Glenglass, Dr Dv.; Thornhill to Elvanfoot, J. T. J. *Annandale*—Burnswark, Lambfairhill, G. Bl.; Hutton of Dryfe, J. Wg.; Queensberry, S.-E.; Chapelhill, Edinburgh Road, J. T. J.; Auchencat burn, Kd., S.-E.; Hartfell, Black's Hope, Capelgill roadside, J. T. J., S.-E.; Grey Mare's Tail, Dobbs Linn, J. Sd.; Craigmichen, J. T. J. *Eskdale*—Wauchope water head, S.-E.; Castle O'er, J. Wn. and R. Bl.; Archie Hill, White Hope Edge, Meikledale, S.-E.

Appears June 12 to 22, J. T. J. On wet roadsides, shingles, in shallow rivulets (from 600 to 2000 feet); fully exposed.

Sedum acre. *Linn*. (Stonecrop).

RECORDS: *Dfs.* and *Kcd.*—P. Gray, 1850 and 1848. *Wgt.*—G. C. Druce, 1883.

LOCALITIES: *Along the shore*—Powfoot to Newbie, Torduff Point, S.-E. *Nithsdale*—Nith side below Dumfries, Hn.; Cowhill, Ad. and S. D. J.; Thornhill, R. A. *Annandale*—Waterbeck, S. E.; Kerr sandbed, J. T. J. *Eskdale*—Glentarras, S.-E.

Appears June 22. On shingles, concrete by shore, old walls, etc.; fully exposed.

VISITORS: Bombus pratorum, Hn.

Sedum rupestre. *Huds.*

RECORD: *Wgt.*—Dr Balfour, 1836.
LOCALITY: Lochnaw, Bl.*

Sempervivum tectorum. *Linn.* (Fooze).

RECORDS: (Escape) *Dfs.*—J. T. Johnstone, 1890. *Kcd.*—J. M'Andrew, 1882. *Wgt.*—Rev. G. Wilson, 1893.
LOCALITIES: House roofs, Moffat, J. T. J.

Ribes grossularia. *Linn.* (Gooseberry).

RECORDS: (Escape) *Dfs.*—P. Gray, 1846. *Kcd.*—J. M'Andrew, 1882. *Wgt.*—G. C. Druce, 1883.
LOCALITIES: A common escape in woods, Dumfries, P. Gr., R. H. M.; Kirkmichael, S. E.; Thornhill, R. A.; Sanquhar, Dr Dv.; Moffat, J. T. J.

Ribes rubrum. *Linn.* (Red Currant).

RECORD: (Escape) *Dfs.* and *Kcd.*—P. Gray, 1846.
LOCALITIES: Not common. *Nithsdale*—Gillie Hill, Harley Bank Bridge, R. H. M.; Cluden Craigs, Routen Brig, P. Gr.; Sanquhar, Dr Dv. *Annandale*—Moffat, J. T. J. *Eskdale*—Along Esk below Langholm, S.-E.

Ribes nigrum. *Linn.* (Black Currant).

RECORDS: (Escape) *Dfs.*—Dr Burgess, 1789. *Kcd.*—J. M'Andrew, 1882. *Wgt.*—G. C. Druce, 1883.
LOCALITIES: *Nithsdale*—Locharbriggs, Amisfield, Dr Br.; Ellisland, P. Gr.

Ribes alpinum. *Linn.*

RECORD: (Escape) *Kcd.*—Charles Scott, 1887.
LOCALITY: Terregles woods, Charles Scott.

Saxifraga oppositifolia. *Linn.*

RECORD: *Dfs.*—Rev. W. Singer, 1843.
LOCALITY: Grey Mare's Tail, W. Sn., W. St., J. Kd., J. Sd., J. T. J., S.-E.; elevation 1000 feet, J. T. J.

Appears April 18 to May 10, J. T. J. On whinstone rocks, well sheltered from wind.

* Requires confirmation.

Saxifraga aizoides. *Linn.*

RECORD: *Dfs.*—Rev. W. Singer, 1843.
LOCALITY: Moffat Hills, W. Sn. and W. Bn.?*

Saxifraga hypnoides. *Linn.* (Ladies' Cushion).

RECORDS: *Dfs.*—Dr Burgess, 1789. *Kcd.—fide* J. M'Andrew, 1882.
Var. sponhemica, *Dfs.*—E. F. Linton, 1889.
LOCALITIES: *Nithsdale*—Merkland, Exc.; Moniaive (reported), J. Cr.; Glenquhargen, J. Wl.; Wanlockhead, J. Sh.; Cample Cleugh, R. A.; Dalveen, J. Sh., S.-E.; Enterkin, Gareland, Dr Dv. *Annandale*—Queensberry, Penbreck, Dr Br., S.-E.; Kinnelhead, S.-E.; Beeftub, J. T. J.; Auchencat Burn, Hartfell Craigs, S.-E.; Wellburn, Beerholm, J. T. J.; Black's Hope, Correifron, W. Ca., J. T. J.; Spoonburn, Whitecoombe, Midlaw Burn (to 2300 feet), Loch Skene, Craigmichen, J. T. J., S.-E. *Eskdale*—Meikledale, S.-E.

Appears May 17 to June 21, J. T. J. On wet mossy rocks or turfs, by mountain streams; in humid atmosphere and wind-sheltered.

VISITORS: Homalomyia, Empis bilineata, Scatophaga stercoraria, Rhamphomyia, sp., S.-E.

Saxifraga granulata. *Linn.*

RECORDS: *Dfs.* and *Kcd.*—P. Gray, 1846. *Wgt.*—J. Gorrie, 1891.
LOCALITIES: *Nithsdale*—Lincluden, C. E. M., F. W. G.; common along Nith and Cluden to Routen Brig, P. Gr., Hn., S.-E.; Moniaive (300 feet), J. Cr.; Cowhill, Ad. and S. D. J.; Auldgirth, Exc.; Tynron, J. Sh. *Annandale*—Common Milke, to three miles up Dryfe, G. Bl. *Eskdale*—Byreburn, Gilnockie, S.-E.; Irvine, C. Y.; occasionally by Esk to half-a-mile above Langholm, S.-E.

Appears April 23, G. Bl.; May 1, S.-E. On moist leaf-mould mixed with alluvium or sand; in shade and well sheltered.

Saxifraga nivalis. *Linn.*

RECORD: *Dfs.*—J. T. Johnstone, 1889.
LOCALITY: Black's Hope (about 2000 feet), J. T. J.
Appears June 22, J. T. J. On wet moss; in shade and shelter of overhanging whinstone rock.

* Doubtful, and requires confirmation.

Saxifraga stellaris. *Linn.*

RECORDS : *Dfs.*—Dr Burgess, 1789. *Kcd.*—J. M'Andrew, 1882.

LOCALITIES : *Nithsdale*—Martown Hill (1650 feet), J. Cr.; Euchan source, Dr Dv. *Annandale*—Queensberry burns, common, Dr Br.; Beeftub, J. T. J.; Hartfell burns, J. T. J., S.-E.; Wellburn and sandbeds, Moffat, J. T. J.; Black's Hope, J. T. J.; Correifron, W. Ca., J. T. J.; Loch Skene, Whitecoombe, Dobb's Linn, J. Sd., J. T. J.

Appears May 18 to June 22, J. T. J. In mud of springs, mossy rocks by small burns ; usually exposed (to 1800 feet).

VISITORS : Empis bilineata, Platychirius clypeatus, albimana, Ascia podagrica, Siphona geniculata, Anthomyia radicum, Hydrotea dentipes, S.-E.

Chrysosplenium oppositifolium. *Linn.*

RECORDS : *Dfs.* and *Kcd.*—P. Gray, 1850. *Wgt.*—G. C. Druce, 1883.

LOCALITIES : *Nithsdale*—Very common by Nith and Cluden, S.-E.; Carruchan, Hn.; Glen, Colonel's Wood, C. E. M.; Sanquhar, Dr Dv. *Annandale*—Kirtle, Milke, very common, Moffat Hills (2000 feet), J. T. J., S.-E. *Eskdale*—Common Kirkandrews, E. Ty.; Penton and Esk very common, S.-E.; Castle O'er, J. Wn. and R. Bl.

Appears March 17 to 28, J. T. J. On wet mud, usually bare of other plants, rocks, walls, springheads, etc.; usually shade and shelter at low altitudes ; fully exposed on hills.

VISITORS : Siphona cristata, geniculata, Microgaster sp., Scatophaga stercoraria, Telephorus bicolor, S.-E.

Chrysosplenium alternifolium. *Linn.*

RECORDS : *Dfs.*—Dr Burgess, 1789. *Kcd.*—P. Gray, 1850.

LOCALITIES : *Nithsdale*—Glen, Grove, F. W. G.; Dalawoodie, R. R.; Isle, S. E.; Scaur, T. A. Br.; Grange, Burnsands, Laggrie, Dr Dv. *Annandale*—Neeze Linn, Kirkmichael Kirk, Dr Br.; Tundergarth, Balgray, Annie's Bridge, G. Bl.; New Mills, Oakrigg, Adam's Holm, J. T. J. *Eskdale*—Kirkandrews, E. Ty.

Appears March 17 to 28, G. Bl., J. T. J. In similar spots to preceding, but limited to shady and sheltered linns below 600 feet.

Parnassia palustris. *Linn.* (Grass of Parnassus).

RECORDS : *Dfs.*—S. W. Carruthers, before 1891. *Kcd.*—P. Gray, 1846. *Wgt.*—G. C. Druce, 1883.

LOCALITIES: *Nithsdale*—Maxwelltown Loch, P. Gr., Hn.; Auldgirth Hill, S.-E.; Drumcork, R. A.; Crawick Kello, Dr Dv. *Annandale*— Limestone Rig, G. Bl.; Hindhill, S. W. Ca.; Beeftub, J. T. J. *Eskdale* —Stennies, Meggat, etc., S.-E.

Appears August 4, J. T. J. On wet peaty hills or whinstone soils; fully exposed.

Drosera rotundifolia. *Linn.* (Sundew).

RECORDS: *Dfs.*—Rev. W. Little, 1834. *Kcd.*—J. Cruickshank, 1836. *Wgt.*—G. C. Druce, 1883.

LOCALITIES: *Nithsdale*—Very common, Newabbey, J. Cru.; Lochar Moss, etc., Hn., S.-E.; Sanquhar, Dr Dv. *Annandale*—Very common Moffat (1000 feet), J. T. J.; *Eskdale*—Archie Hill, Tarras, S.-E.; Castle O'er, J. Wn. and R. Bl.

On wet Sphagnum in peat mosses; fully exposed.

VISITOR: Anthomyids, S.-E.

Drosera longifolia. *Linn.*

RECORDS: *Dfs.* and *Kcd.*—G. N. Lloyd, 1837. *Wgt.*— J. M'Andrew, 1891.

LOCALITIES: *Nithsdale*—Lochar Moss, G. N. Ll., S.-E.; Kirkconnel Moss, G. N. Ll., C. E. M.; roadside, Moniaive (770 feet), J. Cr. *Annandale*—Roundstonefoot Hill, Moffat (1000 feet), J. T. J.

Drosera anglica. *Huds.*

RECORDS: *Dfs.*— G. N. Lloyd, 1837. *Kcd.*—J. Fraser, 1844. *Wgt.*—G. C. Druce, 1883.

LOCALITIES: *Nithsdale*—Lochar Moss, G. N. Ll., F. W. G.; Kirkconnel Moss, P. Gr., S.-E.; Tynron, J. Sh. *Annandale*—Dornock, T. Bl.; Jardine Hall, G. N. Ll.

Myriophyllum spicatum. *Linn.*

RECORDS: *Dfs.* and *Kcd.*—P. Gray, 1850. *Wgt.*—G. C. Druce, 1883.

LOCALITIES: Common in mud of rather sluggish streams in all the valleys.

Myriophyllum verticillatum. Linn.

RECORD: *Dfs.*—J. Sadler, 1858.*
LOCALITY: Loch Skene, J. Sd.

Myriophyllum alternifolium. D. C.

RECORDS: *Dfs.*—Dr F. W. Grierson, 1882. *Wgt.*—G. C. Druce, 1883.

LOCALITIES: *Nithsdale*—Nith, Dumfries, F. W. G. *Eskdale*—Canobie, S.-E.

Hippuris vulgaris. Linn. (Mare's Tail).

RECORDS: *Dfs.*—J. Singer, 1843. *Kcd.*—J. M'Andrew, 1882. *Wgt.*—Rev. G. Wilson, 1893.

LOCALITIES: *Nithsdale*—Lochar Moss, P. Gr.; Mouswald, Tinwald, J. Sn.; Clarencefield, F. W. G.; Fingland Lane (common 1000 feet), J. Cr.; Closeburn, J. Fn. *Annandale*—Kirtlebridge Old Quarry, S.-E.; Earshaig (700 feet), J. T. J.; Eskdalemuir, J. Wn. and R. Bl.

In mud of shallow slowly running streams; sheltered from wind, not specially shaded; probably wind fertilised.

Hydrocotyle vulgaris. Linn.

RECORDS: *Dfs.* and *Kcd.*—P. Gray, 1850. *Wgt.*—G. C. Druce, 1883.

LOCALITIES: Common in all the valleys, Lochar Moss, P. Gr.; Drumcork, R. A.; Townmoor, Sanquhar, Dr Dv.; Moffat Well, J. T. J. *Eskdale*, S.-E.

Appears August 10, J. T. J. Common in marshes (not a sphagnum plant); on both peaty and clayey ground below 800 feet.

Sanicula Europæa. Linn.

RECORDS: *Dfs.*—Dr Davidson, 1886. *Kcd.*—P. Gray, 1850. *Wgt.* —G. C. Druce, 1883.

LOCALITIES: *Nithsdale*—Glen, Hn., C. E. M.; Kirkconnel woods, G. C.; Grove, S.-E.; most linns, Glencairn, J. Cr.; Penpont, Dfs. Herb.; Crawick, Elliock, Dr Dv. *Annandale*—Garpol, Beid Craig, J. T. J.; Corehead, Craigmichen (to 1200 feet), S.-E. *Eskdale*—Canobie to Langholm, abundant, Penton, Bexburn, S.-E.

* Requires confirmation.

FLORA OF DUMFRIESSHIRE.

Appears June 7 to 16. On wet or moist leaf-mould, or leaf-mould mixed with sand or whinstone detritus ; in open, shady, or wind-sheltered glens ; especially common below 900 feet.

VISITORS: Vespa rufa, Syrphus cinctellus, Hydrotea dentipes, Anthomyia radicum, S.-E.

Eryngium maritimum. *Linn.*

RECORDS: *Dfs.*—J. Singer, 1843. *Kcd.*—G. Gordon, 1837. *Wgt.*—Arnott, 1848.

LOCALITIES: Cowans, Arn.; Portwilliam, Hn.; Monreith, J. M'A.; Brighouse, Rosshill, G. N. Ll. and J. M'A., F. W. G.; Kirkcudbright, Exc.; Millstone Quarry, Colvend, J. Fr.; Newbie, J. Sn. and J. Fn.; Southerness, J. M'A.

Along the shore, in sand or shingle ; fully exposed.

Astrantia major. *Linn.*

RECORD : (Escape) *Dfs.*—G. Bell, 1893.
LOCALITY : Poolhouses, Lockerbie, G. Bl.

Cicuta virosa. *Linn.*

RECORDS : *Dfs.*—G. N. Lloyd, 1837. *Kcd.*—J. M'Andrew, 1882.

LOCALITIES : Carlingwark, J. M'A. *Nithsdale*—Lochar Moss, G. C.; Dalswinton, G. C.; Friars' Carse, S.-E. *Annandale*—Æ Water, Castle and Kirk Lochs, Lochmaben, G. N. Ll. and Exc.; ditches and ponds, Lochmaben to Lockerbie, G. Bl.

Grows in shallow stagnant water, usually on mud.

VISITORS: Vespa sylvestris, abundant and sufficient; Bombus lucorum, abundant ; Mimesa Dahlbomi, Panipla vulgaris, Lucilia Cæsar, abundant; Onesia sepulchralis, Borborus sp., Chortophila sp., Hyetodesia incana, Anthomyia radicum, S.-E.

Apium graveolens. *Linn.* (Celery).

RECORDS : (Escape) *Kcd.*—Lot's Wife, Colvend, Rev. J. Fraser, 1882. *Wgt.*—Mull of Galloway, Balfour (1848 ?).

Apium nodiflorum. *Reich.* (Water Parsnip).

RECORDS : 625. Var. *a.*, *Dfs.*—J. T. Johnstone, 1890. *Kcd.* and *Wgt.*—G. Graham, 1836.
Var *b.*, *Dfs.*—Rev. J. Singer, 1843. *Kcd.*—G. N. Lloyd, 1837.

LOCALITIES: *Galloway*—Very common, G. Gr., Bl., Arn. *Annandale*—Loch Skene and Midlaw Burn (1700 to 2000 feet), Earshaig (var. *b.*), Sn.*

Apium inundatum. *Reich.* (Water Parsnip).

RECORDS: *Dfs.*—J. Singer, 1843. *Kcd.*—P. Gray, 1850. *Wgt.*—Arnott, 1848.

LOCALITIES: Dunsky Castle, Arn.; Glenkair, Wed. Herb.; Moniaive, J. Cr.; Maxwelltown Loch, F. W. G., R. H. M.; Lochmaben, J. Sn.; Kirkland, R. A.

Aegopodium podagraria. *Linn.* (Bishop's Weed).

RECORDS: *Dfs.* and *Kcd.*—P. Gray, 1850. *Wgt.*—G. C. Druce, 1883.

LOCALITIES: Very common in all valleys to Sanquhar, Dr Dv.; Moffat, J. T. J.; Langholm, S.-E.; and Eskdalemuir, J. Wn. and R. Bl.

Fairly dry ground, chiefly as a weed in gardens, on waste soil, roadsides, leaf-mould, even cinders and granite; usually wind-sheltered and half-shaded.

VISITORS: Tipbia minuta, R. Sc.; Odynerus spinipes, Ichneumon, sp., Tryphon vulgaris, Siphona geniculata, Anthomyia radicum, S.-E.

Carum verticillatum. *Koch.* (Water Parsnip).

RECORDS: *Dfs.*—Dr Burgess, 1789. *Wgt.* and *Kcd.*—G. Graham, 1836.

LOCALITIES: Very common Wigtown and Kirkcudbright. *Nithsdale*—Ruthwell, Dr Gl.; Mabie Moss, P. Gr.; Maxwelltown Loch, P. Gr., F. W. G., R. H. M.; Glencairn common, J. Cr.; Penpont, Thornhill, Dalveen Pass, W. St.; Drumcork, R. A.; Conrig Bogue, Dr Dv. *Annandale*—Stank Farm, Ruthwell, Dr Br.

Wet, marshy pastures on sandy gravel (? on peat); exposed to sun and wind.

VISITORS: Allantus nothi, Tryphon vulgaris, Pollenia rudis, Hydrotea dentipes, Anthomyia radicum, Siphona cristata, Hyetodesia, sp., Ichneumon, sp., Dolichopodidæ, S.-E.

Carum carui. *Linn.*

RECORDS: (Escape) *Kcd.*—J. M'Andrew, 1882. *Wgt.*—Sir H. Maxwell, 1895.

* This locality is queried by Mr J. T. Johnstone.

Sium angustifolium. *Linn.*

RECORDS: *Dfs.*—Dr Burgess, 1789. *Kcd.*—Arnott, 1844. *Wgt.*—J. M'Andrew, 1882.

LOCALITIES: 1 ridge on high road between Duncow and Kirkmahoe, Dr Br.; Lochmaben, J. Sn.

Pimpinella saxifraga. *Linn.*

RECORDS: *Dfs.* and *Kcd.*—P. Gray, 1850. *Wgt.*—G. C. Druce, 1883.
Var. *c.*, *Dfs.*—Miss Taylor, 1892.

LOCALITIES: Common in all valleys, Sanquhar, Dr Dv.; Black's Hope (2000 feet), J. T. J.; Glencorfe, Blochburn, S.-E.; Glenzier, var. *c.*, E. Ty.; Eskdalemuir, J. Wn. and R. Bl.

Flowers July 15, J. T. J. On dry or rarely wet whinstone rocks, walls; usually wind-sheltered by sides of ravine or by other plants, and usually partly shaded.

VISITORS: Pieris napi, J. C. W.; Allantus nothi, abundant, S.-E., July 18; Chrysogaster splendida, Syrphus ribesii, Orthoneura nobilis, Eristalis tenax, æneus, horticola, Sphærophoria scripta, Syritta pipiens, and 18 other species, J. C. W., l.c., p. 246.

Œnanthe Fistulosa. *Linn.* (Water Dropwort).

RECORDS: *Dfs.*—Dr Burgess, 1789. *Kcd.*—G. N. Lloyd, 1837.

LOCALITIES: Rosshill, G. N. Ll. *Nithsdale*—Between Blackshaw (Caerlaverock) and the side of Lochar water, Dr Br.; Kingholm Merse, J. Sn. *Annandale*—Lochmaben, Th.?

Marshy places, estuarine mud, partly sheltered from wind by other plants, in sun.

Œnanthe pimpinelloides. *Linn.*

RECORDS: 705. Pimpinelloides, *Dfs.*—W. Stevens, 1848. *Kcd.*—G. Graham, 1836. *Wgt.*—J. H. Balfour, 1836.
707. Lachenalii, *Dfs.*—Dr F. W. Grierson, 1882. *Kcd.* J. M'Andrew, 1882. *Wgt.*—Arnott, 1848.

LOCALITIES: *Along the shore*—Portlogan, Cowans, Arn.; Mull of Galloway, G. M'N. and J. M'K.; Rosshill, Balmae, G. Gr.; Lot's Wife, Glen Luffin, J. Fr.; Glencaple, 705, Hn.; 707, F. W. G., R. H. M.; five miles below Dumfries to confluence with Solway, W. St.

Estuarine mud of salt marshes; exposed to sun.

Œnanthe crocata. *Linn.* (Dead Man's Creesh, J. Shaw).

RECORDS: *Dfs.*—J. Shaw, 1882. *Kcd.*—J. M'Andrew, 1882. *Wgt.* J. M'Andrew, 1890.

LOCALITIES: *Nithsdale*—Nithside, Hn.; Annan, Dumfries road, S.-E.; Cluden, R. H. M.; Cargen burns, S.-E.; Tynron, J. Sh.; Holywood, Ad. and S. D. J.; Thornhill, R. A.   *Annandale*—Powfoot, S.-E.; Kirtle Mouth, S.-E.; Scrogg's Bridge, Tundergarth, Lockerbie, G. Bl.; Æ Water, abundant, S.-E.; very common small rivulets and waters at Moffat, J. T. J.   *Eskdale*—Irvine, C. Y.; Wauchope, S.-E.

Appears June 6 to 20. In mud, alluvium, boulder clay or sandstone soils; usually in shallow streams and ponds six to twelve inches deep; exposed to sun and partly wind-sheltered.

VISITORS: Bombus hortorum, Hn.; Vespa rufa, Andrena coitana, S.-E.; Rhingia rostrata, Hn.; Eristalis pertinax, abundant and sufficient; arbustorum, abundant and sufficient; tenax, abundant; Lucilia Cæsar, abundant; Syritta pipiens, Chrysogaster cemetorum, Anthomyia radicum, Chortophila sp., S.-E.

Æthusa cynapium. *Linn.*

RECORDS: (Escape?) *Dfs.*—Dr Davidson, 1891. *Kcd.*—P. Gray, 1850. *Wgt.*—J. M'Andrew, 1889.

LOCALITIES: *Nithsdale*—Three miles from Dumfries, P. Gr.; Glen, S.-E.; Dalgarnock Kirk, R. A.; Nith Mills, Sanquhar, Dr Dv.   *Annandale*—Ecclefechan, S.-E.; Lockerbie roadside, Smithy-Wamphray, Manse road, Kirkpatrick-Juxta, J. T. J.

Appears July 27. Moist or fairly dry places, cultivated ground, roadsides; usually wind-sheltered by long grass or hedges; in sun.

Ligusticum scoticum. *Linn.*

RECORDS: *Kcd.*—J. M'Andrew, 1882. *Wgt.*—Arnott, 1848.

LOCALITIES: *Along the shore*—Port Float, Arn.; Falboque Bay, Borgue, Newabbey, J. M'A. and F. R. C.

Meum athamanticum. *Jacq.* (Bald Money).

RECORDS: *Dfs.*—J. Shaw, 1882. *Kcd.*—P. Gray, 1850. *Wgt.*—Sir H. Maxwell, 1889.

LOCALITIES: *Nithsdale*—Cargen both sides, head of Glen, Th.; Routen Brig, P. Gr.; Stroquhan, G. C.; Glenquhargen Craig, J. Sh.; Trostan, Knocksting, J. Cr.; Carron, Gateslack, R. A.; Euchan,

Crawick, Kello, Dr Dv. *Annandale*—Whitecoombe, J. T. J. *Eskdale*—Eskdalemuir, J. Wn. and R. Bl.

Flowers June 20. Extends to about 1700 feet. Usually moist whinstone soils, alluvium, roadsides, or gravelly boulder clay; exposed to wind and sun.

Coriandrum sativum. *Linn.* (Coriander).

RECORD: (Escape) *Kcd.*—F. R. Coles, 1884.

LOCALITY: Roadside, Underwood, Tongland, F. R. C.

Crithmum maritimum. *Linn.*

RECORDS: *Kcd.*—G. N. Lloyd, 1837. *Wgt.*—Sibbald, 1674.*

LOCALITIES: *Along the shore*—Portpatrick, Mull of Galloway, Dr Br.; Burrowhead, J. M'A.; St. Ninians, Hn.; Kirkandrews, T. Bl.; Ross, Dr Br., G. N. Ll., G. C.; Torrs Point, J. T. S.; Balmae, G. N. Ll.; Balcary, G. N. Ll.; Castle Hill, Fq., C. E. M.; Port o' Warren, F. W. G., S.-E.; Douglas Hall, P. Gr.; Kirkbean, Arn.

Flowers July. On granite and whinstone rocks facing the sea; fully exposed.

Angelica sylvestris. *Linn.*

RECORDS: *Dfs.*—J. T. Johnstone, 1889. *Kcd.*—P. Gray, 1850. *Wgt.*—G. C. Druce, 1883.

LOCALITIES: *Nithsdale*—Dock wall, Dumfries. Hn.; Wood at Kingholm, S.-E.; Craighope Linn, Hn.; Sanquhar, Dr Dv. *Annandale*—Murrayfield, S.-E.; Moffat, J. T. J. *Eskdale*—Wauchope, Blockburn, S.-E.; Eskdalemuir, J. Wn. and R. Bl.

Flowers August 7. On rather moist leaf-mould, alluvium, and whinstone soils; in shade and wind-sheltered.

VISITORS: Polyommatus phloeas, Vespa sylvestris, Bombus terrestris, Halictus rubicundus, Prosopis brevicornis, Chilosia cestracea, Platychirius peltatus, Syrphus topiarus, Eristalis pertinax, horticola, and ten others, J. C. W., l. c., p. 245.

Peucedanum ostruthium. *Koch.* (Masterwort).

RECORDS: (Escape) *Dfs.*—W. Stevens, 1848. *Kcd.*—J. M'Andrew, 1882.

* "It groweth upon the rocks towards the sea in Galloway."—*Scotia Illustrata.*

LOCALITIES: *Nithsdale*—Carronbridge, W. St.; Tynron, J. Sh.; Glenquhargen, Hn. *Annandale*—Adam's Holm, Meikleholmside, J. T. J.
In flower, June 29.

Heracleum spondylium. Linn. (Cow Parsnip or Hogweed).

RECORDS: *Dfs.*—Dr Davidson, 1882. *Kcd.*—P. Gray, 1850. *Wgt.* J. M'Andrew, 1887.

LOCALITIES: Very common in all the valleys to 1300 feet.

Flowers June 26 to July. By roadsides, railway embankments, on cinders, granite, clay, etc.; usually slightly wind-sheltered and half-shaded.

VISITORS: Apis, Bombus lucorum, S.-E.; Vespa, abundant, R. A.; Chrysogaster nigrinus, Scatophaga stercoraria, Lucilia Cæsar, Ortalis omissa, Anthomyia pluvialis, S.-E.; Syrphus corollæ, nemorum, ribesii, Hn.

Heracleum giganteum.

RECORD: (Escape) *Dfs.*—J. T. Johnstone, 1893.
LOCALITY: Moffat.

Scandix pecten-veneris. Linn. (Venus' Comb, Shepherd's Needle).

RECORDS: (Escape) *Dfs.*—Auldgirth Station, Mrs Thomson, 1893.
In flower June 20. Casual plant in garden, Moffat, J. T. J.

Myrrhis odorata. Scop. (Cicely).

RECORDS: *Dfs.*—Dr Burgess, 1789. *Kcd.* and *Wgt.*—J. M'Andrew, 1882 and 1887.

LOCALITIES: *Nithsdale*—Kirkbean, C. E. M.; Kingholm, Glen, Grove, Irongray, C. E. M., S.-E.; Tynron, Moniaive, common, J. Cr.; Holywood, Dr Br.; Auldgirth, S.-E. *Annandale*—Tundergarth, Lockerbie House, G. Bl.; above Moffat, J. T. J. *Eskdale*—S.-E.

Flowers May 6 to May 18, J. T. J. Usually dry or rather moist roadsides, alluvium or whinstone soils up to 600 feet; as a rule wind-sheltered and half-shaded. It also occurs around some of the old towers, where it is probably a remnant of cultivation.

VISITORS: Allantus nothi, Empis opaca, trigramma, Hyelemyia strigosa, Psila fumitara, Siphona cristata, geniculata, Hydrotea dentipes, Anthomyia radicum, S.-E.

Conopodium denudatum. *Koch.* (Earthnut).

RECORDS: *Dfs.* and *Kcd.*—P. Gray, 1850. *Wgt.*—G. C. Druce, 1883.

LOCALITIES: Very common in all the valleys up to Sanquhar, Dr Dv.; Moffat, J. T. J.; White Hope, S.-E.; and Castle O'er, J. Wn. and R. Bl.

Flowers May 9 to June 4. On dry or moist soil, most common on boulder clay, humus or alluvium, but found on all soils; usually half-shaded and in part wind-sheltered.

VISITORS: Allantus nothi, Hn., S.-E.; Syrphus cinctellus, Syritta pipiens, Platychirius manicatus, Scatophaga lutaria, Siphona geniculata, Meligethes viridescens, Orchisa minor, S.-E.

Chærophyllum sativum. *Linn.*

RECORD: (Escape) *Dfs*—Miss Taylor, 1892.
LOCALITY: Glenzier, Canobie, E. Ty.

Chærophyllum temulum. *Linn.* (Chervil).

RECORDS: *Dfs.*—J. Cruickshank, 1829. *Kcd.*—P. Gray, 1850. *Wgt.*—J. M'Andrew, 1887.

LOCALITIES: *Nithsdale*—Brownhall, J. Cru.; very common near Dumfries, S.-E.; Auldgirth, S.-E. (not given for Sanquhar by Dr Dv.). *Annandale*—Annan and Caledonian Railway, very common, S.-E.; Correifron, Coates Hill, J. T. J. *Eskdale*—Canobie, very common, S.-E.

On pretty dry roadsides, cinders of railways, sandy gravel, etc.; in full exposure to wind and sun.

VISITORS: Siphona cristata, geniculata, Hyetodesia incana, Sepsis cynipsea, Eristalis arbustorum, S.-E.

Chærophyllum sylvestre. *Linn.*

RECORDS: *Dfs.*—Dr Davidson, 1886. *Kcd.*—J. M'Andrew, 1882. *Wgt.*—G. C. Druce, 1883.

LOCALITIES: *Nithsdale*—Very common Dumfries, J. M'A., S.-E.; Sanquhar, Dr Dv. *Annandale*—Dryfe, G. Bl.; Black's Hope, Correifron, J. T. J. *Eskdale*—Penton Linn, Woodslee, etc., S.-E.

Appears April 15, G. Bl.; May 4 to June 4, J. T. J. On dry or moist leaf-mould, roadsides, railways; usually half-shaded and part wind-sheltered.

Chærophyllum anthriscus. *Lam.*

RECORDS : *Kcd.*—P. Gray, 1850.* *Wgt.*—J. M'Andrew, 1893.

Caucalis nodosa. *Sm.*

RECORD : (Escape) *Wgt.*—J. M'Andrew, 1888.
LOCALITY : Portwilliam, J. M'A.

Caucalis anthriscus. *Huds.* (Hedge Parsley).

RECORDS : *Dfs.* and *Kcd.*—P. Gray, 1850. *Wgt.*—Rev. G. Wilson, 1893.

LOCALITIES: *Nithsdale*—Caerlaverock, Hn.; near Dumfries, M. J. H., P. Gr.; Sanquhar, Dr Dv. *Annandale*—Moffat, J. T. J. *Eskdale*—Langholm, S.-E.

Flowers July 7. On cindery waste ground, roadsides partly sheltered and shaded by hedges, etc.

VISITORS: Satyrus janira, Simæthis fabriciana, Halictus, Platychirius albimanus, Anthomyia radicum, Hyelemyia strigosa, and nine others, J. C. W.

Daucus carrota. *Linn.* (Carrot).

RECORDS : *Dfs.* and *Kcd.*—P. Gray, 1850. *Wgt.*—Arnott, 1848.

LOCALITIES : Common along the shore. Portpatrick, R. A.; Dunskey, Arn.; Portwilliam, St. Ninian's, Hn.; Ross, G. C.; Annan, Kirtlemouth, S.-E., also G. & S.W. Railway; Holywood, Ad. and S. D. J.; Rashbriggs road, R. A.; Sanquhar, Dr Dv.

On shingle by the shore, waste cindery ground inland; fully exposed.

Conium maculatum. *Linn.* (Hemlock).

RECORDS: *Dfs.* and *Kcd.*—P. Gray, 1850. *Wgt.*—G. C. Druce, 1883.

LOCALITIES : *Nithsdale*—Kingholm, S.-E.; Milldamhead, F. W. G.; Mouswald, Hn., S.-E.; Cummertrees, S.-E.; Auldgirth Station, J. Fn.; Morton Mains, R. A.; Sanquhar Castle, Dr Dv. *Annandale*—Scroggs Mill, abundant, G. Bl.; Beattock, Cornal Tower, J. T. J.

Flowers July 26. On waste ground, cinders or alluvium; usually wind-sheltered but sunny places. (Very likely an escape).

---

* Probably confusion of names.

VISITORS: Bombus lucorum, Hn.; Vespa, abundant and sufficient, Allantus nothi, Syritta pipiens, Hyelemyia strigosa, Telephorus fulvus, S.-E.

Hedera helix. *Linn.*

RECORDS: *Dfs.* and *Kcd.*—P. Gray, 1850. *Wgt.*—J. M'Andrew, 1890.

LOCALITIES: Very common on trees and ruins in all the valleys to about 1100 feet altitude.

Cornus sanguinea. *Linn.* (Dogwood).

RECORD: (Planted) *Dfs.*—*e.g.*, Blackwood Pond, S.-E.; Moffat, J. T. J.

Symphoricarpos racemosus. *D. C.*

RECORD: (Escape) *Dfs.*—G. Bell, 1893.
LOCALITIES: Thornhill, R. A.; Milke near Scroggs, G. Bl.

Adoxa moschatellina. *Linn.*

RECORDS: *Dfs.* and *Kcd.*—P. Gray, 1850. *Wgt.*—J. M'Andrew, 1889.

LOCALITIES: *Nithsdale*—Lincluden, C. E. M., and common Cluden, S.-E.; Grove, Hn.; Moniaive, J. Cr.; Holywood, Ad. and S. D. J.; Crawick Woods, Dr Dv. *Annandale*—Annan, Th.; Auchenbraith, S.-E.; Milke, G. Bl.; Moffat (common from 300 to 2300 feet), Black's Hope, J. T. J. *Eskdale*—Common by Esk to Langholm, Ewes Water, S.-E.; Castle O'er, J. Wn. and R. Bl.

Flowers March 17 to April 24. Usually moist leaf-mould, roadsides, or rarely sandy soil, bare of other plants, and sheltered by hedges or craigs; in full shade or half-shaded.

Sambucus nigra. *Linn.* (Elder).

RECORDS: *Dfs.* and *Kcd.*—P. Gray, 1850. *Wgt.*—J. M'Andrew, 1882.

LOCALITIES: Commonly planted, Portrack, Ak.; Sanquhar, Dr Dv.; Moffat, J. T. J.; Eskdalemuir, J. Wn. and R. Bl.

Deep rich soil in shade, A. M.

Sambucus ebulus. *Linn.* (Dwarf Elder, Bour Tree).

RECORDS: *Dfs.*—Dr Burgess, 1789. *Kcd.*—G. N. Lloyd, 1837.

LOCALITIES: *Nithsdale*—Near Dumfries, Caerlaverock Road, Dr Br.; Netherwood and Observatory, P. Gr.; Belzees, Tinwald, Dr Br.; Holywood, J. Sn.; Carronbridge, R. A.; Saw Mill, Elliock, Dr Dv. *Annandale*—Lochmaben, G. N. Ll.; Turnmoor Mill, Dryfe, J. Sn.

Prefers moist soils, W. Do.

VISITORS: Bombus lucorum, Halictus cylindricus, R. A.; Pollenia rudis, Hyetodesia incana, Anthomyia radicum, S.-E.; Pieris brassicæ, rapi, Vanessa Urticæ, atalanta, Satyrus janira, Characas graminis, Xylophasia polyodon, R. A.

Viburnum opulus. *Linn.* (Guelder Rose).

RECORDS: *Dfs.*—Dr Davidson, 1882. *Kcd.*—P. Gray, 1844. *Wgt.* —J. M'Andrew, 1889.

LOCALITIES: *Nithsdale*—Mavis Grove, Glen, P. Gr.; Carnsalloch, Castle-Douglas road, Hn.; Routen Brig, S.-E.; Drumlanrig, Scaur, R. A.; South Mains, Holm Walk, Dr Dv. *Annandale*—Ecclefechan, S.-E.; very common and established, Dryfe and Milkc, G. Bl.; Raehills, Lochwood, Middlegill, Frenchland, and Wellburn, J. T. J. *Eskdale*— Very common all along Esk, Byreburn, E. Ty., S.-E.; Eskdalemuir, J. Wn. and R. Bl.

VISITORS: Platychirius albimana, Eristalis pertinax, S.-E.

Lonicera periclymenum. *Linn.* (Honeysuckle).

RECORDS: *Dfs.* and *Kcd.*—P. Gray, 1850. *Wgt.*—J. M'Andrew, 1887.

LOCALITIES: *Nithsdale*—Near Dumfries, P. Gr.; Nithside, S.-E.; Cowhill, Ad. and S. D. J.; Dunscore, F. W. G.; Sanquhar, Dr Dv. *Annandale*—Common Moffat, J. T. J. *Eskdale* — Liddel Bridge, Wauchope, Bexburn, S.-E.; Castle O'er, J. Wn. and R. Bl.

Flowers May (1893); July 11 (1891). Usually on leaf-mould in shade and shelter.

VISITORS: Bombus hortorum, s., S.-E., J. C. W.; lapidarius, s., Hn.; pratorum, c.p., Hn.

Galium cruciata. *Scop.* (Bedstraw).

RECORDS: *Dfs.* and *Kcd.*—P. Gray, 1850. *Wgt.*—G. Wilson, 1893.

LOCALITIES: *Nithsdale*—S.-E.; Mavis Grove, Dr Gl.; Moniaive, J. Cr.; Tynron, J. Sh. *Annandale*—From Annan to Moffat very common, J. T. J.; Dryfe, Milke, G. Bl., S.-E. *Eskdale*—Very common everywhere, Wauchope, Langholm (to 1000 feet), S.-E.; Eskdalemuir, J. Wn. and R. Bl.

Flowers April 17, G. Bl.; May 5 to 15, J. T. J. Prefers water-holding soils, alluvials, boulder clay, roadside, humus, occasionally sand or shingle of shores and rivers, carboniferous sandstones, old walls ; exposed to wind and sun or shaded.

VISITORS : Platychirius clypeatus, albimana, Syrphus ribesii, cinctellus, Eristalis pertinax, Syritta pipiens, Helophilus frutetorum, Hyetodesia incana, Dolichopidæ, S.-E.

Galium verum. *Linn.* (Bedstraw).

RECORDS : *Dfs.* and *Kcd.*—P. Gray, 1850. *Wgt.*—G. C. Druce 1883.

LOCALITIES : *Nithsdale*—Common Troqueer, C. E. M.; Dumfries, F. W. G.; Auldgirth, etc., S.-E.; Brachead, Kirkconnel, Dr Dv. *Annandale*—Very common by shore at Newbie, Annan, Kirtlemouth, S.-E.; Beattock, Wamphray, S.-E.; Moffat, J. T. J. *Eskdale*—Glenzier, E. Ty.; Glencorf Burn, Wauchope, S.-E.; Castle O'er, J. Wn. and R. Bl.

Flowers June 12 to 27. Dry rather than moist alluvium, shingles of shore and rivers, roadsides, somewhat sandy holms, granite ; exposed or half-shaded and usually wind-sheltered.

VISITORS : Syrphus ribesii, abundant and sufficient ; Syritta pipiens, abundant and sufficient ; Hyelemyia sp., S.-E.; Siphona cristata, Hydrotea dentipes, Musca corvina, Hn.

Galium palustre. *Linn.*

RECORDS : *Dfs.* and *Kcd.*—P. Gray, 1850. *Wgt.*—G. C. Druce, 1883.
  b. Elongatum, *Dfs.*—Dr Grierson, 1882.
  c. Witheringii, *Dfs.* and *Kcd.*—J. M'Andrew, 1882.
      Wgt.—G. C. Druce, 1883.

LOCALITIES : *Nithsdale*—Cargen, C. E. M.; Grove Hills, S.-E.; Holywood, Ad. and S. D. J.; Roniston, S.-E.; Closeburn, Dr Gl.; Sanquhar, Dr Dv. *Annandale*—Annan town, Kirtle, S.-E.; Lochmaben, P. Gr. (var. *c.*), J. M'A.; Scroggs, S.-E.; Evan Water, and to a mile above Moffat on Annan, S.-E. *Eskdale*—Sark, Blockwell, Glencorfe, Bexburn (to 1400 feet), Archie Hill, S.-E.

Flowers June 13 to 16, J. T. J. Marshy places, on sandstone, alluvium, boulder clay, etc.; usually partly wind-sheltered by long herbage ; in sun.

VISITORS : Syrphus ribesii, Platychirius peltatus, Siphona cristata, Anthomyia radicum, S.-E.

Galium uliginosum. *Linn.*

RECORDS: *Dfs.*—J. M'Andrew, 1882. *Kcd.*—J. Singer, 1843. *Wgt.*—J. M'Andrew, 1893.

LOCALITIES: *Nithsdale*—By Nith, Th. *Annandale*—Lochmaben, J. M'A.; Bcld Craig Burn, S.-E.; Moffat, J. T. J.

Flowers August 10. Wet places (in rocks?); shaded and sheltered.

Galium saxatile. *Linn.*

RECORDS: 751. *Dfs.* and *Kcd.*—P. Gray, 1850. *Wgt.*—G. C. Druce, 1883.
   752. Sylvestre, *Dfs.*—J. Sadler, 1857. *Kcd.*—J. M'Andrew, 1882.

LOCALITIES: 751 very common in all valleys to Sanquhar, Dr Dv.; Hartfell and Coombe Craig (to 2500 feet), J. T. J.; White Hope and Causeway Grain (to 1600 feet), S.-E. 752, Grey Mare's Tail, J. Sd., E. F. L.

Flowers June 1 to 10; 752, July. Dry places on roadsides, peat of higher hills, till, shingles, etc.; usually fully exposed to wind and sun; rarely shaded by pine or other trees. 752 in a very wind-sheltered spot and more humid atmosphere.

VISITORS: Syrphus ribesii, Hn.; cinctellus, abundant, S.-E.; Platychirius albimana, abundant, S.-E.

Galium mollugo. *Linn.*

RECORDS: *Dfs.*—Dr Davidson (1883?). *Kcd.*—J. Fraser, 1843.
   749. Erectum, *Dfs.*—G. Bell, 1893.

LOCALITIES: *Nithsdale*—Virgin Hall, R. A.; Euchan Road, Barr Cottages, Dr Dv. *Annandale*—Annan Road, 749, Th.; Caledonian Railway, Lockerbie to Nethercleugh, 749, G. Bl. *Eskdale*—Old Gretna, S.-E.; roadside near Langholm, S.-E.

Pretty dry, cindery banks or roadsides; wind-sheltered, but in full sunlight.

VISITORS: Perineima nassata, Syritta pipiens, abundant, Chlorosia formosa, Hyetodesia incan, Empis, sp., Spilogaster, sp., S.-E.

Galium boreale. Linn.

RECORDS: *Dfs.*—Dr Burgess, 1789. *Kcd.*—G. N. Lloyd, 1837. *Wgt.*—G. C. Druce, 1883.

LOCALITIES: *Nithsdale*—Martington Ford, near Dumfries, P. Gr.; Blackwood, Dr Br.; Waterside, R. A.; Sanquhar, common, Dr Dv. *Annandale*—Black's Hope, Correifron, Grey Mare's Tail (reaching 2400 feet), Coombe Craigs, J. T. J.

Flowers from July 13 to 27. Usually on dry whinstone rocks in full exposure.

Galium aparine. Linn.

RECORDS: *Dfs.* and *Kcd.*—P. Gray, 1850. *Wgt.*—G. C. Druce 1883.

LOCALITIES: Very common in all the valleys reaching Sanquhar, Dr Dv.; Moffat, J. T. J.; Castle O'er, J. Wn. and R. Bl.

Flowers from July 3 to 27. Usually rather moist ground by roadsides, shingles and sand of rivers and shore; partly shaded and windsheltered.

VISITORS: Vespa sylvestris (perhaps accidental), Anthomyia radicum, abundant, Ichneumon, sp., S.-E.

Galium tricorne. With.

RECORD: (Escape) *Dfs.*—Sanquhar, Dr Davidson.

Asperula odorata. Linn.

RECORDS: *Dfs.* and *Kcd.*—P. Gray, 1850. *Wgt.*—G. C. Druce, 1883.

LOCALITIES: *Nithsdale*—Common Kirkbean, Colonel's Wood, Hn.; Chainston, S.-E.; Cowhill, Ad. and S. D. J.; Blackwood Linn, Exc.; Sanquhar, Dr Dv. *Annandale*—Common Annan, Hn.; very common Æ Water, Kirkmichael, S.-E.; Milke and Dryfe, common, S.-E.; Wellburn, Craigieburn, J. T. J. *Eskdale*—Very common Kirkandrews, E. Ty.; Canobie, Tarras, Bexburn, and Langholm woods, S.-E.; Castle O'er, J. Wn. and R. Bl.

Appears April 23, G. Bl.; flowers from May 4 to 27, J. T. J. Usually on wet humus (often bare of other plants) of woods and linns; in complete shade and shelter.

Asperula taurina. *Linn.*

RECORDS: (Escape) *Dfs.*—Putts, Egypt Bridge, May 3, J. T. Johnstone, 1890.

Sherardia arvensis. *Linn.*

RECORDS: *Dfs.* and *Kcd.*—P. Gray, 1850. *Wgt.*—G. C. Druce, 1883.

LOCALITIES: *Nithsdale*—Dumfries, P. Gr.; Newton, S.-E.; Auldgirth, Exc.; Thornhill, R. A.; Sanquhar, Dr Dv. *Annandale*—Common by shore, S.-E.; Annan, Hn.; Cummertrees, Ak.; Moffat, common, Archbank, J. T.].

Appears April 28, G. Bl.; May 15 to June 5, J. T. J. Pretty dry places, on sandy gravelly soil, cinders of stations, light holmland; in full exposure.

Valeriana dioica. *Linn.*

RECORDS: *Dfs.*—J. Singer, 1843. *Kcd.*—J. Fraser, 1843.

LOCALITIES: *Nithsdale*—Rare Mouswald, Tinwald, J. Sn. *Annandale*—Not common Carterton, G. Bl.; Peter's Moss, Wamphray, J. T. J.; Kirkpatrick-Juxta, F. A. H.; Evan Water and Garpol, J. T. J. and J. Sd.; Beld Craig, Brackenside Burn, Cornal Tower, J. T. J., S.-E.; Capplegill, J. T. J.; Selcoth (800 feet), Crofthead, S.-E.; Carterton Corrie, G. Bl. *Eskdale*—Common second railway bridge from Langholm, Blackknowe Burn, Stennies Water, Meikledale, S.-E.

Appears May 1, G. Bl.; June 5, J. T. J. On wet slopes, till, or rocky whinstone soils; partly wind-sheltered by long grass; in full sun.

Valeriana officinalis. *Linn.*

RECORDS: *Dfs.* and *Kcd.*—P. Gray, 1850. *Wgt.*—G. C. Druce, 1883.

LOCALITIES: *Nithsdale*—Cargen, common, C. E. M.; Dumfries, P. Gr., Hn.; Glen, M. J. H.; Cowhill, Ad. and S. D. J.; Sanquhar, Dr Dv. *Annandale*—Common about Annan and Kirtle mouth by shore, S.-E.; Springkeld, S.-E.; Milke, G. Bl.; Wamphray to Beattock, by Annan, S.-E.; Moffat, common (to 1000 feet), Beeftub, J. T. J. *Eskdale*—Sark, Liddel bridge, Bentpath road, one mile above Langholm, Bexburn, S.-E.

Flowers June 28 to July 5, J. T. J. Usually on wet leaf-mould, boulder clay or holms, shingles of shore and rivers; in shade and usually part wind-sheltered in long grass.

VISITORS: Apis, abundant; Bombus lucorum, Vespa, sp., S.-E.;
Eristalis tenax, Hn.; pertinax, arbustorum, Syrphus ribesii, S.-E.

Valeriana pyrenaica. *Linn.*

RECORDS: *Dfs.*—W. Stevens, 1848. *Kcd.*—P. Gray, 1848. *Wgt.*—Sir H. Maxwell, 1889.

LOCALITIES: *Nithsdale* — Jardington, Routen Brig, and other places along Cluden, P. Gr., Hn., S.-E.; Drumlanrig, W. St. *Annandale*—Granton, Greenhill Cottages, J. T. J. *Eskdale*—Kirkandrews Wood, E. Ty.; Irvine, and two places between Canobie and Langholm, right bank of Esk, S.-E. and C. Y.

On wet humus or sandstone, in full shade and shelter.

VISITORS: Siphona geniculata, very abundant, S.-E.

Valerianella olitoria. *Moench.*

RECORDS: *Dfs.*—P. Gray, 1850. *Kcd.*—J. M'Andrew, 1882. *Wgt.*—J. M'Andrew, 1889.

LOCALITIES: *Along the shore*—Southerness to Arbigland, Th., Exc.; Annan to Kirtle mouth, S.-E. *Nithsdale*—Holywood, Ad. and S. D. J.

On sand or shingles, partly shaded by other plants or fully exposed.

Valerianella dentata. *Poll.*

RECORDS: *Dfs.*—J. Singer, 1843. *Kcd.*—F. R. Coles, 1882. *Wgt.*—G. C. Druce, 1883.

Var. *b. Wgt.*—G. Macnab, 1837.

LOCALITIES: Dee, F. R. C.; Tinwald, J. Sn.

Dipsacus sylvestris. *Linn.*

RECORDS: (Escape) *Dfs.*—Dr Davidson, 1882. *Kcd.*—(Arnott?), 1844. *Wgt.*—Miss Hannay, 1893.

LOCALITIES: Garliestown, Hn. *Annandale*—Milton Woods, J. T. J.

Scabiosa succisa. *Linn.*

RECORDS: *Dfs.* and *Kcd.*—P. Gray, 1850. *Wgt.*—G. C. Druce, 1883.

LOCALITIES: *Nithsdale*—Lochanhead, F. W. G.; Blackwood, Dumfries, P. Gr.; Cample, Thornhill, Hn.; Sanquhar, Dr Dv. *Annandale*—Very common Evan Water, Beeftub, to 1200 feet, Crofthead and Craigmichen Scaurs, S.-E. *Eskdale*—Gilnockie, S.-E.; Castle O'er, J. Wn. and R. Bl.

Flowers June 19 to July 7, J. T. J. In wet or boggy pastures, on whinstone soils or boulder clay; exposed to sun and wind.

VISITORS: Vanessa atalanta, Polyommatus phlœas, Pieris napi (also J. C. W.), Brassicæ, Charæas graminis, Hydræcia nictitans, R. A.; Bombus muscorum, Hn., R. A., J. C. W.; lucorum, R. A.; pratorum, R. A., J. C. W.; terrestris, J. C. W.; Apathus campestris, R. A., J. C. W.; Allantus nothi, Vespa vulgaris, R. A.; Halictus rubicundus, cylindricus, J. C. W.; Eristalis tenax, J. C. W., R. A.; intricarius, J. C. W.; arbustorum, R. A.; Criorhina oxyacantha, Morellia hortorum, Siphona geniculata, S.-E.; Melanostoma scalare, Syrphus balteatus, Helophilus pendulus, J. C. W.; Hyetodesia incana, Chortophila, Homalomyia, Hn.

Scabiosa columbaria. *Linn.*

RECORD: *Dfs.*—S. W. Carruthers, 1885.

LOCALITIES: *Nithsdale*—Cowhill, Ad. and S. D. J. *Annandale*— S. W. Ca. *Eskdale*—Near the border, Canobie, E. Ty.

Scabiosa arvensis. *Linn.*

RECORDS: *Dfs.*—P. Gray, 1850. *Kcd.*—J. Fraser, 1843. *Wgt.*— G. Wilson, 1893.

LOCALITIES: *Nithsdale*—Kirkbean, Dr Dv.; Portland Place Nursery, Glen, Hn.; Dalskairth, Terregles, Hn., P. Gr.; Cowhill, Ad. and S. D. J.; Tynron, J. Sh.; Drumbuie, Dr Dv. *Annandale*—Netherplace, Linghill, Norwood, and Douglas Hall Bridge (Caledonian railway), G. Bl. *Eskdale*—Canobie, E. Ty.; Railway Bridge, Liddel, S.-E.

Pretty dry places on alluvium, cindery railway banks, roadsides; in sun, but wind-sheltered by slopes of river banks, cuttings or hedges.

Eupatorium cannabinum. *Linn.*

RECORDS: *Dfs.*—P. Gray, 1850. *Kcd.*—P. Gray, 1848. *Wgt.*— G. C. Druce, 1883.

LOCALITIES: *Along the shore*—The Ross, G. C.; Port o' Warren, St., S.-E.; Port Ling, Heugh o' Laggan, P. Gr.; Kirkbean, Dr Gl.; Dock Wall, Dumfries, Hn., C. E. M.

On granite soil at base of cliffs; often partly wind-sheltered.

Aster tripolium. *Linn.*

RECORDS: *Dfs.* and *Kcd.*—P. Gray, 1850. *Wgt.*—Dr Arnott, 1848.

LOCALITIES: *Along the shore*—Port Logan, Port Gill, Arn.; Port-William, Hn.; Southerness, G. C.; New Quay to Caerlaverock, P. Gr., F. W. G., Hn., C. E. M., S.-E.

Marshy estuarine mud ; exposed to sun and wind.

VISITORS: Polyommatus phlœas, J. C. W.; Bombus lucorum, Hn.; lapidarius, pratorum, muscorum, terrestris, Eristalis pertinax, arbustorum, æneus, tenax, horticola, Lucilia cornicina, Calliphora sepulchralia, J. C. W.; Sarcophaga carnaria, Hn.; Syrphus corollæ, Siphona geniculata, Pollenia rudis, Anthomyia radicum, Notiphila cinerea, S.-E.; Meligethes æneus, J. C. W. (and 5 anthomyids, l.c., p. 239).

Solidago virgaurea. *Linn.* (Golden Rod).

RECORDS: *Dfs.* and *Ked.*—P. Gray, 1850 and 1848. *Wgt.*—G. C. Druce, 1883.

LOCALITIES: *Nithsdale*—Cluden Mills, Hn., C. E. M., S.-E.; by Nith, F. W. G., Hn., C. E. M.; Moniaive, J. Cr.; Sanquhar, Dr Dv. *Annandale*—Very common Garpol, S.-E.; Beeftub, S.-E.; Wellburn, J. T. J.; Saddleyoke, W. Ca.; Loch Skene (to 2000 feet), S.-E.; Craigmichen, S.-E. *Eskdale*—Gilnockie Bridge and Langholm road, S.-E.; Castle O'er, J. Wn. and R. Bl.

Appears June 13 to 27, J. T. J. On dry rocks, whinstone, granite, sandstone or boulder clay ; exposed to sun or shaded and always partly wind-sheltered in glens or linns.

VISITORS: Apis, Hn., S.-E.; Apathus quadricolor, Hn.; Onesia sepulchralis, Morellia hortorum, abundant and sufficient ; Siphona geniculata, Hydrotea dentipes, Hyetodesia incana, S.-E.

Bellis perennis. *Linn.* (Daisy).

RECORDS: *Dfs.* and *Ked.*—P. Gray, 1850. *Wgt.*—G. C. Druce, 1883.

LOCALITIES : Very common in all the valleys, reaching 1400 feet at White Hope Edge.

Flowers often on January 1. Most common on fairly dry sandy alluvium or gravel, but on all soils except peat ; usually fully exposed in short turf to wind and sun.

VISITORS : Eristalis pertinax, Siphona geniculata, abundant ; Anthomyia radicum, S.-E.

Filago germanica. *Linn.*

RECORDS: *Dfs.*—Dr Gilchrist, 1867. *Ked.*—P. Gray, 1850. *Wgt.* —G. C. Druce, 1883.

LOCALITIES : *Nithsdale*—Tinwald, Dr Gl. *Annandale*—Between Powfoot and Newbie, J. Fn.; Cummertrees, Ak.; Raehills, Exc.*

On dry banks.

* All these require confirmation.

Filago minima. *Fr.*

RECORDS: *Dfs.*—J. Shaw, 1882. *Kcd.*—J. M'Andrew, 1882. *Wgt.*—C. C. Bailey, 1883.

LOCALITIES: *Nithsdale*—Locharbriggs, F. W. G.; Holywood Station, S.-E.; Nith Bridge, Thornhill, J. Cr., R. A.; Tynron, J. Sh. *Annandale*—Powfoot to Newbie, J. Fn.; Holms, Evan Water, J. T. J. *Eskdale*—Tarras Roads, S.-E.

Appears about July 7, J. T. J. On dry sand, gravel or stony roadsides, often shingles of rivers; in full exposure to sun and wind.

Gnaphalium silvaticum. *Linn.* (Cudweed).

RECORDS: *Dfs.* and *Kcd.*—P. Gray, 1850. *Wgt.*—G. C. Druce, 1883.

LOCALITIES: *Nithsdale*—Ruthwell, Dr Gl.; Galla Hill, Terregles, F. W. G.; Holywood, Cowhill, Ad. and S. D. J.; Castle-Douglas road, C. E. M.; Laught Wood, R. H. M.; Sanquhar, Dr Dv. *Annandale*—Common Lockerbie, G. Bl.; Tundergarth, S.-E.; Gardenholm, Granton, fields, Archbank, J. T. J. *Eskdale*—Canobie, E. Ty.; Castle O'er, J. Wn. and R. Bl.

Appears July 14 to August 12. In pastures or cornfields; usually on boulder clay exposed to sun; partly wind-sheltered by herbage.

VISITORS: Anthomyia radicum, abundant, S.-E.

Gnaphalium uliginosum. *Linn.*

RECORDS: *Dfs.* and *Kcd.*—P. Gray, 1850. *Wgt.*—G. C. Druce, 1883.

LOCALITIES: *Nithsdale*—Ruthwell, Dr Gl.; Lochanhead, F. W. G.; Dumfries, P. Gr.; Cowhill, Ad. and S. D. J.; Thornhill, R. A.; Sanquhar, Dr Dv. *Annandale*—Ecclefechan, S.-E.; Poldean to Newbigging, J. T. J.; and damp roadsides Moffat, J. T. J. *Eskdale*—Kirkandrews, E. Ty.

Appears about July 7. On damp roadsides, part wind-sheltered.

Antennaria dioica. *R. Br.*

RECORDS: *Dfs.* and *Kcd.*—P. Gray, 1850. *Wgt.*—G. C. Druce, 1883.

LOCALITIES: *Nithsdale*—Lochanhead, F. W. G.; Grove, S.-E.; Merkland, Thornhill, R. R.; Closeburn, Dr Gl.; Sanquhar, common, Dr Dv. *Annandale*—Common Eskrig, G. Bl.; Kinnelhead, S.-E.; Black's Hope and Whitecoombe (to 2500 feet), J. T. J. *Eskdale*—Castle O'er, J. Wn. and R. Bl.

Appears May 15 to 17, J. T. J. Common in dry pastures over 500 feet, usually shallow soil in short grass over whinstone rocks and fully exposed; rarely well sheltered.

Inula crithmoides. *Linn.*

RECORDS : *Kcd.*—Dr Burgess, 1789. *Wgt.*—Dr Arnott, 1848.

LOCALITIES : Mull, Arn , J. M'A.; Creetown, Exc.; Arbigland, Dr Br.

Pulicaria dysenterica. *Gærtn.*

RECORDS: (Escape) *Wgt.*—Mr Maughan. *Flora Scotica*, 1789.

LOCALITIES : Mull, Mr Maughan ; Monreith Bay, J. M'Andrew.

Bidens cernua. *Linn.*

RECORDS : (Escape) *Dfs.*—J Singer, 1843. *Kcd.*—G. N. Lloyd, 1837. *Wgt.*—J. M'Andrew, 1889.

LOCALITIES : *Nithsdale*— Common Loch, Dabton Loch, R. A. ; *Annandale*—Lochar, G. C. ; Dornock, T. Bl. ; Lochmaben, J. Sn.; Jardine Hall, S. W. Ca.

Bidens tripartita. *Linn.*

RECORDS : *Dfs.*—G. N. Lloyd, 1837. *Kcd.* and *Wgt.* — J. M'Andrew, 1889.

LOCALITIES : Lochar Moss, P. Gr.; Southend, Castle Loch, Lochmaben, G. N. Ll., J. Sn., and Exc.

Chrysanthemum leucanthemum. *Linn.*

RECORDS: *Dfs.* and *Kcd.*—P. Gray, 1850. *Wgt.*—G. C. Druce, 1883.

LOCALITIES : Very common in all the valleys, reaching 2000 feet at Correifron.

Appears May 31 to June 5, J. T. J. Perhaps commonest on sandy or light fairly dry soil in full exposure (not peat).

VISITORS : Calliphora erythrocephala, Cynomyia mortuorum, Hyctodesia basalis, Lucilia Cæsar, abundant, S.-E. Also three unnamed species.

Chrysanthemum segetum. *Linn.*

RECORDS: *Dfs.* and *Kcd.*—P. Gray, 1850. *Wgt.*—Dr Macnab, 1836.

LOCALITIES: *Nithsdale*—Common Ruthwell, Dr Gl.; Dumfries, F. W. G.; Cowhill, Ad. and S. D. J.; Sanquhar, Dr Dv. *Annandale*—Very common near sea, Annan, S.-E. *Eskdale*—Gretna Green, S.-E.

On corn fields, dry sandy or gravelly soil, shingles of shore, cinders of railways, boulder clay, waste ground; fully exposed to wind and sun.

VISITORS: Allantus nothi, Eristalis arbustorum, abundant, Lucilia Cæsar, Platychirius clypeatus, Hyctodesia basalis, Hydrotea dentipes, Anthomyia radicum, S.-E.

Chrysanthemum parthenium. *Pers.*

RECORDS: (Escape) *Dfs.*—Near Kirkandrews, Miss E. Taylor, 1891. *Kcd.*—J. M'Andrew, 1882.

In corn fields.

Matricaria inodora. *Linn.*

RECORDS: 823. Inodora, *Dfs.*—P. Gray, 1850. *Kcd.* – Dr Greville? *Wgt.*—J. H. Balfour (1836?)
b. Salina, *Dfs.*—G. F. Scott-Elliot, 1891.
824. Maritima, *Kcd.*—Mrs Gilchrist-Clark, 1890. *Wgt.* —G. C. Druce, 1883.

LOCALITIES: *Nithsdale*—Holywood, Ad. and S. D. J.; Auldgirth, Exc.; Thornhill, R. A.; Sanquhar, Dr Dv. *Annandale*—Common on shore Annan to Powfoot, Torduff, var *b.*, S.-E.; Moffat, common, J. T. J.

Appears May 27 to June 28, J. T. J. On shingle stones or bare places by shore, inland, chiefly on cinders and waste ground, railways; fully exposed to wind and sun.

VISITORS: Polyommatus phlœas, Choreutis myllerana, Simæthis fabriciana, Halictus cylindricus, rubicundus, Sphecodes affinis, Prosopis brevicornis, Odynerus pictus, J. C. W.; Allantus nothi, Hn.; Syritta pipiens, Sphærophoria, menthrastii, Eristalis nemorum, S.-E.; tenax, pertinax, J. C. W.; Ascia podagrica, J. C. W.; Syrphus ribesii, Musca corvina, Hydrotea dentipes, Ichneumon, sp., S.-E.; Lucilia Cæsar, Siphona cristata, Hn.; Anthomyia radicum, J. C. W. (and twelve other kinds), l.c., p. 237.

Matricaria chamomilla. *Linn.*

RECORD: (Escape) *Dfs.*—Sanquhar, Dr Davidson, 1882 (?)

Anthemis cotula. *Linn.*

RECORD: (Escape) *Dfs.*— Bankhead Coal-pit on railway, Dr Davidson, 1886.

Anthemis arvensis. *Linn.*

RECORD: (Escape) *Kcd.*—Balmae, J. Fraser, 1841-4.

Anthemis nobilis. *Linn.*

RECORD: (Escape) *Kcd.*—Rev. J. Fraser, 1843.
LOCALITIES: Meiklewood, Tongland, F. R. C.; Kirkcudbright, J. Fr.

Anthemis tinctoria. *Linn.*

RECORD: (Escape) *Dfs.*—Miss Hannay, 1893.

Achillea ptarmica. *Linn.*

RECORDS: *Dfs.* and *Kcd.*—P. Gray, 1850. *Wgt.*—G. C. Druce, 1883.
LOCALITIES: *Nithsdale*—Ruthwell, F. W. G.; dockwall, Dumfries, S.-E.; Nithside, Hn, C. E. M.; Cargenwater, Hn.; Sanquhar, Dr Dv. *Annandale*—Annan, common, S.-E.; Moffat, common, J. T. J. *Eskdale* —Very common Langholm, C. Y.; Wauchope, S.-E.; Castle O'er, J. Wn. and R. Bl.

Appears June 30, J. T. J. On rather wet or dry alluvium, boulder clay, granite, roadsides, shingles; usually in sun and windy places.

VISITORS: Eristalis arbustorum, Hydrotea dentipes, Hn.

Achillea millefolium. *Linn.* (Yarrow).

RECORDS: *Dfs.* and *Kcd.*—P. Gray, 1850. *Wgt.*—G. C. Druce, 1883.
Var. villosa, *Wgt.*—Miss Hannay, 1893.
LOCALITIES: *Nithsdale*—Common but sparse to Sanquhar, Dr Dv. *Annandale*—Moffat, common, J. T. J. *Eskdale*—Common Liddel, Tarras and Esk to Castle O'er, J. Wn. and R. Bl., S.E.

Appears May 30 to June 18, J. T. J. Most common moist alluvium, roadsides, railways, shingles; in full sun and partly wind-sheltered.

VISITORS: Pieris rapæ, napi, Polyommatus phloeas, Hydrœcia nictitans, Simæthis fabriciana, Choreutis myllerana, J. C. W.; Allantus nothi, Hn., S.-E.; Melanostoma mellina, Morellia hortorum, Hn.; Pollenia, S.-E.; Syrphus balteatus, J. C. W.; ribesii, S.-E.; Eristalis tenax, pertinax, J. C. W.; Helophilus pendulus, Hn.; Hæmatophora pluvialis, Hn.; Siphona cristata, Hyelemyia strigosa, Hydrotea dentipes, Tenthredo ater, Sarcophago, S.-E.; Syritta pipiens, Oliviera lateralis, J. C. W. Also 12 Anthomyids, Coleoptera, etc., l.c., p. 238.

Tanacetum vulgare. *Linn.* (Tansy).

RECORDS: *Dfs.* and *Kcd.*—P. Gray, 1850. *Wgt.*—G. C. Druce, 1883.

LOCALITIES: *Nithsdale*—Dumfries, P. Gr.; Moniaive, J. Cr.; Old Barr, Sanquhar, Dr Dv. *Annandale*—Moffat (garden escape), J. T. J. *Eskdale*—Castle O'er, J. Wn. and R. Bl.

Probably an escape established in shady sheltered places.

Artemisia maritima. *Linn.*

RECORDS: *Kcd.* and *Wgt.*—Dr Graham, 1836.

LOCALITIES: *Along the shore*—Port Yerrick, J. M'A.; Borgue, F. R. C.; St. Mary's Isle, G. Gr., G. N. Ll.; Cruggleton Castle, East Burrow Head, G. Gr.

Artemisia vulgaris. *Linn.*

RECORDS: *Dfs.* and *Kcd.*—P. Gray, 1850. *Wgt.*—J. M'Andrew, 1887.

LOCALITIES: *Nithsdale*—Southwick, Mrs St.; New Quay, P. Gr., Hn.; Sanquhar, Dr Dv. *Annandale*—Annan water side, Well Road, Archbank, J. T. J. *Eskdale*—Castle O'er, J. Wn. and R. Bl.

Flowers September 6. Along sheltered roadsides; exposed to sun.

Tussilago farfara. *Linn.* (Colt's Foot).

RECORDS: *Dfs.* and *Kcd.*—P. Gray, 1850. *Wgt.*—G. C. Druce, 1883.

LOCALITIES: *Nithsdale*—Dumfries, Hn., P. Gr.; Tynron, J. Sh. *Annandale*—Common by Æ, S.-E.; Milke (to 300 feet), S.-E.; Beattock Hill summit, S.-E.; Moffat, J. T. J. *Eskdale*—Tarras, Gilnockie to Langholm, S.-E.; Castle O'er, J. Wn. and R. Bl.

Appears March 5 to 23, S.-E. On dry, sandy alluvium, sandstones, granite and railway cinders; fully exposed to sun but often wind-sheltered.

Tussilago petasites. *Linn.*

RECORDS: *Dfs.* and *Kcd.*—P. Gray, 1850. *Wgt.*—G. C. Druce, 1883.

LOCALITIES: *Nithsdale*—Dumfries, P. Gr., Hn.; Moniaive, J. Cr.; Sanquhar, Dr Dv. *Annandale*—Annan, Th.; Dumcrieff, J. T. J.; Milke, S.-E. *Eskdale*—Canobie, Langholm, S.-E.; Castle O'er, J. Wn. and R. Bl.

Appears March 29. Up to 300 feet on mossy whinstone rocks, stony alluvium, less common on sandy soil; usually wind-sheltered.

Tussilago petasites × farfara.

RECORD: *Dfs.*—Dr Burgess, 1789.

LOCALITY: Eskbank, on borders of Annandale, near Netherby, Dr Br.

Senecio vulgaris. *Linn.* (Groundsel).

RECORDS: *Dfs.* and *Kcd.*—P. Gray, 1850. *Wgt.*—G. C. Druce, 1883.

LOCALITIES: Very common everywhere.

Blooms all the year, J. T. J. Usually in gardens, cultivated fields, roadsides, and waste ground.

VISITOR: Siphona cristata.

Senecio viscosus. *Linn.*

RECORDS: *Dfs.*—J. T. Johnstone, 1890. *Kcd.*—G. Gordon, 1836.

LOCALITIES: *Nithsdale*—Auldgirth, Exc. *Annandale*—Middlegill, Slaughter-house to Railway Bridge, Moffat, Beattock Station, J. T. J., S.-E.

Appears July. Cinders of railways, waste ground; in full exposure.

Senecio silvaticus. *Linn.*

RECORDS: *Dfs.* and *Kcd.*—P. Gray, 1850. *Wgt.*—G. C. Druce, 1883.

LOCALITIES: *Nithsdale*—Southwick, Mrs Stew.; Dumfries, P. Gr.; Island, Cluden Mill, Ak.; Sanquhar, Dr Dv. *Annandale*—Annan to

Powfoot shore, S.-E.; Wellburn, Blacklaw Burn, Old Carlisle Road, J. T. J.

Appears June 28, J. T. J. On shingles, bare red trap; usually half-sheltered.

VISITOR: Siphona geniculata, S.-E.

Senecio aquaticus. *Huds.*

RECORDS: *Dfs* and *Kcd.*—P. Gray, 1850. *Wgt.*—G. C. Druce, 1883.

LOCALITIES: *Nithsdale*—Cluden Mills, Cargen, Glen, S.-E.; Sanquhar, Dr Dv. *Annandale*—Torduff, Kirtle, S.-E.; common Moffat, J. T. J. *Eskdale*—Common Esk, Glenzier, E. Ty., S.-E.; Castle O'er, J. Wn. and R. Bl.

Appears June 22, J. T. J. On wet alluvial patches by rivers.

VISITORS: Syrphus ribesii, Hyetodesia basalis, incana, Hydrotea dentipes, Anthomyia radicum, Siphona geniculata, S.-E.

Senecio Jacobea. *Linn.* (Ragwort).

RECORDS: *Dfs.* and *Kcd.*—P. Gray, 1850. *Wgt.*—G. C. Druce, 1883.

Var. Flosculosus, *Wgt.*—G. C. Druce, 1883.

LOCALITIES: Very common in all the valleys.

Appears June 17, J. T. J. On old shingles by sea, holmlands, railway cinders; in full exposure to wind and sun.

VISITORS: Bombus, sp., S.-E.; Halictus rubicundus, subfasciatus, R. Sc.; Andrena denticulata, Apathus quadricolor, Eristalis arbustorum, Helophilus frutetorum, Syritta pipiens, Morellia hortorum, Hydrotea dentipes, Nysson dimidiatus, Telephorus fulvus, S.-E.

Senecio sarracenicus. *Linn.*

RECORDS: *Dfs.*—J. Wightman, 1893. *Kcd.*—Mr Maughan, 1789. *Wgt.*—Sir H. Maxwell, 1889.

LOCALITIES: Hedges near Longtown, Cumberland, Mr Jackson, 1796. *Annandale*—Cocklet's Dryfe, J. A. Wg. (Castle-Douglas, New-Galloway, Mg.; Borgue, Kirkmabreck, J. Fr. All Kirkcudbright).

Doronicum pardalianches. *Linn.* (Leopard's Bane).

RECORDS: *Dfs.*—Dr Burgess, 1796. *Kcd.*—J. Fraser, 1844.

LOCALITIES: *Nithsdale*—Dalswinton woods, Exc.; Drumlanrig, W.

St.; Shaw's Bridge, Thornhill, R. A. *Annandale*—Springkeld, abundant, S.-E.; Hoddam Castle, Dr Br.; New Mills, Greenhill Cottages, left bank Birnock, J. T. J. *Eskdale*—Irvine, C. Y.

Appears May 2 to June 2, J. T. J. On beech humus; in full shade and shelter.

Arctium lappa. *Linn.* (Burdock).

RECORDS: *Dfs.* and *Kcd.*—P. Gray, 1850. *Wgt.*—G. C. Druce, 1883.
851. Majus, *Kcd.*—J. M'Andrew, 1888.
853. Minus, *Dfs.*—Dr F. W. Grierson, 1882. *Kcd.* and *Wgt.*—C. C. Newbould, 1883.
854. Intermedium, *Dfs.*—J. Fingland, 1887. *Wgt.*— G. C. Druce, 1883.

LOCALITIES: *Nithsdale*—Ruthwell, 853, F. W. G.; Dumfries, P. Gr.; Caerlaverock, 854, M. J. H.; Glen, Hn.; Cowhill, Ad. and S. D. J.; Elliock Saw Mill, 853, Dr Dv.; Glencairn, J. Cr. *Annandale*—Kirtlebridge, J. T. J.; Beattock to Wamphray, by Annan, S.-E.; Craigieburn (500 feet), J. T. J. *Eskdale*—Kirkandrews, E. Ty.; Glentarras, Gilnockie bridge, above Langholm, S.-E.; Castle O'er, J. Wn. and R. Bl.

Appears July 27, J. T. J. Usually on dry, sandy or clayey alluvium, roadsides, boulder clay, granite; in sun, but usually part wind-sheltered.

VISITORS: Bombus muscorum, Hn., S.-E.; lucorum, abundant; lapidarius, Derhamellus, Anthidium manicatum, S.E.

Serratula tinctoria. *Linn.* (Sawwort).

RECORDS: *Dfs.*—D. Bell, 1837. *Kcd.*—G. N. Lloyd, 1837.

LOCALITIES: *Along the shore*—Nunnery to Senwick, Borgue, Mgh.; along Dee from Kirkcudbright to Kingour, Loch Ken, Mgh., G. N. Ll., Fq., J. M'A.; Portling, Douglas Hall, J. M'A. *Eskdale*—D. Bell.

Saussurea alpina. *D. C.*

RECORD: *Dfs.*—Dr Walker, 1762-1783.

LOCALITIES: *Annandale*—Very common Black's Hope, W. St., J. T. J.; Whitecoombe on both sides, Midlaw and above Loch Skene, Dr Wl., J. T. J., S.-E. (from 1750 to 2000 feet).

Appears July 1 to 19, J. T. J. On pretty moist broken whinstone soil; part shaded and wind-sheltered by narrowness of ravines, S.-E.; or on dry exposed rocks as left side Black's Hope, J. T. J.

Carduus marianus. *Linn.* (Milk Thistle).

RECORD: (Escape) *Wgt.*—Creetown, J. M'Andrew, 1882.

Carduus acanthoides. *Linn.*

RECORDS: (Escape) *Dfs.*—Excursion Natural History Society, 1882. *Kcd.*—G. C. Druce, 1883. *Wgt.*—J. M'Andrew, 1889.

LOCALITIES: *Wigtown*—Along the shore, Port Yerrick, J. M'A. *Kirkcudbright*—Ross, G. C.

Carduus pycnocephalus. *Jacq.*

RECORDS: (Escape) *Dfs.*—Annan to Kirtlemouth, G. F. Scott-Elliot, 1891. *Kcd.*—J. M'Andrew, 1885. *Wgt.*—G. C. Druce, 1883.

Carduus lanceolatus. *Linn.* (Spear Thistle).

RECORDS: *Dfs.*—J. H. Balfour (1836?). *Kcd.*—J. M'Andrew, 1882. *Wgt.*—G. C. Druce, 1883.

LOCALITIES: Common in all the valleys to 900 feet.

Appears about July 10. On rather dry roadsides, holms, corn fields; in sun, but usually part wind-sheltered.

VISITORS: Bombus hortorum, Hn., S.-E.; lucorum, abundant, S.-E.; Megachile willughbiella, Platychirus clypeatus, abundant; Syrphus ribesii, Eristalis intricarius, Empis livida, Sericomyia borealis, Hyetodesia basalis, Vanessa Urticæ, S.-E.

Carduus palustris. *Linn.* (Marsh Thistle).

RECORDS: *Dfs.* and *Kcd.*—P. Gray, 1850. *Wgt.*—G. C. Druce, 1883.

LOCALITIES: Common in all the valleys to at least 800 feet.

Appears June 13 to 26, J. T. J. On wet rather than dry holms, boulder clay, roadsides, shingles of sea shore, cinders, etc.; in sun and wind.

VISITORS: Vespa, sp.; Bombus pratorum, reg. and sufficient, S.-E.; hortorum, Hn.; Platychirus, sp.; Empis livida, S.-E.

Carduus arvensis. *Curt.* (Creeping Thistle).

RECORDS: *Dfs.* and *Kcd.*—P. Gray, 1850. *Wgt.*—G. C. Druce, 1883.

LOCALITIES: Common in all the valleys to 1800 feet, Correifron.

On pretty dry or wet sandy and gravelly fields, boulder clay, alluvium, whinstone soils, etc.; in full sun and wind exposure.

VISITORS: Bombus pratorum, Allantus nothi, S.-E.; Halictus cylindricus, Morellia hortorum, Hn.; Onesia sepulchralis, Platychirus clypeatus, Hydrotea dentipes, Siphona geniculata, Anthomyia radicum, Ablyteles cerinthius, S.-E.

Carduus heterophyllus. *Linn.* (Melancholy Thistle).

RECORDS: *Dfs.*—A. Sibbald, 1820. *Kcd.*—Rev. J. Fraser, 1843.

LOCALITIES: *Nithsdale*—Glenquhargen, J. Sh., Hn., S.-E.; Euchan, Glen, Crawick, Nith, Sanquhar, Dr Dv., R. A.; Thornhill, R. A. *Annandale*—Balgray, Dryfe, G. Bl.; Jardine Hall, Th.; Lochmaben, C. E. M.; Evan Water, Beerholm, Nethermurthat, Crofthead, Annan, J. T. J. *Eskdale*—Junction Glencorfe and Wauchope, Bilholm, Lynholm, Sb., S.-E.; Glenshanna, S.-E.; Castle O'er, J. Wn. and R. Bl.; Eskdalemuir Kirk, S.-E. Elevation up to 1000 feet.

Appears June 22 to July 4, J. T. J. On moist holmland (close to streams) or humus; usually part shaded and half wind-sheltered by banks, etc.

VISITORS: Bombus muscorum, abundant and sufficient; lucorum, abundant and sufficient; Rhingia rostrata, abundant, S.-E.

Onopordon acanthium. *Linn.*

RECORD: (Escape) *Dfs.*—Miss J. Wilson and Mr R. Bell (Castle O'er, 1892)?

Carlina vulgaris. *Linn.*

RECORDS: *Dfs.*—J. Shaw, 1882. *Kcd.* and *Wgt.*—G. N. Lloyd, 1837.

LOCALITIES: *Along the shore*—Portpatrick, J. H. Bl.; Port Logan, G. N. Ll., Arn.; Mullfarm, G. N. Ll.; St. Ninians, Whithorn, Hn., St.; Balcary, G. N. Ll.; Port o' Warren, Port Ling, St.; Auchencairn, J. C. W. *Inland* — Clarencefield, W. Hg.; Waulkmill Farm, J. M'A.; Euchan Glen, J. Sh., Dr Dv.

Dry, grassy slopes.

Centaurea nigra. *Linn.*

RECORDS: *Dfs.* and *Kcd.*—P. Gray, 1850. *Wgt.*—G. C. Druce, 1883.

LOCALITIES: Common in all the valleys to 600 feet.

On moist or dry roadsides, old wall tops, boulder clay, holms, cinders, granite, etc.; in sun but usually wind-sheltered by banks or hedges.

VISITORS: Apis, J. C. W.; Bombus lapidarius, Hn., S.-E.; muscorum, Hn., S.-E.; hortorum, S.-E.; terrestris, Scrhimshiranus, pratorum, J. C. W.; Apathus quadricolor, Hn.; Anthidium manicatum, J. C. W.; Satyrus janira, Hn., J. C. W.; Rhingia rostrata, abundant, Syrphus balteatus, S.-E., J. C. W.; Eristalis æneus, tenax, pertinax, J. C. W.; Sarcophaga, sp., Morellia hortorum, Pollenia rudis, S.-E.; Argynnis aglaia, Pieris napi, rapæ, Polyommatus phlœas, Vanessa Urticæ, and eight others, J. C. W.

Centaurea cyanus. *Linn.*

RECORDS: (Escape) *Dfs.*—R. Armstrong, 1888. *Kcd.*—J. M'Andrew, 1882. *Wgt.*—Rev. J. Gorrie, 1893.

LOCALITY: Dumfries Station, Th.; Auldgirth, Rashbriggs, R. A.

Tragopogon pratensis. (Goat's Beard).

RECORDS: *Dfs.*—J. Singer, 1843. *Kcd.*—Rev. J. Fraser, 1882.

LOCALITIES: *Nithsdale*—Merschead, Kirkbean, J. Fr.; Mouswald, Tinwald, J. Sn. *Annandale*—Broomhouses, to one mile from Nethercleugh, Caledonian Railway, and Lockerbie, to one mile towards Lochmaben, G. Bl.; Kirtlebridge, Caledonian Railway, S.-E.

On dry slopes (chiefly cinders and stones); in sun, wind-sheltered by hay and banks.

VISITORS: Anthomyia radicum, S.-E.

Leontodon hispidus. *Linn.* (Hawkbit).

RECORDS: *Dfs.*—Dr Davidson, 1890. *Kcd.*—J. M'Andrew, 1882.

LOCALITIES: *Nithsdale*—Very common about Dumfries on Nith and Cluden, S.-E.; very common Sanquhar, Dr Dv. *Annandale*—Common on shore, Annan, S.-E.; common Moffat, J. T. J.; and Grey Mare's Tail, S.-E. *Eskdale*—Tarras, Mosspaul, and to 1400 feet Causey Grain, S.E.

Appears June 4 to 20, J. T. J. On dry, rarely wet holmlands, estuarine flats, boulder clay, whinstone soils; usually in short turf exposed to sun and partly wind-sheltered by growing in valleys or ravines.

VISITORS: Apis mellifica, Bombus, sp., Anthomyia radicum, S.-E.

Leontodon autumnalis. *Linn.*

RECORDS: *Dfs.*—(Teste, A. Bennett). *Wgt.*—G. C. Druce, 1883, *b.* pratensis, *Dfs.*—Dr Davidson, 1886.

LOCALITIES: *Nithsdale*—Very common Cowhill, Ad. and S. D. J.; source of Euchan, Glenglass (var. *b.*), Dr Dv. *Annandale*—Very common Coates Hill, Moffat, S.-E.; Annan, Beattock, J. T. J.; Evan, S.-E.; Correifron (2000 feet), J. T. J. *Eskdale*—Along Esk to above Langholm, S.-E.

Appears August 11, J. T. J. On dry or moist holmlands, roadsides, boulder clay, whinstone soils, river shingles, etc.; in full sun and not very windy places.

VISITORS : Lycæna icarus, Simæthis Fabriciana, Crambus, Bombus terrestris, muscorum, Halictus rubicundus, Platychirius manicatus, Syrphus ribesii, Sphærophoria scripta, Brachyopa, bicolor, Sericomyia borealis, Eristalis æneus, tenax, pertinax, Sciara, Trichopthicus cunctans, Anthomyia radicum, Hydrellia griseola, Sitones puncticollis, Calocoris fulvomaculatus, bipunctans, Miris lævigatus, Acocephalus, J. C. W.

Leontodon hirtus. *Linn.*

RECORDS : *Dfs.* and *Kcd.*—Rev. J. Fraser (before 1882). *Wgt.*—J. M'Andrew, 1886.

LOCALITIES : *Along the shore*—Portpatrick, J. M'A.; Glenstocking, Rockliffe, Glenluffin, J. Fr., Fq.; Merse, J. Fr.

Hypochæris radicata. *Linn.* (Cat's Ear).

RECORDS : *Dfs.* and *Kcd.*—P. Gray 1850. *Wgt.*—G. C. Druce, 1883.

LOCALITIES : Very common in all the valleys up to 1200 feet.

Appears June 14 to 27. On dry or moist holms, estuarine alluvials, roadsides, shingles of rivers, granite, etc.; in sun and part wind-sheltered by long grass.

VISITORS : Anthomyia radicum, Hydrotea dentipes, S.-E.

Sonchus arvensis. *Linn.* (Sow-Thistle).

RECORDS : *Dfs.* and *Kcd.*—P. Gray, 1850. *Wgt.*—G. C. Druce, 1883.

LOCALITIES : *Nithsdale*—Near Dumfries, P. Gr.; Sanquhar, Dr Dv. *Annandale*—Cummertrees, Ak.; Moffat, J. T. J. *Eskdale*—Little Tarras, Gilnockie, Burnfoot, S.-E. (not above 800 feet).

Appears August 25, J. T. J. On rather moist boulder clay, holm-lands (corn fields, turnips, etc.), shingles; in sun but part wind-sheltered by corn or long grass, waste ground, J. T. J.

VISITORS : Morellia hortorum, Hydrotea dentipes, Dolichopods, S.-E.

Sonchus oleraceus. *Linn.*

RECORDS : 1009. Oleraceus, *Dfs.*—P. Gray, 1850. *Kcd.*—J. M'Andrew, 1882. *Wgt.*—G. C. Druce, 1883.

1010. Asper, *Dfs.*—Dr Davidson, 1886. *Kcd.*—F. R. Coles, 1882. *Wgt.*—G. C. Druce, 1883.

LOCALITIES: *Nithsdale*—Not common Sanquhar, Dr Dv. *Annandale*—Common on shore, Annan to Kirtle mouth, Hoddam, S.-E.; Murrayfield, G. Bl.; Poldean, Hydropathic, Moffat, J. T. J. *Eskdale*—Langholm, S.-E.; Castle O'er, J. Wn. and R. Bl.

Appears July 3, J. T. J. On fairly dry roadsides, waste ground, boulder clay, shingles of shore and rivers, cinders, old walls, etc.; usually partly shaded and sheltered by dykes, hedges, etc.

VISITORS: Syrphus ribesii, Morellia hortorum, Platychirius clypeatus, Siphona geniculata, Anthomyia radicum, S.-E.

Taraxicum dens-leonis. *D. C.* (Dandelion).

RECORDS: *Dfs.* and *Kcd.*—P. Gray, 1850. *Wgt.*—G. C. Druce, 1883.

c. Palustre, *Dfs.*—Dr Davidson, 1886.

LOCALITIES: Common in all the valleys to 2000 feet.

Appears March 26 to April 24, J. T. J. Var. *a.* on pretty dry roadsides, holms, sandy soil, whinstone rocks, old walls, etc.; in sun or half-shaded, exposed to wind or part sheltered. Var. *c.*, wet hill pastures, Dr Dv.

VISITORS: Empis opaca, Hyctodesia jucana, Siphona cristata, Anthomyia radicum, S.-E.; Halictus rubicundus, villosulus, Andrena albicans, Wilkella, R. Sc.

Crepis virens. *Linn.*

RECORDS: *Dfs.*—Dr Davidson, 1886. *Kcd.*—J. M'Andrew, 1882. *Wgt.*—G. C. Druce, 1883.

LOCALITIES: Common in all the valleys to about 800 feet.

Appears June 7 to 17, J. T. J. On moist or dry railway banks, roadsides, whinstone soils, holms, shingles of shore, etc.; in sun but wind-sheltered by long grass or banks.

VISITORS: Allantus nothi, Andrena bicolor, coitana, furcata, Platychirius manicatus, Morellia hortorum, Anthomyia pluvialis, radicum, Homalomyia, Onesia sepulchralis, Pollenia, sp., Hydrotea dentipes, S.-E.; Telephorus fulvus, Hn.

Crepis biennis. *Linn.*

RECORD: (Escape) *Dfs.*—J. Shaw, 1892.

LOCALITY: Oatfield, Holmhill, Tynron, J. Sh.

Crepis hieracioides. *Jacq.*

RECORDS: *Dfs.* Rev. E. F. Linton, 1889. *Kcd.*—F. R. Coles, 1883.
LOCALITIES: *Annandale*—Correifron, Grey Mare's Tail, Craigmichen Scaur (to 1000 feet), E. F. L. and J. T. J.
Appears July 28 to August 25, J. T. J. On moist whinstone rocks, half shaded and well sheltered from wind in narrow corries.

Crepis paludosa. *Moench.*

RECORDS: *Dfs.*—J. Sadler, 1858. *Kcd.*—F. R. Coles, 1882. *Wgt.* —G. C. Druce, 1883
LOCALITIES: *Nithsdale*— Very common to Sanquhar, Dr Dv.; Tynron, J. Sh. *Annandale*—Common Wamphray, S.-E.; very common in linns (Beld Craig) and corries (Grey Mare's Tail). *Eskdale*—Common Black Knowe Burn, Blochburn, S.-E.; Castle O'er, J. Wn. and R. Bl. (to at least 1800 feet).
Appears May 28, S.-E.; to June 19, J. T. J. On wet holms, shingles, boulder clay, whinstone rocks in shade or half shaded; sheltered from wind by long grass or in narrow valleys.
VISITORS: Allantus nothi, Platychirius clypeatus, Chrysogaster metallina, sp., Helophilus frutetorum, Eristalis nemorum, Pollenia rudis, S.-E.

Cichorium intybus. *Linn.* (Chicory).

RECORD: (Escape) *Dfs.*—T. Brown or J. Wilson, 1882.
LOCALITIES: *Nithsdale*— Locharbriggs, J. Wl.; Auchenessnane, T. Br.; Lochside, Newton, S.-E.
In gravelly, dry fields, S.-E.

Hieracium. (Group Pilosella).

RECORDS: 892. Pilosella, *Dfs.* and *Kcd.*—P. Gray, 1850. *Wgt.*— G. C. Druce, 1883.
893. Aurantiacum (escape) *Dfs.*—J. Wightman, 1892.
LOCALITIES: 892 very common in all the valleys to 1600 feet. Appears May 31, J. T. J. 893, *Annandale*—Railway, Lockerbie, J. Wg., G. Bl., S.-E.; Moffat, J. T. J.; Tundergarth, G. Bl.

892 on hard, dry, stony ground, cinders of railways (also 893), whinstone and trap rocks covered by shallow soil, light soils exposed to sun and wind in short turf.

Hieracium. (Group Alpina Genuina).

RECORD: 896. holosericum, *Ked.*—J. M'Andrew, 1885.
LOCALITY: Milyea (2000 feet), J. M'A.

Hieracium. (Group Alpina Nigrescentia).

RECORDS: 903. Nigrescens, *Dfs.*—J. T. Johnstone, 1890.
912. Centripetale, *Dfs.*—E. F. Linton, 1893.

LOCALITIES: *Annandale*—Black's Hope, 903, J. T. J.; Correifron, 903, 912, J. T. J., E. F. L.; Midlaw Burn, Selcoth, 912, E. F. L.; Lochanburn, 912, J. T. J.

Appears July to August, J. T. J.

Hieracium. (Group Cerinthoidea).

RECORDS: 917. Callistophyllum, *Dfs.*—E. F. Linton, 1893.
920. Iricum, *Dfs.*—Dr Burgess, 1789.
923. Langwellense, *Dfs.*—E. F. Linton, 1893.

LOCALITIES: *Nithsdale*—Elliock Bridge, 920, Dr Dv. *Annandale*—Black's Hope, 920, 923, J. T. J., E. F. L.; Correifron, Grey Mare's Tail, 920, Dr Br., J. Sd., J. T. J.; Selcoth, 923, E. F. L.; Midlaw Burn, 917, E. F. L.

Appears July to August. Somewhat sheltered and in moist atmosphere on whinstone rocks.

Hieracium. (Group Oreadea).

RECORDS: 929. Schmidtii, *Dfs.*—J. T. Johnstone, 1893.
935. Rubicundum, *Dfs.*—E. F. Linton, 1890.
938. Argenteum, *Dfs.*—E. F. Linton, 1889.
939. Nitidum, *Dfs.*—J. T. Johnstone, 1892.
940. Sommerfeltii, *Dfs.*—E. F. Linton, 1893.
942. Onosmoides.
  b. Buglossoides, *Dfs.*—E. F. Linton, 1893.
943. Saxifragum, *Dfs.*—Mr Backhouse, 1850.*

LOCALITIES: *Annandale*—Moffat, 929, 935, J. T. J., E. F. L.; Crofthead, 938, J. T. J.; Beeftub, 938, J. T. J.; Black's Hope, 938, 943, 940, 942, J. T. J., E. F. L.; Correifron, 938, 943, J. T. J.; Grey Mare's Tail, 938, 942, E. F. L.; 943, Bk.; Selcoth, 942, E. F. L.; Andrew's Whinnie, 939, J. T. J.

Appears July to August. On dry rarely wet whinstone rock, 938; mudstones, 939; usually exposed, sometimes wind-sheltered.

---

* According to Mr Linton, Backhouse's form is the preceding plant onosmoides, *b*. buglossoides.

Hieracium. (Group Vulgata).

RECORDS: 945. Stenolepis.
b. anguinum, *Dfs.*—W. R. Linton, 1893.
952. Murorum, *Dfs.*—P. Gray, 1850.
k. ciliatum, *Dfs.*—J. T. Johnstone, 1893.
w. sarcophyllum, *Dfs.*—E. F. Linton, 1893.
953. Euprepes, *Dfs.*—E. F. Linton, 1893.
958. Cæsium.
c. pallidum, *Dfs.*—J. Sadler, 1858.
964. Duriceps micracladium, *Dfs.*—E. F. Linton, 1893.
967. Vulgatum, *Dfs.*—W. Carruthers, 1864. *Kcd.*—P. Gray, 1850. *Wgt.*—G. C. Druce, 1883.
969. Stenophyes, *Dfs.*—W. R. Linton, 1893.
971. Angustatum, *Dfs.*—E. F. Linton, 1893.
974. Diaphanoides, *Dfs.*—G. F. Scott-Elliot, 1892.
977. Gothicum, *Dfs.*—J. T. Johnstone, 1891.
978. Sparsifolium, *Dfs.*—J. T. Johnstone, 1890.
679. Rigidum *g.* tridentatum, *Dfs.*—Dr Davidson, 1886. *Kcd.*—J. M'Andrew, 1885.

LOCALITIES: *Nithsdale*—Newton, Dalveen Pass, S.-E.; Thornhill, 958, S.-E.; Carron Glen, Sanquhar, 958, J. Sd., S-E.; Dr Dv.; Euchan, 979 *g.*, Dr Dv. *Annandale*—Beeftub, 978, J. T. J.; Lockerbie, 958, 974, S.-E.; Beld Craig, 958, S.-E.; Croft Head, 958, 964, Wellburn, 958, J. T. J.; Black's Hope, 969, W. R. L.; Selcoth, 964, J. T. J.; Lochanburn, 964, J. T. J.; Duff of Kinnel, 978, J. T. J.; Moffat, 952, *k. w.*; 945, J. T. J.; Craigmichen Scaur, 978, J. T. J. *Eskdale*—14 miles north of Langholm, 971, S.-E.; Gilnockie, 958 *c.*, S.-E.; Harperwhat, Billholm, 967, S.-E.

Appears, 952 and 967, May 30, S.-E.; remainder July to August, J. T. J. The commonest forms on sandstones are 967, 974, and 958 *c.*; on trap-rocks, 967 and 974; on boulder clay, 958 *c.*; on whinstones, 967, 971, and 978; most are exposed or half-shaded (967).

VISITORS: (Dryfe and Lochanburn, May). Empis billineata, Eristalis pertinax, Syrphus albostriatus, Anthomyia radicum, S.-E.

Hieracium. (Group Prenanthoides).

RECORD: 987. Prenanthoides, *Dfs.*—E. F. Linton, 1889.

LOCALITIES: *Nithsdale*—Carserig, Elliock Sawmills, Dr Dv. *Annandale*—Grey Mare's Tail, E. F. L.; Correifron, J. T. J. *Eskdale*—Canobie, S.-E.

On moist rocks, half-shaded and wind-sheltered in long grass.

Hieracium. (Group Foliosa).

RECORDS: 989. Strictum* *Dfs.*—E. T. Linton, 1891.
990. Corymbosum, *Dfs.*—Dr Davidson, 1886. *Ked.*— J. M'Andrew, 1885.
991. Auratum, *Dfs.*—Dr Davidson, 1886.
992. Crocatum, *Dfs.*—W. Carruthers, 1864. *Ked.*— *Fide* J. M'Andrew, 1844.

LOCALITIES: *Nithsdale* — Southerness, 992, J. M'A. ; Ryehill of Nith, 992, Dr Dv.; Sanquhar, 991, Dr Dv. *Annandale* — Garpol, 992, W. Ca., J. T. J.; Rowantree Grain, 989, J. T. J.; Frenchland burn, 989, 992, 991, J. T. J.; Cornal Tower burn, 989, 992 ; Gallows Hill, 989, J. T. J.; Black's Hope, Spoonburn, 989, 992, J. T. J. *Eskdale*— Burnfoot, 992, S.-E.

Appears August 1. Usually in shaded and moist places on sandstones, boulder clay, etc.

Hieracium. (Group Sabauda).

RECORDS: 994. Boreale, *Dfs.*—Dr Davidson, 1886. *Ked.*—J. T. Syme, 1836. *Wgt.*—G. C. Druce, 1883.

LOCALITIES : *Nithsdale*—Holywood, S.-E. ; Sanquhar, Dr Dv. *Eskdale*—Canobie, Langholm, S.-E.

On sandstones, granite, leaf mould ; in shade and wind-sheltered.

Hieracium. (Group Umbellata).

RECORDS: *Dfs.*—J. Shaw, 1882. *Ked.*—Dr Arnott, 1844. *Wgt.*— J. M'Andrew, 1891.

LOCALITIES: *Nithsdale*—Southerness, J. M'A.; Knockenjig Ford, Dr Dv.; Tynron, J. Sh. *Annandale*—Alton, J. T. J. *Eskdale*—Gilnockie, S.-E.

Appears July to August. On sandstone walls, part shaded and sheltered.

Lapsana communis. *Linn.* (Nipplewort).

RECORDS: *Dfs.* and *Ked.*—P. Gray, 1850. *Wgt.*—G. C. Druce, 1883.

LOCALITIES: *Nithsdale* —Caerlaverock, F. W. G.; very common by Nith and Cluden, S.-E.; Sanquhar, Dr Dv. *Annandale*—Solway Bridge,

---

* This appears to be now considered a variety of Crocatum.

S.-E.; Tundergarth, G. Bl.; Moffat, J. T. J. *Eskdale*—Langholm, Wauchope, S.-E.; Castle O'er, J. Wn. and R. Bl.

Appears June 11 to 14. On dry humus, roadsides, waste ground, boulder clay, whinstone soils, gravelly soils; in shade or half shaded; usually wind-sheltered.

VISITORS: Andrena albicans, Syrphus ribesii, Siphona geniculata, Hyelemyia strigosa, Anthomyia radicum, S.-E.

Lobelia dortmanna. *Linn.*

RECORDS: *Dfs.*—G. N. Lloyd, 1837. *Kcd.*—G. Gordon, 1837. *Wgt.*—G. C. Druce, 1883.

LOCALITIES: *Nithsdale*—Loch Kindar, G. Go., P. Gray, E. M. C.; Lotus Loch, Hn., P. Gr.; Keir; Loch Urr (700 feet), Morton Loch, J. Sh. *Annandale*—Lochmaben, J. Sn., G. N. Ll.

In pretty deep quiet lakes with a muddy bottom.

Jasione montana. *Linn.* (Sheep's Bit).

RECORDS: *Dfs.* and *Kcd.*—P. Gray, 1850 and 1848. *Wgt.*—J. H. Balfour (1836 ?).

LOCALITIES: *Nithsdale*—Very common near Dumfries, S.-E.; Cluden Mill, Hn.; Glen, F. W. G.; Holywood, Portrack, S.-E.; Cowhill, Ad. and S. D. J.; Thornhill sandstone quarries, Hn., R. A.; Sanquhar, Dr Dv. *Annandale*—Seashore at Annan, Torduff, S.-E.; Dryfesdale Cemetery, Bishop's Cleugh, Lockerbie, G. Bl.; Carpol, J. Bl.; Moffat, J. T. J. *Eskdale*—Kirkandrews, E. Ty.

Appears June 22 to 28, J. T. J. On dry, sandy, or stony banks and fields, old walls; in shade or exposed to sun; usually in short grass and exposed to wind.

VISITORS: Pieris napi, J. C. W.; Bombus lucorum, Hn.; Halictus, sp., S.-E.; Platychirius clypeatus, peltatus, albimana, Siphona geniculata, Anthomyia radicum, S.-E.

Campanula trachelium. *Linn.*

RECORDS: (Escape) *Dfs.*—Miss Hannay, 1893. *Kcd.*—F. R. Coles, 1883.

LOCALITIES: Tongland, F. R. C.; Holywood, Hn., S.-E.

Campanula latifolia. *Linn.*

RECORDS: *Dfs.*—J. Singer, 1843. *Kcd.*—P. Gray, 1844. *Wgt.*—J. M'Andrew, 1893.

LOCALITIES: *Nithsdale*—Kirkbean, Exc.; Newabbey, Hn.; Tinwald, J. Sn.; Jardington, Martington, F. W. G.; Dock wall, Hn.; Lincluden, Grove, P. Gr., M. J. H.; above Routen Brig on Cairn, S.-E.; Moniaive, J. Cr.; Tynron, J. Sh., R. A; Clauchries, R. A.; Euchan, Newark, Polskeoch, Crawick, Dr Dv. *Annandale*—Annan, J. Sn., J. T. J.; Tundergarth Linns, Lamonbie Mill, G. Bl.; Wamphray, J. T. J. *Eskdale*—Kirkandrews, E. Ty.; very abundant, Bilholm to Bentpath, S.-E.

Appears August 12, J. T. J. On rather moist humus or alluvium; in shade or half-shaded and well sheltered from wind.

VISITOR: Bombus pratorum (on newly opened flowers), S.-E.

Campanula rapunculoides. *Linn.*

RECORDS: (Escape) *Dfs.*—G. Bell, 1893. *Kcd.*—F. R. Coles, 1883.

LOCALITIES: Tongland, F. R. C.; Castle Loch near Lockerbie road, G. Bl.; Dalton, J. Wg.

Campanula rotundifolia. *Linn.* (Bluebell).

RECORDS: *Dfs.* and *Kcd.*—P. Gray, 1850. *Wgt.*—G. C. Druce, 1883.

LOCALITIES: *Nithsdale*—Very common to Sanquhar, Dr Dv. *Annandale*—Very common (to 2000 feet), Loch Skene, S.-E. *Eskdale*—Common Wauchope, road to Bentpath, S.-E.; Castle O'er, J. Wn. and R. Bl.

Appears June 20 to July 5. On pretty dry whinstone soils, roadsides, boulder clay, etc.; usually half-shaded and part wind-sheltered or quite exposed.

VISITORS: Vanessa Urticæ, J. C. W.; Bombus terrestris, J. C. W.; lucorum, S.-E.; Andrena gwynana, R. Sc.; Siphona cristata, Hyelemyia strigosa, Dolichopids, S.-E.; Anthomyia radicum, S.-E., J. C. W.; Meligethes, J. C. W.

Campanula persicifolia. *Linn.*

RECORD: (Escape) *Dfs.*—J. Singer, 1843. (Johnstone parish).

Vaccinium myrtillus. *Linn.* (Blaeberry).

RECORDS: *Dfs.*—W. Keddie, 1854. *Kcd.*—P. Gray, 1848 *Wgt.*—G. C. Druce, 1883.

LOCALITIES: *Nithsdale*—Common Dalskairth, Terregles, P. Gr.; Craighope Linn, Scaur Water, S.-E.; Sanquhar, Dr Dv. *Annandale*—Gallows Hill, Moffat, Kd.; Hartfell and Whitecoombe (to 2400 feet),

S.-E. *Eskdale*—Common to Castle O'er, J. Wn. and R. Bl., and Mosspaul, S.-E.

Appears May 12 to 17, J. T. J. On wet mossy rocks or humus of woods and linns; in shade and shelter or fully exposed on peat and whinstone rocks, both at low and high altitudes.

VISITORS: Bombus terrestris, muscorum, abundant and sufficient, S.-E.

Vaccinium uliginosum. *Linn.*

RECORD: *Dfs.*—J. T. Johnstone, 1891.

LOCALITIES: *Annandale*—(1800 to 2000 feet), Whitecoombe, J. T. J. *Eskdale*—Castle O'er, J. Wn. and R. Bl. (specimen not seen).

On soil near damp whinstone rocks. Have never seen it in flower or fruit, only in foliage, J. T. J.

Vaccinium vitis idæa. *Linn.*

RECORDS: *Dfs.*—J. Singer, 1843. *Kcd.*—J. M'Andrew, 1882.

LOCALITIES: *Nithsdale*—Trostan Hill (1250 feet), J. Cr.; Sanquhar district, Dr Dv. *Annandale*—Raehills, J. Sn.; Queensberry, R. A.; Hartfell, Black's Hope and Whitecoombe (not uncommon from 800 to 2400 feet), S. W. Ca., J. T. J., S.-E.

Appears June 17 to 22, J. T. J. On dry exposed whinstone rocks, covered with short turf.

VISITORS: Rhamphomyia, sp.; Empis, sp., S.-E.

Vaccinium oxycoccos. *Linn.*

RECORDS: *Dfs.*—Rev. J. Little, 1834. *Kcd.*—P. Gray, 1850. *Wgt.* —G. C. Druce, 1883.

LOCALITIES: *Nithsdale*—Glensone, J. Wl.; Kirkconnel, E. M. C.; Lochar Moss, F. W. G.; Harv., Lotus, Hn.; Terregles, P. Gr.; Irongray, C. E. M.; common Glencairn, J. Cr.; Drumcork, Scaur, R. A.; Townmoor, Sanquhar, Dr Dv. *Annandale*—Templand, S.-E.; Breygillhead, etc., Moffat, S. W. Ca. and J. T. J.; Beeftub, Kd. *Eskdale*— The Flow, S.-E.; Langholm, C. Y. Elevation 250 to 900 feet.

Appears May to June. On sphagnum becoming peat; fully exposed.

VISITORS: Siphona cristata, geniculata, S.-E.

Arctostaphylos ulva-ursi. *Spreng.* (Bearberry).

RECORDS: *Dfs.*—J. Singer, 1843. *Kcd.*—Rev. J. Fraser, 1843.

LOCALITY: *Annandale*—Whitecoombe (2400 feet), J. T. J.; Corriefron (2200 feet), J. T. J.

In fruit May 28, 1893, J. T. J.; in good flower June 3, 1894. On dry whinstone rocks; fully exposed. Grows in matted masses where found.

Andromeda polifolia. *Linn.*

RECORDS: *Dfs.*--Dr Burgess, 1789. *Kcd.*—P. Gray, 1848. *Wgt.*—G. C. Druce, 1883.

LOCALITIES: *Nithsdale*—Solway Moss, Dr Br.; Lochar Moss, Dr Gl.; Kirkconnel Moss, T. Bl., E. M. C.; Maxwelltown Loch, Terregles, P. Gr.; Black Loch, Dr Dv. *Annandale*—Shillahill Bog, W. M. H. M.; Dalfibble, Stanemoor, Dr Br.; Spedlings, J. Wg.; Templand Moss, W. M. H. M.; Johnstone, J. Sn. *Eskdale*—T. B. Bl., *fide* S. W. Ca.

Appears May. On wet sphagnum peat; fully exposed.

Erica vulgaris. *Linn.* (Ling).

RECORDS: *Dfs.* and *Kcd.*—P. Gray, 1850. *Wgt.*—G. C. Druce, 1883.

b. Incana, *Dfs.*—J. Sadler, 1858. *Kcd.*—J. M'Andrew, 1882.

LOCALITIES: Very common in all the valleys at all altitudes. Usually moist ground, peat, whinstone soils, etc.; fully exposed to wind and sun.

VISITORS: Polyommatus phlœas, Peronea aspersana, J. C. W.; Apis, abundant, S.-E., J. C. W.; Bombus muscorum, pratorum, Scrimschiranus, terrestris, Platychirius albimanus, manicatus, Sericomyia borealis, J. C. W., and four Anthomyids, etc.

Erica cinerea. *Linn.* (Heather).

RECORDS: *Dfs.* and *Kcd.*—P. Gray, 1850. *Wgt.*—G. C. Druce, 1883.

LOCALITIES: Very common in all the valleys.

Appears June 20 to 29, J. T. J. On moist or dry peat; fully exposed.

VISITORS: Apis, S.-E., J. C. W.; Bombus lucorum, S.-E.; terrestris lapidarius, muscorum, pratorum, latreillelus, J. C. W.; Colletes succincta, R. Sc.; Apathus campestris, Platychirius albimanus, J. C. W.; Eristalis pertinax, S.-E.

Erica tetralix. *Linn.* (Cross-leaved Heather).

RECORDS: *Dfs.* and *Kcd.*—P. Gray, 1850. *Wgt.*—G. C. Druce, 1883.

LOCALITIES: Very common in all the valleys.

Appears June 15 to 29, J. T. J. On moist or dry peat; fully exposed.

VISITORS: Apis, J. C. W.; Bombus lucorum, S.-E.; muscorum, hortorum, J. C. W.; Micropalpus vulpnus, S.-E.; Platychirius peltatus, Hn.

Pyrola rotundifolia. *Linn.*

RECORD: *Kcd.*—P. Gray, 1850?

Pyrola media. *Sw.* (Wintergreen).

RECORDS: *Dfs.*—Rev. W. Little, 1834. *Kcd.*—J. Fraser, 1843.

LOCALITIES: *Nithsdale*—Dalskairth, P. Gr.; Lochanhead, J. Fr.; Cargen Glen, C. E. M.; Merkland, Exc.; Sanquhar, Dr Dv. *Annandale*—Lochmaben, C. E. M.; Garpol, Lochanburn, Cornal Tower burn, Wellburn, Selcoth, J. T. J.; Whitecoombe, J. T. J. and S.-E. (2400 feet, growing along with Arctostaphylos uva-ursi). *Eskdale*—Stuart's Wood (under 900 feet), E. Ty.

Appears June 21 to July 12. On pretty dry humus; in shade and shelter.

Pyrola minor. *Sw.*

RECORDS: *Dfs.*—J. Shaw, 1882. *Kcd.*—P. Gray, 1850. *Wgt.*—G. C. Druce, 1883.

LOCALITIES: *Nithsdale*—Solway Moss, S.-E.; Glen, S.-E., F. W. G.; Holywood, S.-E.; Druidhall Mill, Penpont, J. Sh. and T. Br.; Thornhill, Redpaths, Drum, Templand, Trigony, Morton, R. A.; Mennock, Glendyne, Craigdarroch, Dr Dv. *Annandale*—Lochmaben, Bk., S.-E.; wood near Dryfesdale Cemetery, Torwood, G. Bl.; Brackenside, S.-E.; Beld Craig, Garpol, J. Sd., J. T. J.; Crofthead, Cornal Tower, Wellburn, Selcoth, J. T. J. *Eskdale*—Solway Moss (under 900 feet), E. Ty.

Appears June 21 to July 12, J. T. J. In dry or wet humus or whinstone, often peaty (often rather bare); in full shade and shelter.

VISITORS: Bombus, sp.; Chortophila, sp.; Lithocolletis, sp.; Adrastus limbatus, S.-E.

Pyrola secunda. *Linn.*

RECORDS: *Dfs.*—J. Singer, 1843. *Kcd.*—J. M'Andrew, 1883.

LOCALITIES: *Nithsdale*—Long Wood, Exc. *Annandale*—Garpol, Beld Craig, J. Sd., J. Bl.; Black's Hope, and five of the corries of Moffat Water (1100 to 1200 feet), Johnstone, Duff Kinnel, J. T. J.

Appears July 13. On dry whinstone rocks with very little soil; in sun but partly wind-sheltered in narrow corries under the influence of the humid atmosphere from the burns.

Primula veris. *Linn.*

RECORDS: 1061. Acaulis (primrose), *Dfs.* and *Kcd.*—P. Gray, 1850. *Wgt.*—G. C. Druce, 1883.
1062. Veris (cowslip), *Dfs.*—J. Shaw, 1882. *Kcd.*—J. Cruickshank, 1836. *Wgt.*—Rev. W. W. Newbould, 1883.
1063. Elatior (escape?) (oxlip), *Dfs.*—Miss Adams and Miss S. D. Johnstone, 1890. *Kcd.*—Rev. J. Fraser, 1882.

LOCALITIES: *Nithsdale*—1061 common to the limit of peat haggs; 1062, Arbigland, C. E. M.; Tynron, J. Sh.; Cowhill, Ad. and S. D. J.; 1063, Cowhill, Ad. and S. D. J.; 1062, Dabton, Kirkbog, R. A. *Annandale*—1061 common to 1000 feet; 1062, Balgray, Milke, G. Bl.; Mill Meadows, J. T. J. *Eskdale*—1061 very common.

Appears March 11, G. Bl.; March 31 to April 15, J. T. J. 1061 on dry or moist humus, whinstone, boulder clay, more rarely light soils, sandstones and alluvium; in half-shaded, fully shaded or exposed, sheltered or windy places. 1062 and 1063 are probably escapes established in good soil, well-shaded and sheltered.

VISITORS: (Primrose) Bombus hortorum, reg. and sufficient, S.-E.

Lysimachia vulgaris. *Linn.* (Loosestrife).

RECORDS: *Dfs.*—J. Little, 1834. *Kcd.*—P. Gray, 1848. *Wgt.*—J. M'Andrew, 1893.

LOCALITIES: *Nithsdale* — Clarencefield, Th.; Cargenwater, S.-E.; Maxwelltown Loch, P. Gr., J. H. Bl.; Friars' Carse, Dr Gl.; ditch, Gateside, Saw Mill, Sanquhar, Dr Dv. *Annandale*—Castle Loch, Lochmaben, Mg., J. Sn., Th.; Murrayfield, G. Bl.; Johnstone, J. Lt., J. Sn. *Eskdale*—Irvine, C. Y.

In shallow water on mud sheltered from wind.

Lysimachia nummularia. *Linn.* (Moneywort).

RECORDS: *Dfs.*—Dr Burgess, 1789. *Kcd.*—See J. M'Andrew. *Wgt.*—J. M'Andrew.

LOCALITIES: *Nithsdale*—Broomlands, Dr Gl.; Cludenbank, Th., Hn., C. E. M.; Dabton, Kirkbog, R. A. *Annandale*—Cleugh, called Neese Linn, about half a mile from Kirkmichael Church, Dr Br.; Garpol, Beld Craig, J. Bl.; Eskrig, near Dumfries Road, G. Bl.

Lysimachia ciliata. *Linn.*

RECORDS: (Escape) *Dfs.*—R. Armstrong, 1888. *Kcd.*—R. H. Masterman, 1891.

LOCALITIES: Waterside farm wood, Kirkconnel Lodge, R. H. M.; near Dumfries, Hn.; Waterside, Morton, R. A.

Lysimachia nemorum. *Linn.* (Wood Pimpernel).

RECORDS: *Dfs.*—Dr Davidson, 1886. *Kcd.*—P. Gray, 1850. *Wgt.* —G. C. Druce, 1883.

LOCALITIES: *Nithsdale*—Common Glencaple, Hn.; common by Cluden and Nith, S.-E.; Scaur Water, Exc.; Craighope Linn, Hn.; Sanquhar, Dr Dv. *Annandale*—Common Dryfe, G. Bl.; common Moffat, J. T. J.; Queensberry and Correifron (to 1700 feet), S.-E. *Eskdale*— Very common below Langholm, S.-E., Hn.; Castle O'er, J. Wn. and R. Bl.; to 1500 feet, White Hope, S.-E.

Appears May 18 to June 7, J. T. J. On damp humus, clay, roadsides, shingle, whinstone rocks; usually in shade and wind-sheltered (in drains in the hills).

VISITORS: Borborus equinus, May 26; Dolichopods S.-E.

Trientalis Europea. *Linn.*

RECORD: *Dfs.*—A. Hutton.

LOCALITIES: *Eskdale*—(Requires confirmation) A. Ht. *Nithsdale*— Holywood, W. Bn.?

Glaux maritima. *Linn.*

RECORDS: *Dfs.* and *Kcd.*—P. Gray, 1850. *Wgt.*—G. C. Druce, 1883.

LOCALITIES: *Along the shore*—Very common from half a mile below Kingholm Mill to the border, Hn., C. E. M., J. Wl., S.-E.; Arbigland, Th.

Appears June, J.-E. Chiefly on barer parts of estuarine mud flats fully exposed, also on shingle and rarely on sand.

Anagallis arvensis. Linn.

(Poor Man's Weatherglass, Pimpernel).

RECORDS: 1075. *Dfs.* and *Kcd.*—P. Gray, 1850. *Wgt.*—G. C. Druce, 1883.
1076. Cœrulea, *Kcd.*—Sir Mark Stewart, 1843. *Wgt.* —G. Wilson, 1893.

LOCALITIES: *Along the shore*—Portwilliam, Rosshill, Hn.; Southwick (also 1076), P. Gr., J. H. Bl.; Priestside, S.-E.; common Annan to Gretna, S.-E. *Nithsdale*—Dumfries, F. W. G.; Cowhill, Ad. and S. D. J.; Glencairn, J. Cr.; Capenoch, R. A.; Sanquhar, Dr Dv. *Annandale*—Scroggs, G. Bl.; gardens (not in fields), J. T. J. *Eskdale*—Bilholm, J. Wn. and R. Bl.

Dry shingles of shore, boulder clay, garden soil, roadsides, holmlands; in sun and wind.

Anagallis tenella. Linn.

RECORDS: *Kcd.*—P. Gray, 1848. *Wgt.*—Arnott, 1848.

LOCALITIES: *Along the shore*—Portpatrick, J. H. Bl.; Moat, Dunsky, Arn., P. Gr.; Rosshill, G. C.; Garliestown, Hn.; Innerwell, Hn.; Laggan Hill, Hn.; Colvend, Port Ling, P. Gr., Fq., C. E. M.; Port o' Warren, St., S.-E.; Solway coast near mouth of Nith, W. St.; Kirkconnel, M. W.

Appears June, S.-E. In marshy estuarine or inland bogs; in sun, but often part wind-sheltered.

Centunculus minimus. Linn.

RECORDS: *Dfs.*—J. Cruickshank, 1836? *Kcd.*—F. R. Coles, 1883.
LOCALITY: Near Dumfries, J. Cr.*

Samolus valerandi. Linn.

RECORDS: *Dfs.*—Dr Burgess, 1789. *Kcd.*—P. Gray, 1848. *Wgt.* —Dr Arnott, 1848.

LOCALITIES: *Along the shore*—Common Portpatrick, R. R.; Killiness, Arn.; Whithorn, Portwilliam, Garliestown, Hn.; Ross, G. N. Ll., G. C.; Port o' Warren, S.-E.; Carbelly, Newabbey, Dr Gl.; Glencaple

* Both much require confirmation.

Quay, P. Gr., F. W. G., Hn., C. E. M., Th.; Brow, S.-E.; Priestside, Ruthwell, Dr Br., S.-E.

Appears July, S.-E. On wet, peaty, or muddy ground by the sea; usually in sun and wind.

Pinquicula vulgaris. *Linn.* (Butterwort).

RECORDS: *Dfs.* and *Kcd.*—P. Gray, 1850. *Wgt.*—G. C. Druce, 1883.

LOCALITIES: *Nithsdale*—Common Glen, Holywood, Hn.; Cowhill, Ad. and S. D. J.; Auldgirth, S.-E.; Glenquhargen, Exc.; Sanquhar, Dr Dv. *Annandale*—Very common Hartfell and hills, Moffat Water, J. T. J., S.-E. *Eskdale*—Tarras Water, Glencorfe, S.-E.; Castle O'er, J. Wn. and R. Bl. (ascends to 2000 feet).

Appears May 24 to June 22, J. T. J. On wet moss, moss-covered stones or bare soil (peat, whinstone, boulder clay, trap); in sun and scarcely wind-sheltered.

Pinquicula lusitanica. *Linn.*

RECORDS: *Kcd.*—J. Fraser, 1843. *Wgt.*—Dr Arnott, 1843.
LOCALITIES: Loch Dee, Drumbuie, Barrhead, Balmaclellan, J. M'A.

Utricularia vulgaris. *Linn.* (Bladderwort).

RECORDS: 1208. Vulgaris, *Dfs.*—P. Gray, 1846. *Kcd.*—P. Gray, 1850. *Wgt.*—J. M'Andrew, 1890.
1209. Neglecta, *Dfs.*—J. Corrie, 1891. *Kcd.*—F. R. Coles, 1885.

LOCALITIES: Barscraigh, Nunton, 1209, F. R. C., S.-E.; Lochar Moss, P. Gr.; Blackstone (850 feet), 1093, J. Cr.

In water of shallow ditches and ponds, in peat bogs, on mud of rich organic character.

Utricularia minor. *Linn.*

RECORDS: *Dfs.*—N. A. Dalzell, 1836. *Kcd.*—Dr Burgess, 1789. *Wgt.*—Sir H. Maxwell, 1889.

LOCALITIES: *Nithsdale*—Lochar Moss, near Racks, F. W. G.; East side Black Loch, about a mile S.W. from Kirkconnel, Dr Br.; Blackstone (857 feet), J. Cr. *Annandale*—Applegarth, J. Sn. *Eskdale*—Solway Moss, N. A. D.

Utricularia intermedia. *Hayne.*

RECORDS: *Dfs.*—J. Fingland, 1887. *Kcd.*—F. R. Coles, 1883.
LOCALITY: *Nithsdale*—Loch Urr (700 feet), Girharrow, J. Cr.

Utricularia bremii. *Heer.*

RECORD: *Wgt.*—J. M'Andrew, 1890? (doubtful).

Vinca minor. *Linn.*

RECORDS: (Escape) *Dfs.*—P. Gray, 1850. *Kcd.*—J. M'Andrew, 1882.
LOCALITIES: *Nithsdale*—Upper Newton, S.-E.; Castle-Douglas Road, Hn.; Cowhill, Ad. and S. D. J.; Dalswinton Woods, Exc. *Annandale*—Old Well Road, J. T. J. *Eskdale*—Kirkandrews, E. Ty.

Appears March 20 to May 1. Escape fully naturalised; in sheltered half-shaded places on good soil.

Vinca major. *Linn.*

RECORD: (Escape) *Kcd.*—Mrs Thompson, 1893.
LOCALITY: Roadside, Colvend.

Fraxinus excelsior. *Linn.* (Ash).

RECORDS: *Dfs.*—P. Gray, 1850. *Kcd.*—J. M'Andrew, 1882.
LOCALITIES: *Nithsdale*—Common Dumfries, S.-E.; Thornhill, R.A.; Ashcleugh, Dr Dv. *Annandale*—Craigieburn, J. T. J.; Beld Craig, F. W. G.

Prefers rich deep soil and exposure, A. M.

Ligustrum vulgare. *Linn.* (Privet).

RECORDS: *Dfs.* and *Kcd.*—P. Gray, 1850. *Wgt.*—J. M'Andrew, 1883.
LOCALITIES: Common hedge plant everywhere.

Prefers shade and shelter, A. M.

Erythræa centaurium. *Linn.* (Centaury).

RECORDS: 1086. Centaurium, *Dfs.*—P. Gray, 1850. *Kcd.*—G. N. Lloyd, 1837. *Wgt.*—Dr Arnott, 1848.
 b. Capitata Koch, *Kcd.*—Mrs Stewart, 1893. *Wgt.*—J. M'Andrew, 1890.

1088. Littoralis, *Dfs.*—W. Stevens, 1848. *Kcd.*— Dumfries Herbarium, 1866. *Wgt.*—J. H. Balfour, 1836.
1089. Pulchella, *Dfs.*—J. Cruickshank, 1836.

LOCALITIES: *Along the shore*—Portpatrick, 1086, 1088, Arn., J. H. Bl.; Kirkmaiden, 1086, *b.*, J. M'A.; Rerrick, Rosshill, G. N. Ll.; Southwick, 1086 *b.*, St.; Douglas Hall, Hn.; mouth of Nith on each side, 1086, 1088, 1089, J. Cru., W. St., P. Gr.; Kingholm Mill, S.-E.; Glencaple, Hn.; 1089, C. E. M.; Caerlaverock, and common Annan to Kirtle mouth, S.-E.; Torduff, S.-E.

Appears on shingle, estuarine alluvium, boulder clay, etc.; in sun, usually against cliffs.

VISITORS: Eristalis intricarius, Empis livida, Siphona geniculata, Hyelemyia, sp., S.-E.

Gentiana campestris. *Linn.* (Gentian).

RECORDS: *Dfs.*—S. W. Carruthers, 1890. *Kcd.*—P. Gray, 1848. *Wgt.*—G. C. Druce, 1883.

LOCALITIES: *Nithsdale*—Laught Wood, R. H. M.; Dalskairth Hill, P. Gr., R. H. M.; Glen, Hillhead, R. H. M.; Mabie, F. W. G.; Moniaive, J. Cr. (500 to 550 feet); Blacknest, Kirkbride, Newton, R. A.; Euchan, Barrmoor, Dr Dv. *Annandale*—Valencines near Garpol, Alton, J. T. J.; Greygill Head, S. W. Ca.; Correifron, footpath to Craigmichen Scaurs, six to eight miles along Selkirk road, Birkhill, J. T. J. (from 500 to 850 feet).

Appears July 27 to September 6, J. T. J. On pastures of whinstone soil; fully exposed to wind and sun.

Menyanthes trifoliata. *Linn.* (Buckbean).

RECORDS: *Dfs.*—P. Gray, 1850. *Kcd.*—J. Cruickshank, 1836. *Wgt.*—G. C. Druce, 1883.

LOCALITIES: *Nithsdale*—Maxwelltown Loch, Hn.; Somervile House, J. Cru.; Newton and by Cluden, S.-E.; Maryfield, Th., S.-E.; Merkland, Exc.; Girharrow (850 feet), Loch Urr (700 feet), J. Cr.; Dabton, Morton, R. A.; Black Loch, Dr Dv. *Annandale*—Common Archbank Moor, Parks, Meikleholmside Hill, Lochhouse, Chapel, Earshaig Lakes, etc., J. T. J. *Eskdale*—Castle O'er, J. Wn. and R. Bl.

Appears May 24 to 26, J. T. J. Amongst rushes in alluvial or peaty marshes in full exposure.

VISITORS: Apis abundant and sufficient; also Bombus spp., S.-E.

Polemonium cœruleum. *Linn.*

RECORDS: *Dfs.*—R. H. Masterman, 1891. *Kcd.*—P. Gray, 1850. *Wgt.*—G. C. Druce, 1883.

LOCALITIES: *Nithsdale*—Newabbey Road, Calton's Loaning, P. Gr.; Caerlaverock, R. H. M.; Friars' Carse Loch, S.-E.

An escape establishing itself in marshy fields by hedges, etc.

Convolvulus arvensis. *Linn.*

RECORDS: (Escape) *Dfs.*--Dr Davidson, 1886. *Kcd.*—J. M'Andrew, 1882. *Wgt.*—J. M'Andrew, 1889.

LOCALITIES: Rockcliffe, Th.; Locharbriggs, Hn.; Sanquhar? Dr Dv.

Convolvulus tricolor.

RECORD: (Escape) *Dfs.*—Miss Hannay, 1892.

LOCALITY: Dumfries Station, Hn.

Convolvulus sepium. *Linn.*

RECORDS: *Dfs.*—Dr Burgess, 1789. *Kcd.*—P. Gray, 1848. *Wgt.*—J. M'Andrew, 1883.

LOCALITIES: *Nithsdale*—Conheath, G. Go., P. Gr.; Southerness, Exc.; Cluden Mill, C. E. M.; Moniaive by river, J. Cr.; Thornhill, Hn.; Cample, Rashbriggs, R. A. *Annandale*—Hedge on side of burn below Stank House, Dr Dv.; Broom farm, G. Bl.

An outcast or escape along the shore, or by hedges, etc.

VISITORS: Bombus hortorum, Apathus quadricolor, Empis livida, Hn.

Convolvulus soldanella. *Linn.*

RECORDS: (Escape) *Dfs.*—G. F. Scott-Elliot, 1892. *Kcd.*—J. M'Andrew, 1882. *Wgt.*—Dr Arnott, 1848.

LOCALITIES: *Along the shore*—Monreith, J. M'A.; Killiness, Cowans, Arn.; Southerness, J. Fr.; Cummertrees, Gretna Green, S.-E.

On sandy shores in full exposure.

Cuscuta epilinum. *Weihe.*

RECORD: (Escape) *Dfs.*—On flax near Dumfries, G. N. Lloyd, 1837.

FLORA OF DUMFRIESSHIRE. 121

Cuscuta epithymum. *Linn.*

RECORD: (Escape) *Kcd.*—Dr Burgess, 1789.
LOCALITY: Castle-Douglas, Dr Br.

Mertensia maritima. *Don.* (Smooth Gromwell).

RECORDS: *Kcd.*—Rev. G. M'Conachie, 1882. *Wgt.*—J. H. Balfour, 1836.
LOCALITIES: *Along the shore*—West Tarbet, J. H. Bl.; Whiteport Bay, Rerrick, G. M'C.; Kirkmaiden, J. M'A.

Pulmonaria officinalis. *Linn.*

RECORD: (Escape) *Dfs.*—G. Bell, 1892.
LOCALITIES: Catch Hall Loaning, Lockerbie, G. Bl.; Bilholm, Langholm, J. Wn. and R. Bl.
Appears April 4, G. Bl.

Lithospermum arvense. *Linn.* (Gromwell).

RECORDS: (Escape) *Dfs.*—J. M'Andrew, 1882. *Kcd.*—J. Fraser, 1843.
LOCALITY: Near Dumfries in cultivated fields, J. M'A.

Lithospermum officinale. *Linn.* (Gromwell).

RECORDS: (Escape) *Kcd.*—P. Gray, 1848. *Wgt.*—J. H. Balfour, 1843.
LOCALITIES: *Along the shore*—Dundrennan Abbey, Fq., J. Fr.; Luce Abbey, J. H. Bl.; Southwick, P. Gr.; Tongland Bridge, F. R. C.

Myosotis palustris. *With.* (Forget-Me-Not).

RECORDS: 1115. Palustris, *Dfs.* and *Kcd.*—P. Gray, 1850. *Wgt.*— G. C. Druce, 1883.
    b. strigulosa, *Dfs.*—Dr F. W. Grierson, 1882. *Wgt.*—G. C. Druce, 1883.
    1114. Cœspitosa, *Dfs.*—W. Stevens, 1848. *Kcd.*— J. M'Andrew, 1882. *Wgt.*—G. C. Druce, 1883.
    1116. Repens, *Dfs.*—W. Stevens, 1848. *Kcd.*—F. R. Coles, 1882. *Wgt.*—G. C. Druce, 1883.

LOCALITIES : *Nithsdale*—1115, common Maxwelltown Loch, P. Gr.; Cargen, Hn.; Glencaple, Hn.; Cowhill, Ad. and S. D. J.; Crawick, Connelbush, 1114, Sanquhar, Dr Dv. *Annandale*—1115, common in lower part Annan, S.-E. (also 1114, Hn.); Lockerbie, G. Bl.; Wamphray Water, 1114, J. T. J.; Craigmichen Scaurs, 1114, J. T. J.; Selcoth, 1114, J. T. J.; Grey Mare's Tail, 1114 and 1116, W. St., E. F. L., J. T. J. (to 2000 feet at Loch Skene, J. T. J.). *Eskdale*—Penton, S.-E.; Castle O'er, J. Wn. and R. Bl.

Appears, 1114, July 26 to August 26, J. T. J.; 1115, May 26, J. T. J. 1115 in ditches, usually in shallow water ; usually sheltered. 1114 and 1116 in springs in mud in the hills, damp roadsides ; both exposed to wind and sun.

VISITORS : Platychirius clypeatus, albimana, Hn.; Helophilus frutetorum, Hydrotea dentipes, Chortophila, sp., S.-E.

Myosotis sylvatica. *Linn.*

RECORDS : *Dfs.* and *Kcd.*—P. Gray, 1850. *Wgt.*—Miss Hannay, 1893.

LOCALITIES : *Nithsdale*—Glen, Hn., P. Gr. ; Cowhill, Ad. and S D. J.; Cluden Woods, S.-E. ; Blackwood Linn, Exc. *Annandale*— Dryfe, G. Bl. *Eskdale*—Pretty common by Esk, Gilnockie, S.-E.

Gathered June 28, S.-E. On moist humus ; in shade and shelter.

VISITORS : Platychirius clypeatus, Hn.

Myosotis arvensis. *Roth.*

RECORDS : *Dfs.*—P. Gray, 1850. *Kcd.*—J. M'Andrew, 1882. *Wgt.* —G. C. Druce, 1883.

b. Umbrosa, *Dfs.*—Dr Davidson, 1886. *Kcd.*—J. M'Andrew, 1882.

LOCALITIES : *Nithsdale*—Very common Glencaple, Hn.; Glen, S.-E.; Cowhill, Ad. and S. D. J.; Sanquhar (also var. *b.*), Dr Dv. *Annandale* —Springkeld, S.-E.; Milke, G. Bl.; Auchencas (var. *b.*), S.-E.; Craigieburn, S.-E., etc.; Moffat, J. T. J. *Eskdale*—Gilnockie, Glencorfe, S.-E.; Castle O'er, J. Wn. and R. Bl.

Appears May 4 to 7, J. T. J. On dry or damp roadsides, manure heaps ; var. *b.*, humus, waste ground, gardens, etc.; usually in shade and well wind-sheltered.

VISITORS : Anthomyidæ, many, S.-E.

Myosotis collina. *Hoffm.*

RECORDS: *Dfs.*—Dr F. W. Grierson, 1882. *Kcd.*—Rev. J. Fraser, 1843.

LOCALITIES: *Nithsdale*—Kirkmahoe, F. W. G.

Myosotis versicolor. *Pers.*

RECORDS: *Dfs.* and *Kcd.*—P. Gray, 1850. *Wgt.*—C. C. Bailey, 1883.

LOCALITIES: Common in all the valleys to about 800 feet.

Appears May 7 to 15, J. T. J. On dry gravelly or sandy fields, cinders of railways, whinstone soils (arable); in full exposure to sun and wind.

VISITORS: Sepsis cynipsea, Anthomyidæ and Dolichopodidæ, abundant, S.-E.

Lycopsis arvensis. *Linn.* (Bugloss).

RECORDS: (Escape) *Dfs.*—Dr Gilchrist, 1862. *Kcd.*—P. Gray, 1850. *Wgt.*—Miss Hannay, 1893.

LOCALITIES: *Along the shore*—Arbigland, Exc.; Clarencefield, Old Quay, Dr Gl.; Colvend, P. Gr.; Cummertrees, Ak. (I do not know Dr Davidson's locality).

Anchusa sempervirens. *Linn.* (Alkanet).

RECORDS: *Dfs.*—Rev. W. Bennet, 1890. *Kcd.*—P. Gray, 1846. *Wgt.*—J. M'Andrew, 1893.

LOCALITIES: *Nithsdale*—Kirkbean, Th., Hn.; Rosehall, Calton's Loaning, P. Gr.; roadside, Amisfield, J. Wg.; Cowhill, Ad. and S. D. J.; Sanquhar Castle, Dr Dv. *Annandale*—Hutton Manse, J. Wg.; Kirkpatrick-Juxta Manse, W. Bn., J. T. J.; Dumcrieff, Heatheryhaugh, W. Bn.

Roadsides in sun; but partly wind-sheltered by hedges, etc.

Symphytum officinale. *Linn.* (Comfrey).

RECORDS: *Dfs.*—J. Sadler, 1858. *Kcd.*—P. Gray, 1846. *Wgt.*—Rev. G. Wilson, 1893.

b. Patens, *Dfs.*—Dr F. W. Grierson, 1882. *Wgt.*—G. C. Druce, 1883.

LOCALITIES: *Nithsdale*—Common by Nith and Cluden, near Dumfries, Hn., S.-E.; Jardington, var. *b.*, F. W. G.; Irongray roadside *(b.)*, R. H. M.; Routen Brig, S.-E.; Maxwelton House, Th.; Moniaive by river, J. Cr.; Tynron, J. Sh.; Templand Bridge, R. A.; Braeheads, Mansepool, Kirkconnell, Dr Dv. *Annandale*—Common lower part Annan, Th., S.-E.; Lochmaben, Exc.; Garpol, Beld Craig, J. Sd.; Moffat, common, J. T. J., S. W. Ca.; Beeftub Road, S.-E. *Eskdale*—Ewes Water, S.-E.; Castle O'er, J. Wn., and R. Bl.

Appears May 7 to June 4, J. T. J. In moist or wet holmlands of river banks; half-shaded or in sun, and usually sheltered by banks or long grass.

VISITORS: Bombus muscorum, lucorum, pratorum, Hn., S.-E.

Symphytum asperrimum. *Bab.*

RECORDS: (Escape) *Dfs.*—Dr Davidson, 1886. *Wgt.*—G. C. Druce, 1883.

LOCALITY: Auchengruith, Sanquhar, Dr Dv.

Symphytum tuberosum. *Linn.*

RECORDS: (Escape) *Dfs.*—Dr Davidson, 1886. *Kcd.*—J. M'Andrew, 1882. *Wgt.*—J. M'Andrew, 1893.

LOCALITIES: *Nithsdale*—Templand Bridge, R. A.; North Kirkconnel Station, Dr Dv. *Annandale*—Dumcrieff, J. T. J.

Echium vulgare. *Linn.* (Viper's Bugloss).

RECORDS: (Escape) *Dfs.*—J. Singer, 1843. *Kcd.*—J. Fraser, 1843. *Wgt.*—G. C. Druce, 1883.

LOCALITIES: *Nithsdale*—Solway Moss, Herb.; Tinwald, J. Sn.; Tynron, J. Sh.; Auldgirth? J. Fn. *Annandale*—Moffat, W. Ca.

Borago officinalis. *Linn.* (Borage).

RECORDS: (Escape) *Dfs.*—Mrs Carthew-Yorstoun, 1886. *Kcd.*—J. M'Andrew, 1882. *Wgt.*—G. C. Druce, 1883.

LOCALITIES: *Annandale*—Kerr, Crockspool, J. T. J. *Eskdale*—Irvine, C. Y.

Hyoscyamus niger. *Linn.*

RECORD: (Escape) *Wgt.*—Rev. J. Gorrie, 1891.

LOCALITY: Garliestown Bay, Hn.

VISITOR: Bombus muscorum, Hn.

Solanum Dulcamara. *Linn.* (Bittersweet).

RECORDS: *Dfs.*—Dr Burgess, 1789. *Kcd.*—P. Gray, 1846. *Wgt.* —G. C. Druce, 1883.

LOCALITIES: *Nithsdale*—Cargen Pow, Hn., C. E. M., J. M'A.; Castle-Douglas Road, C. E. M., Th.; Kingholm, P. Gr.; Glencaple, Hn.; Nith Bridge, Hn.; Lincluden Abbey, P. Gr.; Cowhill, Ad. and S. D. J.; Dunscore, F. W. G.; Backwater Marsh, 330 feet, J. Cr.; Rashbriggs Round Plantation, R. A. *Annandale*—Beside burn below Stank House, Dr Br.; Powfoot to Newbie, S.-E., Shillahill, G. Bl.; Holm and Beerholm, Annan Water, Middlegill, J. T. J.; Beattock, S.-E.

On broken pretty dry alluvial banks, boulder clay, shingle; in sun and partly wind-sheltered.

Solanum nigrum. *Linn.* (Nightshade).

RECORD: *Wgt.*—G. Graham, 1836.

LOCALITIES: *Along the shore*—Portwilliam to Glenluce, J. H. Bl.; Sandhead, Portwilliam, G. Gr.

Atropa belladona. *Linn.* (Deadly Nightshade).

RECORD: (Escape) *Dfs.*—Dr Burgess, 1789.

LOCALITIES: Among ruins of Abbey of Holywood, Dr Br.; Lochmaben Castle, J. Sn.

Datura stramonium. *Linn.*

RECORDS: (Escape) *Wgt.*—Miss Hannay, 1893. *Kcd.*—J. C. Willis, 1894.

LOCALITIES: Garliestown, Hn.; Auchencairn, J. C. W.

Orobanche major. *Linn.* (Broomrape).

RECORDS: *Dfs.*—Sir W. Jardine, 1837. *Kcd.*—P. Gray, 1846.

LOCALITIES: *Nithsdale*—Locharbriggs, J. Wl.; Kirkconnel Avenue, Harper; Harleybank, Cluden Craigs, P. Gr.; White Bridge, Ru.; Dalawoodie, R. R.; Cluden Mills, Hn.; Auldgirth, F. W. G.; Blackwood, Harv.; Thornhill, J. M'A. *Annandale*—Jardine Hall, Sir W. J.

On roots of broom; full shade and shelter.

Orobanche rubra. *Linn.*

RECORDS: *Kcd.* and *Wgt.*—Rev. J. Fraser, 1843.

Lathræa squamaria. *Linn.* (Toothwort).

RECORDS: *Dfs.*—T. Brown, 1882. *Kcd.*—P. Gray, 1846.

LOCALITIES: *Nithsdale*—Terregles, Dr Gl.; Grove, P. Gr.; Scaurwater, T. Br. *Annandale*—Craigieburn Wood, J. T. J.

Appears April, 1894. On hazel or elm roots, in humus; well-shaded and sheltered.

Verbascum thapsus. *Linn.* (Mullein).

RECORDS: *Dfs.*—Dr Davidson, 1886. *Kcd.*—J. Fraser, 1843. *Wgt.*—G. C. Druce, 1883.

LOCALITIES: *Along the shore*—Garliestown, Hn.; Rough Island, Th.; Ross Hill, G. C.; Orchardton, J. M'A. *Inland*—Maxwelltown Station, Hn., C. E. M.; Caerlaverock, J. Wl.; Dumfries, J. Sn.; Cowhill, Ad. and S. D. J.; Thornhill, R. A.; Sanquhar, Dr Dv. *Annandale* —Jardine Hall, J. Wg. *Eskdale*—Kirkandrews, E. Ty.

Antirrhinum majus. *Linn.*

RECORD: (Escape) *Wgt.*—Mr Farquharson, 1873.

LOCALITY: Dundrennan Churchyard.

Linaria vulgaris. *Mill.*

RECORDS: *Dfs.*—P. Gray, 1850. *Kcd.*—P. Gray, 1846. *Wgt.*— G. C. Druce, 1883.

LOCALITIES: *Nithsdale*—Near Dumfries, P. Gr.; Cargen, Hn.; Glencaple, Hn., M. J. H.; Glen, F. W. G.; Tinwald, Locharbriggs, Hn.; Cluden Bridge, S.-E.; Moniaive, J. Cr.; Redpaths, Nithbank, R. A.; G. & S.-W. Railway, Sanquhar, Dr Dv. *Annandale*—Common by shore, Whinnyrigg to Browhouses, Caledonian Railway, Lockerbie, Lochmaben, Hn., S.-E. *Eskdale*—Gretna Green, S.-E.; Kirkandrews, E. Ty.; Railway at Canobie, S.-E.

Appears June, S.-E. On dry cindery railway banks, boulder clay of cliffs, roadside banks, shingle, etc.; in full sun and wind exposure.

VISITORS: Apis (stealing); Bombus hortorum, reg. and sufficient, lucorum (efficient?), S.-E.

Linaria minor. *Desf.*

RECORDS: *Dfs.*—Dr F. W. Grierson, 1882. *Kcd.*—F. R. Coles, 1883.

LOCALITIES: *Annandale*—Common along the Caledonian Railway Line, Lockerbie, Lochmaben to Wamphray, Beattock, Moffat, F. W. G., S.-E. *Eskdale*—Gretna Green, S.-E.
Confined to dry cindery soil of railway tracks.

Linaria cymbalaria. *Mill.*

RECORD: (Escape) *Dfs.*—J. T. Johnstone, 1890. *Kcd.*—J. M'Andrew, 1882. *Wgt.*—J. M'Andrew, 1893.
LOCALITIES: *Nithsdale*—Old wall near Castledykes, M. J. H., Troqueer Churchyard, C. E. M. and on walls. *Annandale*—Sandbed, Moffat, J. T. J.

Linaria purpurea. *Linn.*

RECORD: (Escape) *Wgt.*—G. C. Druce, 1883.

Scrophularia nodosa. *Linn.* (Figwort).

RECORDS: *Dfs.* and *Kcd.*—P. Gray, 1850. *Wgt.*—G. C. Druce, 1883.
LOCALITIES: *Nithsdale*—Very common by Cargen, Cluden, Cairn and Nith, S.-E., C. E. M.; Sanquhar, Dr Dv. *Annandale*—Kirtle mouth, Annan, Milke, Wamphray to Moffat, common, S.-E., J. T. J. *Eskdale*—Gretna Green, Glenzier, Sark, Liddel, Esk to Bentpath, and Castle O'er, S.-E., J. Wn., and R. Bl.
Appears June 11 to 14, J. T. J. On wet, damp, or fairly dry river banks, usually alluvial, boulder clay, or humus, also shingles and gravelly soil; usually half-shaded or fully shaded, or in sun partly wind-sheltered by banks.
VISITORS: Bombus muscorum, Hn., S.-E.; hortorum, S.-E.; Vespa sylvestris, abundant; Allantus Scrophulariæ, S.-E.

Scrophularia aquatica. *Linn.*

RECORDS: *Dfs.*—J. Singer, 1843 (?) *Kcd.*—J. Fraser, 1882. *Wgt.*—J. M'Andrew, 1890.
a. Balbisii, *Kcd.*—Dr F. W. Grierson, 1882. *Wgt.*—J. M'Andrew, 1890.
LOCALITY: Annan, J. Sn. (Lot's Wife, var *a.*, F. W. G.).

Scrophularia vernalis. *Linn.*

RECORDS: (Escape) *Dfs.*—Dr Burgess, 1789. *Kcd.*—G. N. Lloyd, 1837.

LOCALITY: Hoddam Castle, Dr Br.

Mimulus luteus. *Linn.* (Monkey Flower).

RECORDS: (Escape) *Dfs.*—W. Stevens, 1848. *Kcd.*—J. M'Andrew, 1882. *Wgt.*—J. M'Andrew, 1886.

LOCALITIES: *Nithsdale*—Dockwall, P. Gr., Hn., F. W. G.; Moat House, Exc.; Cluden Mills, S.-E.; Ewanston (370 feet), Caitloch (429 feet), J. Sh., J. Cr.; Cowhill, Ad. and S. D. J.; Friars' Carse to Auldgirth, S.-E.; Dunscore Road, Auldgirth, Th.; Thornhill, Cample, R. A.; Craighope Linn, Scaur, Hn.; Drumlanrig, W. St.; Cumnock, J. Fn. *Annandale*—Scroggs Mill, Abigailburn, G. Bl.; East Kirkpatrick-Juxta Kirk, F. A. H.; Adam's Holm Kerr, J. T. J.; Frenchland Burn, W. Bn. *Eskdale*—Eskdalemuir Kirk, S.-E.

Moist shingles, trap, holmland by rivers and burns, roadsides; half-shaded and well-sheltered.

Digitalis purpurea. *Linn.* (Foxglove, Bloody Fingers, J. Shaw).

RECORDS:—*Dfs.* and *Kcd.*—P. Gray, 1850. *Wgt.*—G. C. Druce, 1883.

LOCALITIES: Very common in all valleys by rivers, linns, etc., to 1100 feet.

Appears from June 7 to 20, J. T. J. On pretty dry slopes, on sandy soil, sandstones, holms, shingles, boulder clay (rarer on whinstone); half-shaded or exposed; usually wind-sheltered.

VISITORS: Bombus hortorum, reg. and sufficient, S.-E., J. C. W.; muscorum, terrestris, J. C. W.

Veronica serpyllifolia. *Linn.*

RECORDS: *Dfs.* and *Kcd.*—P. Gray, 1850. *Wgt.*—C. C. Bailey, 1883.

b. humifusa, *Dfs.*—J. T. Johnstone, 1891.

LOCALITIES: *Nithsdale*—Jardington, Midnunnery, S.-E., Kirkmahoe, Hn.; Sanquhar, Dr Dv. *Annandale*—Ecclefechan, Milke, Craigboar, S.-E.; Wellroad Moffat, Whitecoombe (2000 feet), var. *b.*; J. T. J.

Eskdale—Penton, Canobie, Gilnockie, Langholm, White Hope (1400 feet), S.-E.

Appears April 15 to May 20, J. T. J. On dry, rarely moist sods on old walls, sandy soil, holms, roadsides; usually fully exposed in short turf.

VISITOR : Anthomyia radicum, S.-E.

Veronica officinalis. *Linn.*

RECORDS: *Dfs.* and *Kcd.*—P. Gray, 1850. *Wgt.*—G. C. Druce, 1883.

LOCALITIES: *Nithsdale*—Cargen, F. W. G; Glen Hills, above Grove, Maxwelltown Station, Cluden Valley, Th., S.-E.; Sanquhar, Dr Dv. *Annandale*—Tundergarth, Kirkmichael, Beeftub, S.-E.; Moffat, Black's Hope, J. T. J. *Eskdale*—Burnghaell's Head, common by Esk to Langholm, Wauchope, S.-E.; Castle O'er, J. Wn. and R. Bl.; White Hope, Causey Grain (1400 feet), S.-E.

Appears June 7 to 25, J. T. J. On dry, rarely moist whinstone rocks, old walls, shingles, roadsides, boulder clay, cinders of railways, etc.; in sun or shaded (often under beech), usually quite exposed in short turf or bare places.

VISITOR : Bombus lucorum, S.-E.

Veronica anagallis. *Linn.*

RECORDS: *Dfs.* and *Kcd.*—P. Gray, 1850. *Wgt.*—G. C. Druce, 1883.

LOCALITIES: *Nithsdale*—Dalskairth, P. Gr.; Cargen and Tributaries, F. W. G., M. J. H., Hn., S.-E.; Maxwelltown Loch, F. W. G. *Annandale*—Kirtlebridge Quarry, S.-E.; Kerr, along ditches from Putts to Hydropathic, Moffat; Annan Water, J. T. J.

Appears June 22 to July 5, J. T. J. Usually in sluggish steams from six inches to a foot deep, part sheltered in ditches, etc.

Veronica beccabunga. *Linn.* (Brooklime).

RECORDS: *Dfs.* and *Kcd.*—P. Gray, 1850. *Wgt.*—G. C. Druce, 1883.

LOCALITIES: Very common in all the valleys to at least 1000 feet.

Appears June 1 to 14, J. T. J. In wet ditches, by roadsides, shingles, etc.; usually in sun and part wind-sheltered.

VISITORS: Platychirius clypeatus, very abundant, peltatus, albimana, Rhingia rostrata, Empis bilineata, Siphona geniculata, Hydrotea dentipes, Hyetodesia incana, S.-E.

Veronica scutellata. *Linn.*

RECORDS: *Dfs.*—Professor Balfour, 1863. *Kcd.*—P. Gray, 1850. *Wgt.*—A. Bennet, 1882.

LOCALITIES: *Nithsdale*—Merse, J. M'A., S.-E.; New Loch, R. A.; Ulzieside Fold, Grange mill dam, Dr Dv. *Annandale*—Carterton, Carrick, G. Bl.; Castle Loch and Halleaths, G. Bl., S.-E.; Beld Craig, J. H. Bl.; Beattock hill and meadows, J. T. J.; Adam's Holm, Minnygap, J. T. J.; Wellburn, Carr, J. T. J.; Grey Mare's Tail (to 900 feet), J. T. J.

Appears May 26 to June 21, J. T. J. In marshy ground, ditches, etc.; usually part shaded and sheltered by long grass, etc.

Veronica montana. *Linn.*

RECORDS: *Dfs.*—Dr Singer, 1843. *Kcd.*—P. Gray, 1846.

LOCALITIES: *Nithsdale*—Mavisgrove, P. Gr.; Colonel's Wood, Hn.; along Cargen and Glen, Hn., S.-E.; Irongray Manse Linn, P. Gr.; Tinwald, Dr Sn.; along Nith, Lincluden, S.-E.; Cowhill, Ad. and S. D. J.; Auldgirth Linn, Exc. *Annandale*—Dalebank Wood, Annan, Dr Sn.; Dryfe, Tundergarth, G. Bl.; Garpol, Beld Craig, J. T. J. *Eskdale*—Scotch Dyke, Penton. Tarras, Byreburn, Esk below and above Langholm, S.-E.

Appears April 24, G. Bl.; June 5 to 16, J. T. J. On wet humus, usually over whinstone or alluvium; in shade or half-shaded and windsheltered.

VISITORS: Platychirius albimana, Syrphus ribesii, Borborus equinus, Hyelemia strigosa, Anthomyia radicum, Hyetodesia, sp., reg. and abundant; Chortophila, sp., S.-E.

Veronica chamædrys. *Linn.* (Bird's Eye Speedwell).

RECORDS: *Dfs.* and *Kcd.*—P. Gray, 1850. *Wgt.*—G. C. Druce, 1883.
Var. Pilosa, *Dfs.*—G F. Scott-Elliot, 1890.

LOCALITIES: *Nithsdale*—Common to Sanquhar, Dr Dv. *Annandale*—Very common Beeftub (1300 feet), S.-E. *Eskdale*—Very common to Castle O'er, J. Wn., R. Bl.; White Hope, Causey Grain Head (1500 feet), S.-E.

Appears April 30 to May 5. On moist or dry roadsides, holms, shingles, humus, whinstone soils, clay, cinders of railways; usually in sun and part wind sheltered by long grass.

VISITORS: Bombus pratorum, Hn.; Ascia podagrica, Anthomyia sulciventris, radicum, Chortophila, sp , S.-E.

Veronica hederæfolia. Linn.

RECORDS: *Dfs.* and *Kcd.*—P. Gray, 1850. *Wgt.*—J. M'Andrew, 1886.

LOCALITIES: *Nithsdale*—Ruthwell, M. J. H.; Newabbey Road, Hn.; Broomlands, C. E. M.; Auldgirth Station, Exc.; Nith Mills, Dr Dv. *Annandale*—Ruthwell shore, M. J. H.; Douglas Bridge, Lockerbie, G. Bl.

Appears April 19, G. Bl. On dry roadsides, garden or bare waste ground, cinders of railways; half-shaded or in sun and usually wind-sheltered.

VISITOR: Orchisa minor, S.-E.

Veronica agrestis. Linn.

RECORDS: 1167. Polita, *Dfs.*—P. Gray, 1876. *Wgt.*—G. C. Druce, 1883.
1168. Agrestis, *Dfs.* and *Kcd.*—P. Gray, 1850. *Wgt.*—G. C. Druce, 1883.

LOCALITIES: *Nithsdale*—Glen, S.-E.; Sanquhar, Dr Dv. *Annandale*—Cummertrees, Ak.; Æ Water, Th.; Vicarland, Annan Water, J. T. J.; Crooks, 1167, J. T. J. *Eskdale*—Canobie, S.-E.

Appears February 25 to September, J. T. J. Common weed in gardens, waste ground, cinders, sandy holms; usually exposed.

Veronica buxbaumii. Ten.

RECORDS: (Escape) *Dfs.*—P. Gray, 1846. *Kcd.*—F. R. Coles, 1885. *Wgt.*—C. C. Bailey, 1883.

LOCALITIES: *Nithsdale*—Rosehall, Dumfries, P. Gr. *Eskdale*—Woodslee, S.-E.

Veronica arvensis. Linn.

RECORDS: *Dfs.* and *Kcd.*—P. Gray, 1850. *Wgt.*—G. C. Druce, 1883.

LOCALITIES: *Nithsdale*—Arbigland Beach, Th.; near Dumfries, common, P. Gr.; along G. and S.-W. Railway, S.-E.; Holywood, Hn., Dr Dv. *Annandale*—Kirtlebridge, along Caledonian Railway and Dumfries Branch, common, S. E.; Moffat, J. T. J. *Eskdale*—Canobie, Gilnockie, Langholm Hill, Bexburn, Meggat, Eskdalemuir Kirk, S.-E.

Appears April 22, G. Bl.; May 8 to June 3, J. T. J. On dry cinders of railways, sandy soil, waste ground, turf on old walls, shingles, stony till; usually exposed, on ground bare of other plants.

Bartsia viscosa. *Linn.*

RECORDS: *Kcd.*—P. Gray, 1850. *Wgt.*—Dr Graham, 1836.
LOCALITIES: Dumfries, P. Gr.; Glasserton, Portwilliam, Dr Gr.

Bartsia odontites. *Huds.*

RECORDS: *Dfs.* and *Kcd.*—P. Gray, 1850. *Wgt.*—G. C. Druce, 1883.

LOCALITIES: *Nithsdale*—Dumfries, F. W. G.; Holywood, Cowhill, Ad. and S. D. J.; Sanquhar, Dr Dv. *Annandale*—Ecclefechan, Lochmaben, Lockerbie, G. Bl., S.-E.; common Moffat, J. T. J.; Capelgill (600 feet), J. T. J. *Eskdale*—Glenzier, E. Ty.; Canobie, common, S.-E.: Castle O'er, J. Wn. and R. Bl.

Appears July 31 to September 6, J. T. J. On bare, dry roadsides; fully exposed.

VISITORS: Bombus pratorum, abundant and sufficient, S.-E.

Euphrasia officinalis. *Linn.* (Eyebright).

RECORDS: *Dfs.* and *Kcd.*—P. Gray, 1850. *Wgt.*—G. C. Druce, 1883.
b. gracilis, *Dfs.*—J. T. Johnstone, 1891.

LOCALITIES: *Nithsdale*—Common Jardington, Dalawoodie, S.-E.; Cowhill, Ad. and S. D. J.; Closeburn, Dr Gl.; Sanquhar, Dr Dv. *Annandale*—Very common Beattock Hill, var *b.*, J. T. J. (reaching 2300 feet); Loch Skene, J. T. J. *Eskdale*—Very common Wauchope, White Hope (1400 feet), S.-E.; Castle O'er, J. Wn. and R. Bl.

Appears June 4 to 27, J. T. J. On dry or damp slopes of boulder clay, sand of seashore, holms, whinstone soils, roadsides, granite, peat; usually exposed to wind and sun.

VISITORS: Bombus lucorum, S.-E.; hortorum, Hn.; Allantus nothi, Andrena coitana, Hn.; Helophilus pendulus, Melanostoma mellina, Hn.; Platychirius manicatus, clypeatus, Sericomyia borealis, S.-E.

Rhinanthus cristagalli. *Linn.* (Yellow Rattle).

RECORDS: *Dfs.* and *Kcd.*—P. Gray, 1850. *Wgt.*—G. C. Druce, 1883.

LOCALITIES: Very common in all the valleys, reaching 2300 feet, Loch Skene.

Appears June 1 to 20, J. T. J. On dry or rarely marshy sandy holms, whinstone rocks, granite, along the shore; fully exposed to wind and sun.

VISITORS: Bombus lapidarius, Hn.; lucorum, abundant and sufficient; pratorum, abundant and sufficient, S.-E.

Pedicularis palustris. *Linn.* (Lousewort).

RECORDS: *Dfs.* and *Kcd.*—P. Gray, 1850. *Wgt.*—G. C. Druce, 1883.

LOCALITIES: *Nithsdale*—Very common Lochanhead, F. W. G.; Newton, S.-E.; Holywood, Ad. and S. D. J.; Dunscore, Dr Gl.; Sanquhar, Dr Dv., etc. *Annandale*—Lochmaben, S.-E., Hn.; Moffat, J. T. J. *Eskdale*—The Flow, S.-E.; Castle O'er, J. Wn. and R. Bl. Elevation up to 1200 feet.

Appears May 24 to June 25, J. T. J. In wet peaty marshes; fully exposed.

VISITOR: Bombus lucorum, S.-E.

Pedicularis silvatica. *Linn.*

RECORDS: *Dfs.* and *Kcd.*—P. Gray, 1850. *Wgt.*—G. C. Druce, 1883.

LOCALITIES: *Nithsdale*—Newabbey, Grove, S.-E.; Cowhill, Ad. and S. D. J.; Sanquhar, Dr Dv. *Annandale*—Lochmaben, S.-E.; Moffat (to 1300 feet), J. T. J. *Eskdale*—Castle O'er, J. Wn. and R. Bl.

Appears May 2, G. Bl.; May 24 to June 11, J. T. J. On moist peat marshes, sandy holms, whinstone soils, etc.; exposed to wind and sun.

VISITORS: Bombus pratorum, reg. and sufficient; lucorum, reg. and sufficient, S.-E.

Melampyrum arvense. *Linn.*

RECORD: *Wgt.*—"Var hians," G. C. Druce, 1883.

Melampyrum pratense. *Linn.* (Cow Wheat).

RECORDS: *Dfs.* and *Kcd.*—P. Gray, 1850. *Wgt.*—G. C. Druce, 1883.

d. montanum, *Dfs.*—Dr Balfour, 1856.

LOCALITIES: *Nithsdale*—Glencaple, C. E. M.; Cargen, Glen, Hn., F. W. G.; Woodlands, Routen Brig, S.-E.; Cowhill, Ad. and S. D. J.; Craighope Linn, Hn., R. A.; Carron, R. A.; Euchan Glen, Dr Dv.

Annandale—Kirtlebridge, S.-E.; Gimmonbie, G. Bl.; Black's Hope, Correifron, var. *d.*, J. T. J.; Andrewswhinnie, var *d.*, J. T. J.; Grey Mare's Tail, vars. *a.* and *d.*, Dr Bl., J. T. J. *Eskdale*—Torduff shore, S.-E.; Castle O'er, J. Wn. and R. Bl. Elevation 300 to 2000 feet, J. T. J.

Appears June 14 to 22, J. T. J. On dry, shallow soil over whinstone rocks; usually in shade and wind-sheltered (var *d.* exposed).

VISITORS: Bombus muscorum, abundant and sufficient; Bombus pratorum, abundant and sufficient; Bombus terrestris (biting holes), S.-E.

Melampyrum silvaticum. *Linn.*

RECORD: *Kcd.*—G. N. Lloyd, 1837.

LOCALITY: Queen Mary's Cave, G. N. Ll.

Lycopus Europæus. *Linn.* (Gipsy Wort).

RECORDS: *Dfs.*—J. Cruickshank, 1836. *Kcd.*—P. Gray, 1846. *Wgt.*—Dr Graham, 1836.

LOCALITIES: *Nithsdale*—Tinwald, Dr Sn.; Maxwelltown Loch, P. Gr. *Annandale*—Castle and Halleaths Lochs, Lochmaben, G. Bl., S.-E.

Mentha sylvestris. *Linn.*

RECORDS: 1219. Alopecuroides, *Wgt.*—G. C. Druce, 1883.
1220. Longifolia, *Dfs.*—J. Sadler, 1858. *Kcd.*—Mrs Stewart, 1893.

LOCALITIES: *Nithsdale*—Newabbey, St. *Annandale*—Beld Craig, J. Sd., J. T. J. *Eskdale*—Kirkandrews, E. Ty.

On wet paths, St.

Mentha viridis. *Linn.* (Spear Mint).

RECORDS: (Escape) *Dfs.*—J. T. Johnstone, 1891. *Kcd.*—J. M'Andrew, 1882.

LOCALITY: Riddings by waterside, J. T. J.

Mentha piperita. *Huds.*

RECORDS: (Escape) *Dfs.*—J. Fingland, 1887. *Kcd.*—J. M'Andrew, 1882.

LOCALITY: Ditch glebe, Sanquhar, Dr Dv.

Mentha aquatica. *Linn.*

RECORDS: 1224. Hirsuta, *Dfs.*—P. Gray, 1876. *Kcd.*—F. R. Coles, 1882. *Wgt.*--G. C. Druce, 1883.

LOCALITIES: *Nithsdale*—Dumfries Dock Wall, 1224, P. Gr., Hn., Sanquhar (?), Dr Dv. *Annandale*—Lochmaben, Exc.; Adamsholm, J. T. J. *Eskdale*—Kirkandrews, E. Ty.

Appears August 7, J. T. J. In wet, marshy clay.

VISITORS: Picris napi, Vanessa urticæ, Polyommatus phlœas, J. C. W.; Bombus muscorum, Hn., J. C. W.; Apathus quadricolor, Hn.; campestris, J. C. W.; Halictus rubicundus, J. C. W.; Eristalis tenax, Hn., J. C. W.; æneus, horticola, J. C. W.; Volucella pellucens, and seven Anthomyids and Coleoptera, J. C. W.

Mentha sativa. *Linn.*

RECORDS: 1225. Sativa, *Dfs.*—T. B. Bell, 1882. *Kcd.*—P. Gray, 1850. *Wgt.*—G. C. Druce, 1883.*
 a. rivalis, *Dfs.*—Dr Davidson, 1890.
 1226 Rubra, *Kcd.*—P. Gray, 1850.
 1229. Gentilis, *Dfs.*—Dr Walker, 1789.

LOCALITIES: *Nithsdale*—Near Dumfries, P. Gr.; Sanquhar, 1225 *a.*, Dr Dv. *Annandale*—Lochmaben, Exc.; Adamsholm, J. T. J.; Moffat Water, below Correifron, 1229, Dr Wl. *Eskdale*—T. B. Bl.

Appears August 7.

Mentha arvensis. *Linn.*

RECORDS: *Dfs.*—Dr Davidson, 1886. *Kcd.*—J. M'Andrew, 1882. *Wgt.*—G. C. Druce, 1883.

b. nummularia, *Dfs.*—Dr Davidson, 1886.

LOCALITIES: *Nithsdale*—Lochanhead, F. W. G.; Maxwelltown, Hn.; Dumfries, S.-E., Sanquhar, *b.*, Dr Dv. *Annandale*—Ecclefechan, S.-E.; Archbank fields, J. T. J. *Eskdale*—Canobie, E. Ty.; Gilnockie, S.-E.; Langholm, C. Y.

Appears August 11, J. T. J. On dry corn fields, specially on boulder clay; in part wind-sheltered.

VISITORS: Scatophaga stercoraria, Siphona cristata, Hydrotea dentipes, Lophius albomarginatus, Telephorus fulvus, Hn.

---

* These are all doubtful as records of the sub-species.

Mentha pulegium. *Linn.* (Pennyroyal).

RECORD : (Escape) *Dfs.*—Dr Singer, 1843.
LOCALITY : Holywood, Dr Sn.

Thymus serpyllum. *Fr.* (Thyme).

RECORDS : 1234. *Dfs.* and *Kcd.*—P. Gray, 1850.   *Wgt.*—G. C. Druce, 1883.
1235. Chamædrys, *Wgt.*—G. C. Druce, 1883.

LOCALITIES : *Nithsdale*—Common Nith, Hn.; Cluden Mills, Routen Brig, S.-E.; Glenquhargen, Hn.; Sanquhar, Dr Dv. *Annandale*—Annan, Æ, S.-E.; Evan Water, S.-E.; Kerr, J. T. J.; Beeftub, Black's Hope, and Loch Skene (to 2400 feet), J. T. J., S.-E. *Eskdale*—Langholm, Wauchope, S.-E.; Castle O'er, J. Wn. and R. Bl.

Appears June 14 to 22, J. T. J. On dry, often bare whinstone rocks, old walls, shingles, trap, boulder clay; often half-shaded and sheltered or quite exposed.

VISITORS : Apis, abundant and sufficient; Bombus lucorum, Apathus quadricolor, Calliphora erythrocephala, Micropalpus vulpuria, (Cynomyia mortuorum ?), S.-E.

Origanum vulgare. *Linn.* (Marjoram).

RECORDS : (Escape) *Dfs.*—Dr Singer, 1843.   *Kcd.*—P. Gray, 1850. *Wgt.*—G. C. Druce, 1883.

LOCALITIES : *Nithsdale*—Near Dumfries, P. Gr.; Cowhill, Ad. and S. D. J.; Doocot Knowe, R. A. *Annandale*—Westerhall, Dr Sn.; about a mile along Milke below and above Scroggs, G. Bl.

Calamintha acinos. *Clairv.* (Basil Thyme).

RECORDS : (Escape) *Kcd.*—P. Gray, 1846.   *Wgt.*—J. M'Andrew, 1893.

LOCALITIES : *Nithsdale*—Troqueer, J. Fr.; first mile along Castle-Douglas Road, Glen, P. Gr.; Cargen Bridge, F. W. G.

Calamintha clinopodium. *Benth.* (Basil).

RECORDS : *Dfs.*—J. Wilson, 1882 (?).   *Kcd.*—J. Fraser, 1843.   *Wgt.*—J. M'Andrew, 1890.

LOCALITIES : *Nithsdale*—Scaur, Keir Bridge, R. A. *Annandale*—Annan to Gretna shore, S.-E.; Milke above Scroggs, G. Bl.; Dryfe,

G. Bl.; Wamphray, Craigbeck Bridge, Grey Mare's Tail, Craigmichen Scaurs (500 to 2000 feet), J. T. J. *Eskdale*—Liddel Railway Bridge, S.-E.; Castle O'er, J. Wn. and R. Bl.

Appears August 12 to September 7, J. T. J. On pretty dry slopes of cinders, sandy gravel, stony till, on alluvium; in sun or shade and part wind-sheltered.

VISITOR : Bombus muscorum, S.-E.

Nepeta glechoma. *Benth.* (Ground Ivy).

RECORDS: *Dfs.* and *Kcd.*—P. Gray 1850. *Wgt.*—G. C. Druce, 1883.

LOCALITIES : *Nithsdale*—Caerlaverock, S.-E.; common Nith, Hn., C. E. M.; Grove, Friars' Carse, S.-E.; Tynron (400 feet), J. Sh.; Sanquhar, Dr Dv. *Annandale*—Shieldhill, Milke, S.-E.; Heathery Haugh, J. T. J. *Eskdale*—Gilnockie, S.-E.

Appears April 30, J. Sh.; to May 24, J. T. J. On pretty dry banks by roads, whinstone, sandstones, dykes, humus, etc.; half or quite shaded and wind-sheltered.

VISITOR : Bombus muscorum, abundant and sufficient, S.-E.

Prunella vulgaris. *Linn.* (Self-heal).

RECORDS: *Dfs.* and *Kcd.*—P. Gray, 1850. *Wgt.*—G. C. Druce, 1883.

LOCALITIES : *Nithsdale*—Cluden Bridge, and very common Dumfries, S.-E.; Sanquhar, Dr Dv. *Annandale*—Annan, Kirtlebridge, S.-E.; Moffat, J. T. J. *Eskdale*—Very common everywhere to Castle O'er, J. Wn. and R. Bl.

Appears June 20 to 25, J. T. J. On dry or moist shingles, holms, boulder clay, whinstone soils, granite, etc.; fully exposed.

VISITORS : Bombus muscorum, abundant and sufficient, S.-E., J. C. W.; terrestris, J. C. W.

Scutellaria galericulata. *Linn.* (Scull-cap).

RECORDS : *Dfs.*—J. Sadler, 1858. *Kcd.*—G. N. Lloyd, 1837. *Wgt.* G. C. Druce, 1883.

LOCALITIES : *Nithsdale*—Southerness, Exc.; Brow (Lochar), S.-E.; Lincluden below Abbey, S.-E.; Cowhill, Ad. and S. D. J.; Loch Urr (700 feet), J. Cr.; Friars' Carse Loch, S.-E., Dr Gl.; Laught Road, New

Loch, R. A. *Annandale*—Lochmaben (Castle and Halleaths Lochs), J. T. J., Th., S.-E., G. Bl.; Grey Mare's Tail, J. Sd.

On sandy often shelly or gravelly loch margins ; half-shaded and in part wind-sheltered by grass, etc.

VISITORS: Bombus hortorum, muscorum, S.-E.

Scutellaria minor. *Linn.*

RECORDS: *Kcd.* - G. N. Lloyd, 1837. *Wgt.*—Dr Balfour (1836?).

LOCALITIES: Portwilliam, Dr Bl.; Trostrie Loch, T. Bl.; Conaughty field, Dundrennan, G. N. Ll.; Auchencairn, J. C. W.; Aird's Point, Murbroy, Laggan Hill, J Fr., F. W. G., Th., Hn.

Marrubium vulgare. *Linn.* (Horehound).

RECORDS: (Escape) *Dfs.*—Dr Singer, 1843. *Kcd.*—Natural History Society?

LOCALITY: Holywood, Dr Sn.

Stachys betonica. *Benth.* (Betony).

RECORDS: *Dfs.*--Dr Singer, 1843. *Kcd.*—Mr Maughan, 1789. *Wgt.*—Miss Hannay, 1893.

LOCALITIES: *Nithsdale*—Newabbey, E. M. C.; Caerlaverock, J. Fn.; Glen, Hn.; Holywood, Dr Sn.; Auldgirth, F. W. G., J. Fn.; Scaur, Th.; common Euchan and Euchanmouth to Elliock Bridge, J. Sh., Dr Dv. *Annandale*—Gimmenbie, G. Bl.; Selkirk Road, Moffat, J. T. J.; Beld Craig, Garpol, Grey Mare's Tail, J. Sn. *Eskdale*—Kirkandrews, E. Ty. (under 600 feet).

Appears June 30, G. Bl.; August to September, J. T. J. On pretty dry roadsides or humus; half-shaded and wind-sheltered.

VISITORS: Bombus hortorum, abundant and sufficient, Hn., S.-E.; muscorum, pratorum, Hn.

Stachys silvatica. *Linn.* (Woundwort).

RECORDS: *Dfs.*—Dr Davidson, 1886. *Kcd.*—P. Gray, 1850. *Wgt.* —G. C. Druce, 1883.

LOCALITIES: Very common in all the valleys, reaching 1400 feet Whitehope.

Appears June 24 to 29, J. T. J. On moist leaf-mould, clayey holms, roadsides ; in shade and wind-sheltered.

VISITORS: Bombus muscorum, lucorum, pratorum, S.-E.; Platychirius clypeatus, Hn.

Stachys palustris. *Linn.*

RECORDS: Palustris, *Dfs.* and *Kcd.*—P. Gray, 1850. *Wgt.*—Dr Balfour, 1836.
Palustris × sylvatica, *Dfs.*—J. T. Johnstone, 1889. *Kcd.*—G. N. Lloyd, 1837. *Wgt.*—Dr Arnott, 1848.

LOCALITIES: *Nithsdale*—Kingholm Mill, S.-E.; Cargen, M. J. H.; Dumfries, P. Gr.; Sanquhar, Dr Dv. *Annandale*—Brow-well, F. W. G.; Annan, Halleaths, S.-E.; Lochmaben, Th.; Craigbeck (× sylvatica), J. T. J.; Hydropathic grounds, J. T. J. *Eskdale*—Glenzier, E. Ty.; Castle O'er, J. Wl. and R. Bl

Appears June 22 to 24, J. T. J. In moist holms, by roadsides, cinders of railways; usually in sun and part wind-sheltered by long grass.

VISITORS: Bombus muscorum, abundant and sufficient, Hn., S.-E., J. C. W.; hortorum, Hn., J. C. W.; lucorum, S.-E.; Anthidium manicatum, J. C. W.; Platychirius peltatus, S.-E.; manicatus, albimanus, Melanostoma scalare, Rhingia rostrata, Anthomyia radicum, J. C. W.

Stachys arvensis. *Linn.*

RECORDS: *Kcd.*—F. R. Coles, 1882. *Wgt.*—G. C. Druce, 1883.

Galeopsis tetrahit. *Linn.*

RECORDS: 1260. (Versicolor) *Dfs.*—Sir J. G. Cullum, 1789. *Kcd.*—P. Gray, 1850.
1261. Tetrahit, *Dfs.* — P. Gray, 1850. *Kcd.*—J. M'Andrew, 1882. *Wgt.*—G. C. Druce, 1883.
b. bifida, *Dfs.*—J. T. Johnstone, 1890. *Wgt.*—G. C. Druce, 1883.

LOCALITIES: *Nithsdale*—Glencaple, Hn.; common Dumfries, 1260, 1261, F. W. G., S.-E., Hn., C. E. M., Th.; Old Barr, Greenhead, Wanlockhead, 1260, Dr Dv. *Annandale*—Priestside Flow, Kirtle, abundant, S.-E.; Beld Craig, 1260, J. Sd.; Craigbeck, 1261 *b.*, J. T. J.; Moffat, very common, J. T, J. *Eskdale*—Gretna Green, 1260, Sir J. G. C.; Esk valley, common, S.-E.; Castle O'er J. Wn. and R. Bl. (to 800 feet).

Appears June 30 to July 8, J. T. J. In moist or dry corn fields, on boulder clay, holms, sandstones; usually in sun and wind-sheltered by grass or other plants.

VISITORS: Bombus muscorum, Hn., S.-E., J. C. W.; lucorum, Hn., S.-E.; hortorum, Hn.; terrestris, J. C. W.; Syrphus balteatus, Hn.; Platychirius peltatus, Hn.

Lamium amplexicaule. *Linn.*

RECORDS: 1263. Amplexicaule, *Dfs.*—P. Gray, 1850. *Kcd.*—J. M'Andrew, 1882. *Wgt.*—G. C. Druce, 1883, 1264. Intermedium, *Dfs.*—J. Cruickshank, 1836. *Kcd.* —Field Club Excursion, 1893. *Wgt.*—Dr Graham, 1836.

LOCALITIES: *Nithsdale*—Southerness, also 1264, Exc.; Dumfries, P. Gr.; Brownhall, J. Cru.; Fourmerkland, Th.; Auldgirth Station, 1264, Exc.; Tynron, J. Sh.; Sanquhar, Dr Dv. *Annandale*—Annan Station, S.-E., C. E. M. *Eskdale*—Woodslee Orchard, S.-E.; Irving House, C. Y.

On dry cinders or sandy waste ground ; fully exposed or part wind-sheltered by long grass.

Lamium purpureum. *Linn.*

RECORDS: 1266. Purpureum, *Dfs.* and *Kcd.*—P. Gray, 1850. *Wgt.*—G. C. Druce, 1883.

LOCALITIES: *Nithsdale*—Carsethorn, E. M. C.; Dumfries, P. Gr.; Glencaple, F. W. G.; Newton, S.-E.; Sanquhar, Dr Dv. *Annandale*— Along the shore, common, S.-E.; Scroggs, G. Bl.; Moffat, common, J. T. J. *Eskdale*—Langholm, S.-E.; Castle O'er, J. Wn. and R. Bl.

Appears Jan. 3, G. Bl.; March 23 to May 28, J. T. J. On dry or moist garden soil, waste ground free of other plants, cinders of stations, boulder clay, etc.; fully exposed.

VISITOR: Bombus muscorum, reg. and sufficient, S.-E.

Lamium album. *Linn.*

RECORDS: *Dfs.* and *Kcd.*—P. Gray, 1850. *Wgt.*—Rev. J. Gorrie, 1891.

LOCALITIES: *Nithsdale*—Newabbey, Lincluden, Cargen, Hn.; Dumfries, F. W. G.; Nunwood corner of Terregles Park, S.-E.; Cowhill, Ad. and S. D. J.; Thornhill, R. A.; Carco, Dr Dv. *Annandale*—Lockerbie, Hn.; seven miles north of Annan, S.-E.; Chapel Farm (700 feet), J. T. J. *Eskdale*—Esk, Meggat, S.-E.

Appears May 21 to June 30, J. T. J. In wet or moist ditches by roadsides ; half-sheltered by long grass or hedges.

VISITORS: Bombus muscorum, terrestris hortorum, Hn.; Syrphid, S.-E.

Lamium maculatum. *Linn.*

RECORD : (Escape) *Kcd.*—J. M'Andrew, 1882.
LOCALITY : Grove Road, Hn.

Lamium galeobdolon. *Crantz.*

RECORD : (Escape) *Dfs.*—Miss Ethel Taylor, 1892.
LOCALITY : Kirkandrews, E. Ty.

Teucrium scorodonia. *Linn.* (Wood Sage).

RECORDS : *Dfs.* and *Kcd.*—P. Gray, 1850. *Wgt.*—G. C. Druce, 1883.

LOCALITIES : *Nithsdale*—Very common Cluden Mills, S.-E.; Sanquhar, Dr Dv. *Annandale*—Common Powfoot shore, S.-E.; Catch Hall Loaning, G. Bl.; Moffat, very common, J. T. J. *Eskdale*—Common Langholm, Meikledale, Mosspaul, S.-E. (to 1500 feet).

Appears June 25 to July 3. On dry whinstone rocks, granite, shingles of shore ; usually half-shaded and in part wind-sheltered.

VISITORS : Apis, S.-E.; Apathus campestris, J. C. W.; Bombus muscorum, abundant reg., Hn., J. C. W., S.-E.; hortorum, J. C. W., Hn.; pratorum, abundant and sufficient, S.-E.; terrestris, J. C. W.; Syrphus, sp., S.-E.

Ajuga reptans. *Linn.* (Bugle).

RECORDS : *Dfs.* and *Kcd.*—P. Gray, 1850. *Wgt.*—G. C. Druce, 1883.

LOCALITIES : *Nithsdale*—Common by Nith and Cluden, S.-E.; Carron Glen, C. E. M.; Sanquhar, Dr Dv. *Annandale*—Moffat, very common (to 2000 feet), Correifron, J. T. J. *Eskdale*—Very common by Esk, Tarras, etc. (to 1400 feet), Archie Hill, S.-E.; Castle O'er, J. Wn. and R. Bl.

Appears April 23, J. Bl.; May 15 to 20, J. T. J. On moist humus, holmlands, roadsides, etc.; in shade or half-shaded and usually windsheltered.

VISITORS : Bombus muscorum, abundant and sufficient ; Rhingia rostrata, Platychirius albimana, S.-E.

Ajuga pyramidalis. *Linn.*

RECORD : *Dfs.*—J. T. Johnstone, 1888.
LOCALITY : *Annandale*—Black's Hope, J. T. J.

Appears May 23, J. T. J. On pretty dry grass ledges amongst whinstone rocks, about 1250 feet.

Statice limonium. *Linn.* (Sea Lavender).

RECORDS: 1054. Limonium, *Kcd.* and *Wgt.*—Mr Maughan, 1789. 1055. Rariflora, *Kcd.*—Dr Arnott, 1848. *Wgt.*—Dr Graham, 1836. 1056. Auriculæfolia, *Wgt.*—Dr Balfour, 1843.
 a. occidentalis, *Wgt.*—J. M'Andrew, 1890
 b. intermedia, *Wgt.*—J. M'Andrew, 1890.

LOCALITIES: *Along the shore*—Mull, 1055, Dr Gr.; 1056, J. M'A.; West Tarbet, 1056 *a.*, J. M'A.; Orchardton Bay to Garliestown, 1054, 1055, Dr Bl., J. M'A., Hn.; Burran Point, P. Gr.; Borgue, Ross, J. M'A.; St. Mary's Isle, Kirkcudbright, Mau., G. N. Ll., J. M'A.; Auchencairn Bay, J. M'A., C. E. M.; Rockcliffe, 1055, Bab.

On muddy ground along the shore, Fq.

VISITORS: Bombus hortorum, Melegethes, sp., J. C. W.

Armeria vulgaris. *Willd.* (Thrift).

RECORDS: *Dfs.* and *Kcd.*—P. Gray, 1850. *Wgt.*—G. C. Druce, 1883.

LOCALITIES: *Along the shore*—Very common from Newabbey to Kingholm and thence left bank to Cumberland border, Hn., S.-E.; Loch Skene, J. Sd.

On estuarine mud below level, highest tides; fully exposed.

VISITORS: Bombus pratorum, Syrphus corollæ, Hn.; Anthomyia radicum, Dolichopods, Hilara maura, S.-E.

Plantago major. *Linn.*

RECORDS: *Dfs.* and *Kcd.*—P. Gray, 1850. *Wgt.*—C. C. Bailey, 1883.

LOCALITIES: Very common in all the valleys to 600 feet.

On bare ground by roadsides, shingles at rivers and shore; in full exposure.

Plantago media. *Linn.*

RECORDS: *Dfs.*—P. Gray, 1850. *Kcd.*—J. M'Andrew, 1882.

LOCALITIES: *Nithsdale*—Dumfries, P. Gr. *Annandale*—Torduff Point, S.-E.; Westburn above Clarefoot, J. T. J.

Appears July 25. On stones, concrete, sandy ground, river shingles; fully exposed.

Plantago lanceolata. *Linn.* (Plantain).

RECORDS: *Dfs.* and *Kcd.*—P. Gray, 1850. *Wgt.*—G. C. Druce, 1883.

LOCALITIES: Very common in all the valleys to 1400 feet, Whitehope, S.-E.

Appears April 14, G. Bl. In short turf, bare rocks, river shingles, etc., on all soils except peat, and usually exposed; rarely in shade and wind-sheltered.

VISITORS: Polyommatus icarus (Bombus, sp.; Apis, c. p., also rarely Platychirius, sp., c. p.), S.-E.

Plantago maritima. *Linn.*

RECORDS: *Dfs.*—P. Gray, 1850. *Kcd.*—J. M'Andrew, 1882. *Wgt.* —G. C. Druce, 1883.

LOCALITIES: *Along the shore*—Very common from Kingholm Merse to Cumberland border, F. W. G., S.-E., Hn, E. Ty.

On wet estuarine mud; fully exposed.

Plantago coronopus. *Linn.* (Buck's Horn).

RECORDS: *Dfs.*—P. Gray, 1850. *Kcd.*—J. M'Andrew, 1882. *Wgt.*—G. C. Druce, 1883.

LOCALITIES: *Along the shore*—Three miles from Dumfries, P. Gr.; Annan mouth and occasionally to Gretna Bridge, S.-E.

Bare places on estuarine mud, sandstone quays, shingles; fully exposed.

Littorella lacustris. *Linn.*

RECORDS: *Dfs.*—Dr Burgess, 1789. *Kcd.*—J. M'Andrew, 1882. *Wgt.*—G. C. Druce, 1883.

LOCALITIES: *Nithsdale*—Cluden above Lincluden Abbey, P. Gr.; Morton Mains Loch, R. A. *Annandale*—Castle, Kirk, Mill and Halleaths Lochs, Lochmaben, Dr Br., S.-E.; Loch Skene, Dr Bl., J. T. J.

On wet mud, margins of lochs; often over-flowed or exposed.

Scleranthus annuus. *Linn.* (Knowel).

RECORDS: *Dfs.*—Dr Burgess, 1796. *Kcd.*—P. Gray, 1850. *Wgt.* —G. C. Druce, 1883.

LOCALITIES : *Nithsdale*—Dumfries, F. W. G.; Maxwelltown, Hn.; Racks, Holywood, S.-E.; Kirkmahoe, Hn.; Sanquhar, Dr Dv. *Annandale*—By Annan, Moffat, J. T. J. *Eskdale*—Canobie, E. Ty., S.-E.; Langholm, S.-E.; Broomholm (S. Polycarpos), Dr Br.

Appears May 18, J. T. J. On dry sandy waste ground bare of other plants.

Salicornia herbacea. *Linn.* (Glasswort).

RECORDS : *Dfs.*—Dr Singer, 1843. *Kcd.*—Mr Farquharson, 1873. *Wgt.*—G. C. Druce, 1883.

LOCALITIES : *Along the shore*—Common Southerness, J. M'A.; Corbelly, Newabbey, Dr Gl.; Glencaple, Hn.; Annan mouth, Dr Sn.

On wet estuarine mud ; usually part sheltered from wind.

Suæda maritima. *Dumort.*

RECORDS : *Kcd.*—Mrs Gilchrist-Clark, 1867. *Wgt.*—J. M'Andrew, 1890.

LOCALITIES : *Along the shore*—Wigtown, J. M'A.; Port Yerrick, Eggerness, Hn.; Southwick, St.; Auchencairn, J. C. W.; Brighouse, H. R. C.; Ross, G. C.; Colvend Scaur, Th.; Fq.; Southerness, Th.

On mud of salt marshes, Th.

Salsola kali. *Linn.* (Saltwort).

RECORDS : *Dfs.*—J. Fingland, 1891. *Kcd.*—Mrs Gilchrist-Clark, 1867. *Wgt.*—G. C. Druce, 1883.

LOCALITIES : *Along the shore*—Lag Point, Monreith, J. M'A.; Ross, G. C.; Arbigland to Southerness, C. E. M.; Glencaple, Hn.; Powfoot to Newbie, J. Fn.

Sandy shores.

Chenopodium vulvaria. *Linn.*

RECORD : (Escape) *Kcd.*—Mrs Gilchrist-Clark, 1867.
LOCALITY : Ross.

Chenopodium album. *Linn.*

RECORDS : *Dfs.*—P. Gray, 1850. *Kcd.*—Mrs Gilchrist-Clark, 1867. *Wgt.*—G. C. Druce, 1883.

a. Candicans, *Dfs.*—Dr F. W. Grierson, 1882. *Wgt.*—G. C. Druce, 1883.

b. Viride, *Dfs.*—Dr Davidson, 1890. *Kcd.*—J. M'Andrew, 1882.
c. Paganum, *Wgt.*—G. C. Druce, 1883.

LOCALITIES : *Nithsdale*—Dumfries, P. Gr., and var. *a.*, F. W. G.; Cowhill, Ad. and S. D. J.; common Sanquhar, var. *b.*, Dr Dv. *Annandale*—Waste ground and fields, J. T. J.; var. *b.*, waste ground, J. T. J.

Chenopodium urbicum. *Linn.*

RECORD : (Escape) *Dfs.*—G. F. Scott-Elliot, 1892.
LOCALITY : Wellburn, Moffat, S.-E.

Chenopodium bonus henricus. *Linn.*
(Good King Henry).

RECORDS : *Dfs.*—P. Gray, 1850. *Kcd.*—J. M'Andrew, 1882. *Wgt.*—G. C. Druce, 1883.

LOCALITIES : *Nithsdale*—Dumfries, P. Gr. ; Locharbriggs, Hn.; Auldgirth Station, S.-E.; Druidhall, Penpont, J. Sh.; Templand, Durisdeer, R. A.; Craighope Linn, Hn. *Annandale*—Queensberry Foot, Dr Gl.; Wellburn, Cornal Tower Burn, Craigbeck, Craigieburn, J. T. J. *Eskdale*—Irvine, Langholm, Burnfoot, S.-E.

Appears May 28 to July 5, J. T. J. On waste ground, railway cinders, gardens, river shingles ; fully exposed.

Beta maritima. *Linn.*

RECORDS : *Kcd.*—J. M'Andrew, 1882. *Wgt.*—Dr Arnott, 1848.
LOCALITIES : *Along the shore*—Cowans, Port Logan, Arn.; Wigtown, J. M'A.; Creetown, J. M'A.

Atriplex portulacoides. *Linn.*

RECORDS : *Wgt.*—Dr Balfour, 1843.
LOCALITIES : Mull, Dr Bl., Dr Gr.; Garliestown, Hn.

Atriplex patula. *Linn.* (Orache).

RECORDS : 1307. Littoralis.
  b. serrata (?), *Kcd.* and *Wgt.*—J. M'Andrew, 1888.
1308. Patula, *Dfs.*—J. Fingland (1886?). *Kcd.*—J. M'Andrew, 1882

b. erecta, *Dfs.*—W. Stevens, 1848.   *Wgt.*—Dr Balfour, 1843.
c. angustifolia, *Dfs.*—Dr Davidson, 1886.   *Kcd.* and *Wgt.*—G. C. Druce, 1883.
1309. Hastata, *Dfs.*—W. Stevens, 1848.   *Kcd.*—G. C. Druce, 1883.
1310. Deltoidea, *Kcd.* and *Wgt.*—G. C. Druce, 1883.
1311. Babingtonii, *Kcd.*—J. M'Andrew, 1882.   *Wgt.*—G. C. Druce, 1883.

LOCALITIES : *Nithsdale*—Dumfries, in fields, common, S.-E.; Thornhill and Drumlanrig, 1308 *b.* and 1309, W. St.; Sanquhar, 1308 *c.*, Dr Dv.   *Annandale*—Annan to Kirtle mouth, S.-E.; Powfoot to Newbie, J. Fn.   *Eskdale*—Gretna shore, S.-E.; Burnfoot, S.-E.

Sandy places on shore, fields on light sandy soil, S.-E. (1308 *b.*, dunghills, W. St.).

Atriplex rosea. *Linn.*

RECORDS : *Kcd.* and *Wgt.*—Dr Balfour, 1843.

LOCALITIES : *Along the shore*—Drummore, Cowans, Arn.; Portwilliam, Hn.; Whitethorn, Southerness, Dr Bl.; Glencaple, Hn.

Muddy shingles.

Rumex aquaticus. *Linn.*

RECORDS : *Dfs.*—W. Keddie, 1854.   *Kcd.*—J. M'Andrew, 1882.   *Wgt.*—G. C. Druce, 1883.

LOCALITIES : *Nithsdale*—Sanquhar, Dr Dv.   *Annandale*—Beattock, Kd.; Beld Craig, J. Sd., J. T. J.

Rumex crispus. *Linn.*

RECORDS : *Dfs.* and *Kcd.*—P. Gray, 1850.   *Wgt.*—G. C. Druce, 1883.

LOCALITIES : *Nithsdale*—Common Dumfries, P. Gr.; Sanquhar, Dr Dv.   *Annandale*—Ecclefechan, S.-E.; Moffat, J. T. J.

Dry ditches, pretty common to 600 feet.

Rumex obtusifolius. *Linn.*

RECORDS : *Dfs.*—Dr Davidson, 1886.   *Kcd.*—J. M'Andrew, 1882.   *Wgt.*—G. C. Druce, 1883.

LOCALITY : Very common in all the valleys.

Along roadsides, boulder clay, holms, etc.; fully exposed.

Rumex hydrolapathum. *Huds.*

RECORD: (Escape) *Kcd.*—J. M'Andrew, 1882.
LOCALITY: Lovers' Walk, Carlingwark.

Rumex conglomeratus. *Murr.*

RECORDS: *Kcd.*—F. R. Coles, 1883.* *Wgt.*—G. C. Druce, 1883.*
LOCALITY: None given.

Rumex sanguineus. *Linn.*

RECORDS: *Dfs.*—Dr Gilchrist, 1882. *Kcd.*—J. M'Andrew, 1882. *Wgt.*—Dr Balfour, 1843.
b. viridis, *Dfs.* — Dr Davidson, 1886. *Kcd.* — J. M'Andrew, 1882.
LOCALITIES: Very common in all the valleys. Along the shore, on shingles of rivers, roadsides, sandy holms, etc.; fully exposed to wind and sun.

Rumex alpinus. *Linn.*

RECORD: (Escape) *Dfs.*—W. Stevens, 1848.
LOCALITY: Eccles, Penpont and Closeburn Mills, W. St.

Rumex friesii. *Bab.*

RECORD: (Escape) *Dfs.*—G. F. Scott-Elliot, 1892.
LOCALITY: Bilholm, S.-E.

Rumex acetosa. *Linn.*

RECORDS: *Dfs.* and *Kcd.*—P. Gray, 1850. *Wgt.*—G. C. Druce, 1883.
LOCALITIES: Very common in all the valleys to 1500 feet, Causeway Grain.
Appears May 3, J. T. J. On dry or moist holms by rivers, sandy soil, shingles, old walls; usually in sun or half-shaded, and in part wind-sheltered by long grass, etc.

* Require confirmation.

Rumex acetosella. *Linn.*

RECORDS: *Dfs.* and *Kcd.*—P. Gray, 1850. *Wgt.*—G. C. Druce, 1883.

LOCALITIES: Very common in all the valleys to 2300 feet, Loch Skene.

Appears May 4, J. T. J. On dry whinstone rocks, trap rocks, shingles of rivers, gravelly till, sandy holms, etc.; usually exposed or in part wind-sheltered.

Oxyria reniformis. *Hook.*

RECORD*: *Dfs.*—Dr Singer, 1843.

LOCALITIES: *Annandale*—Hartfell Craigs, Black's Hope, Correifron, J. T. J., S.-E.; Grey Mare's Tail, Dr Sn., Kd., Dr Bl., J. Sd., T. Bl., W. S. H., S. W. Ca.

Appears May 24 to June 22, J. T. J. On dry whinstone rocks in moist atmosphere, from 1700 to 2300 feet; usually in narrow, and therefore wind-sheltered gullies.

Polygonum aviculare. *Linn.* (Knotgrass).

RECORDS: *Dfs.* and *Kcd.*—P. Gray, 1850. *Wgt.*—G. C. Druce, 1883.

LOCALITIES: Very common in all the valleys to 800 feet.

Appears June 30 to July 9. In cultivated fields, by roadsides, cinders of railway stations; fully exposed.

Polygonum maritimum. *Linn.*

RECORDS: Var. Roberti, *Dfs.*- G. F. Scott-Elliot, 1892. *Kcd.*—J. M'Andrew, 1882. *Wgt.*—Dr Graham, 1836.

LOCALITIES: *Along the shore*—Portpatrick, S.-E.; Kirkmaiden, Dr Bl.; Drummore, Cowans, Arn.; Portwilliam, Hn.; Rerrick, G. M'C.; the Brow to Kirtle mouth, Annan, S.-E.

On sand or shingles by sea, waste ground and roadsides; fully exposed.

---

* This is one of the most flagrant examples known to me of "record-making." A record was published as new in 1888, although it has been, at least, seven times published before, and specimens existed in London, Edinburgh, and Dumfries.

Polygonum convolvulus. *Linn.*

RECORDS: *Dfs.* and *Kcd.*—P. Gray, 1850. *Wgt.*—G. C. Druce, 1883.

LOCALITIES: *Nithsdale*—Dumfries, P. Gr.; Maxwelltown, S-E.; Castle-Douglas Road, C. E. M.; Sanquhar, Dr Dv. *Annandale*—Common by shore, S.-E.; Lochmaben, G. Bl.; common Moffat, J. T. J. *Eskdale*—Esk, S.-E.

On sand or shingles of shore and rivers, cinders of railways, arable fields, etc.; usually in part sheltered by herbage or fully exposed.

Polygonum viviparum. *Linn.*

RECORD: *Dfs.*—Dr Singer, 1843.

LOCALITIES: *Annandale*—West side of Hartfell, Black's Hope, Saddleyoke ravine, Correifron, Birkie Cleugh, Whitecoombe, J. T. J.; Midlaw Burn and above Loch Skene, S.-E. (1800 to 2200 feet).

Appears July 19, J. T. J. On moist grass ledges amongst whinstones or mudstones; fully exposed or slightly wind-sheltered.

Polygonum bistorta. *Linn.*

RECORDS: *Dfs.* and *Kcd.*—P. Gray, 1850 and 1846. *Wgt.*—J. M'Andrew, 1893.

LOCALITIES: *Nithsdale*—Cargen Water, Hn.; Woodlands, R. H. M.; Cresswell, Dr Gl.; Guilyhill, R. H. M.; Cowhill, Ad. and S. D. J.; Auldgirth, Hn.; Caitloch (exterminated), J. Cr.; Grovehill, Waterside, Holmhill, R. A. *Annandale*—Douglas Hall Bridge, Lockerbie, G. Bl.; Craiglands, Old Well Road, Kerr, Dyke, J. T. J.

Appears May 28 to June 2, J. T. J. On moist or nearly dry roadsides, railway banks; usually exposed or shaded.

VISITORS: Allantus nothi, abundant; Perineima nassata, Volucella bombylans, Empis tessellata, abundant, S.-E.

Polygonum amphibium. *Linn.*

RECORDS: *Dfs.*—Miss F. A. Hope, 1881. *Kcd.*—P. Gray, 1850. *Wgt.*—G. C. Druce, 1883.
  b. terrestre, *Dfs.*—J. T. Johnstone, 1890. *Kcd.*—J. M'Andrew, 1882.

LOCALITIES: *Nithsdale*—Near Dumfries, P. Gr.; Ruthwell, Lincluden, F. W. G.; Loch Urr (700 feet), J. Cr.; Thornhill, var. *b.*, R. A.; Sanquhar, Dr Dv. *Annandale*—Priestside, S.-E.; Roberthill, Dryfe (also var. *b.*), Caledonian Railway (var. *b.*), G. Bl.; Kirkpatrick-Juxta, F. A. H.; Beattock (var *b.*), J. T. J. *Eskdale*—Glenzier, S.-E.; Castle O'er, J. Wn. and R. Bl.

In water of lochs or sluggish streams with roots in mud. Var. *b.* on waste dry ground, corn fields, etc.

VISITORS : Syrphus, sp.; Dolichopodidæ, Tetanocera, sp., E. Ty.

Polygonum lapathifolium. *Linn.*

RECORDS : *Dfs.*—Dr Davidson, 1886. *Kcd.*—P. Gray, 1850. *Wgt.* —G. C. Druce, 1883.

Var. incana, *Dfs.*—J. T. Johnstone, 1889.

LOCALITIES : Very common in all the valleys to about 900 feet (var. Beattock, J. T. J.).

Appears July 30, J. T. J. On dry soil of cultivated fields, cinders of railways, etc.; fully exposed.

Polygonum persicaria. *Linn.*

RECORDS: *Dfs.* and *Kcd.*—P. Gray, 1850. *Wgt.*—G. C. Druce, 1883.

Nodosum, var. glandulosum (?), J. T. J.

LOCALITY : Very common in all the valleys to 900 feet.

Appears July 30. On dry soil, cultivated fields ; fully exposed.

VISITORS : Hyetodesia incana, Siphona cristata, Hn.

Polygonum lapathifolium × persicaria.

RECORD : *Dfs.*—J. T. Johnstone, 1891.

Polygonum hydropiper. *Linn.* (Waterpepper).

RECORDS: *Dfs.* and *Kcd.*—P. Gray, 1850. *Wgt.*—G. C. Druce, 1883.

LOCALITIES : *Nithsdale*—Dalscone, F. W. G.; Lincluden Holms, Cluden Mill, Hn., S.-E.; Cowhill, Ad. and S. D. J.; New Loch, Rashbriggs, R. A.; Sanquhar, Dr Dv. *Annandale*—Lochmaben, S.-E.; Moffat, common, J. T. J. *Eskdale*—Esk, Liddel, S.-E.

Appears July 30, J. T. J. On wet places of holms or lake margins ; usually fully exposed.

Polygonum minus. *Huds.*

RECORDS: *Dfs.*—J. Cruickshank, 1839. *Kcd.*—P. Gray, 1850.
LOCALITY: *Nithsdale*—Lochar Moss, J. Cru., P. Gr.

Daphne mezereum. *Linn.*

RECORD: (Escape) *Wgt.*—Sir H. Maxwell, 1889.

Hippophae rhamnoides. *Linn.* (Buckthorn).

RECORDS: (Escape) *Dfs.*—Dr Davidson, 1886. *Wgt.*—C. C. Bailey, 1883.
LOCALITY: Morison House, Moffat, J. T. J. (planted).

Euphorbia helioscopia. *Linn.* (Sunspurge).

RECORDS: *Dfs.* and *Kcd.*—P. Gray, 1850. *Wgt.*—G. C. Druce, 1883.

LOCALITIES: *Nithsdale* — Dumfries, F. W. G., P. Gr., Th., Hn.; Holywood, S.-E.; Sanquhar, Dr Dv. *Annandale*—Very common by shore, Annan to Kirtle, S.-E.; Ecclefechan, S.-E.; Lochmaben, C. E. M.; Skellshead, Saughtrees, and on cultivated fields, Moffat, J. T. J. *Eskdale*—Meggat, S.-E.

Appears July 4, J. T. J. On dry, sandy, or clayey fields, shingles of shore, etc.; usually in sun, but in part wind-sheltered.

Euphorbia peplus. *Linn.*

RECORDS: *Dfs.*—Dr F. W. Grierson, 1882. *Kcd.*—J. M'Andrew, 1882. *Wgt.*—G. C. Druce, 1883.

LOCALITIES: *Nithsdale*—Thornhill, R. A.; Sanquhar Station, Dr Dv. *Annandale*—Garden, Moffat, J. T. J.; Lockerbie, G. Bl. *Eskdale*—Langholm, C. Y.

Appears July 2. (Probably an escape established, but in Kirdcudbright "very common," J. M'A.).

Euphorbia exigua. *Linn.*

RECORDS: *Dfs.*—Mrs Thomson, 1893. *Kcd.*—J. M'Andrew, 1882. *Wgt.*—J. M'Andrew, 1889.

LOCALITIES: *Along the shore*—Isle of Farne, J. M'A.; Ross, G. C.;

Kirkcudbright, Mullock Bay, Rerrick, J. M'A.   *Nithsdale*—Auldgirth Station, Th.   *Annandale*—Annan, Th.
On cinders of railway inland, probably through trucks.

Euphorbia segetalis. *Linn.*

RECORDS: *Wgt.*—James Smith (*fide* G. Don), 1802.   *Kcd.*—T. Bell, 1882.
LOCALITY: St. Ninian's Cave, Hn.; Kirkandrews, T. Bl.

Euphorbia paralias. *Linn.*

RECORD: *Wgt.*—J. M'Andrew, 1886.
LOCALITY: Portpatrick, J. M'A.

Euphorbia amygdaloides. *Linn.*

RECORD: (Escape) *Dfs.*—Miss Adams and Miss S. D. Johnstone, 1890.
LOCALITY: Cowhill, Ad. and S. D. J.

Euphorbia portlandica. *Linn.*

RECORD: *Wgt.*—J. M'Andrew, 1890.
LOCALITY?

Mercurialis perennis. *Linn.* (Dogs' Mercury).

RECORDS: *Dfs.* and *Kcd.*—P. Gray, 1850.   *Wgt.*—G. C. Druce, 1883.

LOCALITIES: *Nithsdale*—Very common in deciduous woods to Sanquhar, Hn., S.-E., Dr Dv.   *Annandale*—Very common Milke, Annan, etc., S.-E., J. T. J.   *Eskdale*—Very common by Esk, Tarras, Bexburn, S.-E.; Castle O'er, J. Wn. and R. Bl.
Appears March 20, J. Sh., to April 3, J. T. J. On moist or dry leaf-mould in shade and shelter.
VISITOR: Apis c.p., S.-E.

Empetrum nigrum. *Linn.* (Crowberry).

RECORDS: *Dfs.* and *Kcd.*—P. Gray, 1850.   *Wgt.*—J. M'Andrew, 1886.
LOCALITIES: *Nithsdale*—Girharrow, J. Cr.; Lowthers, Dr Dv.

Annandale—Hindhill, S. W., Ca.; Black's Hope, Whitecoombe, Craigmichen, etc.; common from 450 feet at Archbank to 2690 feet at Whitecoombe, J. T. J. *Eskdale*—Castle O'er, J. Wn. and R. Bl.; Ewesleesdowns, J. Rae.

Appears May 18, J. T. J. On dry whinstone rock; fully exposed, and in peat moss.

Callitriche aquatica. *Sm.*

RECORDS: 636. Verna, *Dfs.*—P. Gray, 1850? *Kcd.*—J. M'Andrew, 1882. *Wgt.*—C. C. Bailey, 1883.

637. Stagnalis, *Dfs.*—J. Sadler, 1858. *Kcd.*—J. M'Andrew, 1882. *Wgt.*—Dr Balfour, 1843.

639. Hamulata, *Dfs.*—Dr Davidson, 1886. *Wgt.*—G. C. Druce, 1883.

b. pedunculata, *Dfs.*—W. Stevens, 1848.

641. Autumnalis, *Dfs.*—Dr Gilchrist, 1882. *Kcd.*—Dr F. W. Grierson, 1882. *Wgt.*—C. C. Bailey, 1883.

LOCALITIES: *Nithsdale*—Dumfries, 636, P. Gr.; Old Quay, Dumfries, 641, Dr Gl.; common Broomrigg, 637, Hn., S.-E.; Closeburn, Townfoot, Thornhill, 641, J. Fn.; common Sanquhar, 637, Dr Dv.; Auchengruith, Sanquhar Reservoir, 639, Dr Dv. *Annandale*—Moffat, 637, J. Sd.; margin Loch Skene, 639, var. *b.*, W. St. *Eskdale*—Hardgrave, L. Smith; common by Esk, S.-E.

In mud of springs and shallow rills by roads, etc.

Urtica urens. *Linn.*

RECORDS: (Probably Escape) *Dfs.*—P. Gray, 1850. *Kcd.*—J. M'Andrew, 1882. *Wgt.*—G. C. Druce, 1883.

LOCALITIES: *Nithsdale*—Kirkbean, Exc.; Nunfield, F. W. G.; Lochar Moss (firs near Racks), S.-E.; Holmhill, J. Sh.; farmyards, Sanquhar, Dr Dv. *Annandale*—Lochmaben, S.-E.; Nethermill, J. T. J. *Eskdale*—1 mile north of Bentpath on Meggat, S.-E.

On rubbish or manure heaps, roadsides; fully exposed or shaded.

Urtica dioica. *Linn.*

RECORDS: *Dfs.* and *Kcd.*—P. Gray, 1850. *Wgt.*—G. C. Druce, 1883.

LOCALITIES : Very common in all the valleys to 1400 feet, S.-E.
On pretty dry roadsides, humus, holms, etc. (not on peat) ; in shade and shelter or quite exposed.

Parietaria officinalis. *Linn.* (Pellitory).

RECORDS: *Kcd.* and *Wgt.*—J. M'Andrew, 1882.
LOCALITY: Colvend (one spot), J. M'A., Dr Dv., Th., Hn.
In crannies of granite rocks on shore ; sheltered.

Humulus lupulus. *Linn.* (Hop).

RECORDS: (Escape) *Dfs.* and *Kcd.*—P. Gray, 1850. *Wgt.*—G. C. Druce, 1883.
LOCALITIES : *Nithsdale*—Thornhill, A. M. ; Dumfries, P. Gr. *Annandale*—Lockerbie, G. Bl. *Eskdale*—Castle O'er, J. Wn. and R. Bl.

Ulmus montana. *Sm.* (Wych Elm).

RECORD: (Planted) *Dfs.*—P. Gray, 1850. *Kcd.*—F. R. Coles, 1885. *Wgt.*—G. C. Druce, 1883.
Prefers dry sandy loam and sunny places, W. Do., A.M.

Ulmus campestris. *Sm.* (Elm).

RECORDS: (Planted) *a.* suberosa, *Dfs.*—Dr Davidson, 1886. *Kcd.* —J. M'Andrew, 1882. *Wgt.*—G. C. Druce, 1883.
Dry sandy loam, A. M.

Myrica gale. *Linn.* (Bog Myrtle).

RECORD: *Dfs.*—G. N. Lloyd, 1837. *Kcd.*—P. Gray, 1846. *Wgt.* —C. C. Bailey, 1883.
LOCALITIES : *Nithsdale*—Criffel, S.-E.; Lochanhead, F. W. G.; Terregles, Maxwelltown Loch, P. Gr.; Drumclyer, S.-E. *Annandale*— Lockerbie, S.-E.; Rigghead, etc., J. T. J. *Eskdale*—Castle O'er, J. Wn. and R. Bl.
Appears April 17, G. Bl.; May 23, J. T. J. In peat bogs ; fully exposed.

Alnus glutinosa. *Linn.* (Alder).

RECORDS: *Dfs.*—P. Gray, 1850. *Kcd.*—J. M'Andrew, 1882. *Wgt.*—G. C. Druce, 1883.
LOCALITIES: Very common by rivers in all valleys to 700 feet.
Prefers moist holms; fully exposed, A. M., W. Do., S.-E.

Betula alba. *Linn.* (Birch).

RECORDS: 1385. Verrucosa, *Dfs.* and *Kcd.*—P. Gray, 1858. *Wgt.*—G. C. Druce, 1883.
1386. Pubescens, *Dfs.*—Dr Davidson, 1886. *Kcd.*—F. R. Coles, 1885. *Wgt.*—G. C. Druce, 1883.
LOCALITIES: Very commonly planted in all valleys to 900 feet.
Open sandy ground, A. M.; moist sunny places, W. Do.

Corylus avellana. *Linn.* (Hazel).

RECORDS: *Dfs.* and *Kcd.*—P. Gray, 1850. *Wgt.*—G. C. Druce, 1883.
LOCALITIES: Very common in woods, chiefly shaded and sheltered; on leaf mould to 800 feet, S.-E.
Prefers dry deep soil; shaded and sheltered places, W. Do., A. M.

Carpinus betulus. *Linn.* (Hornbeam).

RECORD: (Planted) *Dfs.*—Dr Balfour, 1843. *Kcd.*—J. M'Andrew, 1882.

Fagus sylvatica. *Linn.* (Beech).

RECORD: (Planted) *Dfs.* and *Kcd.*—P. Gray, 1850. *Wgt.*—G. C. Druce, 1883.
Prefers dry sunny places, W. Do.; specially calcareous, A. M.

Castanea sativa. *Mill.* (Chestnut).

RECORDS: (Planted) *Dfs.*—J. T. Johnstone, 1890. *Kcd.* and *Wgt.*—J. M'Andrew, 1882.
Prefers dry sheltered places, W. Do.

Quercus robur. Linn.

RECORDS : (Planted) *a.* pedunculata, *Dfs.* and *Kcd.*—J. M'Andrew, 1882. *Wgt.*— G. C. Druce, 1883.
c. sessiliflora, *Kcd.*—J. M'Andrew, 1882. *Wgt.*—G. C. Druce, 1883.
Prefers dry sunny places, W. Do.; deep rich soil and shelter, A. M.

Salix pentandra. Linn.

RECORDS: *Dfs.*—Dr Burgess, 1789. *Kcd.*—P. Gray, 1848. *Wgt.*—G. C. Druce, 1883.
LOCALITIES : *Nithsdale*—Castlefairn, J. Cr.; Leadhills, Dr Br.; Elliock Wood, Burnfoot, Sanquhar, Dr Dv. *Annandale*—Roadsides, Moffat, Dr Br.; fifth milestone old Dumfries Road, J. T. J.; banks Caledonian Railway at Greskine, J. T. J.; Houslack, Annan Water, J. T. J.

Salix fragilis. Linn. (Crack Willow).

RECORDS : *Dfs.*—Dr Davidson, 1886. *Kcd.*—J. M'Andrew, 1882.
b. decipiens, *Dfs.*—J. T. Johnstone, 1891.
LOCALITIES : *Nithsdale*—Elliock Woods, Dr Dv. *Annandale*—Barnhill Bridge, Granton Roadside, Upper Murthat, between railway and road, J. T. J.

Salix alba. Linn.

RECORDS : *Dfs.*—P. Gray, 1850. *Wgt.*—Dr Arnott, 1848.
c. vitellina, *Dfs.*—J. T. Johnstone, 1890.
LOCALITIES : *Nithsdale*—Holmwoods, Dr Dv. *Annandale*—Hydropathic grounds, Moffat, var. *c.*, J. T. J.; Birnock, Auchencastle, J. T. J.

Salix amygdalina. Linn (= triandra).

RECORD : *Dfs.*—Dr Walker, 1762.
b. Hoffmaniana, *Dfs.*—J. T. Johnstone, 1891.
LOCALITIES : *Annandale*—Stitherick Wood, Annan, Dr Wl.; Annan at Putts (with var. *b.*), J. T. J. *Eskdale*—Black Esk, Dr Wl.
Appears May 26, J. T. J.

Salix purpurea. Linn.

RECORDS : Purpurea, *Dfs.*—Dr Walker, 1796. *Kcd.*—J. M'Andrew, 1882. *Wgt.*—Rev. J. Gorrie, 1893.
× Viminalis, *Dfs.*—Dr Walker, 1796.

LOCALITIES : *Nithsdale*—Elliock, Dr Wl.; Nithside, Dr Dv. *Annandale*—Milke, S.-E.; Craigieburn × Dr Wl.; Adam's Holm, Putts, J. T. J. *Eskdale*—Netherby, Dr Wl.

Appears April 25, J. T. J.

Salix viminalis. *Linn.*

RECORDS: Viminalis, *Dfs.*—Dr Davidson, 1886. *Kcd.*—P. Gray, 1850. *Wgt.*—G. C. Druce, 1883.
   a. stipularis, *Dfs.*—J. T Johnstone, 1891.
   × caprea, *Dfs.*—J. T. Johnstone, 1891. *Kcd.*—*Fide* J. M'Andrew, 1882. *Wgt.*—G. C. Druce, 1883.
   d. ferruginea, *Wgt.*—G. C. Druce, 1883.

LOCALITIES : *Nithsdale*—Sanquhar, Dr Dv. *Annandale*—Evanwater holms × caprea, J. T. J.; Annan water, Oakrigside, var. *a.* J. T. J.; Barnhill Bridge, J. T. J. *Eskdale*—Bilholm, S.-E.

Salix caprea. *Linn.*

RECORDS: 1401. Caprea, *Dfs.*—Dr Davidson, 1886. *Kcd.*—J. M'Andrew, 1882. *Wgt.*—Dr Arnott, 1848.
   1399. Cinerea, *Dfs.*—J. T. Johnstone, 1891. *Kcd.*—J. M'Andrew, 1882. *Wgt.*—G. C. Druce, 1883.
   b. aquatica, *Wgt.*—Dr Arnott, 1848.

LOCALITIES : *Nithsdale*—Sanquhar, by Nith, common, 1401, Dr Dv. *Annandale*—Moffat water, 1401, J. T. J.; Annan, 1399, J. T. J. *Eskdale*—Mosspaul, 1401, S.-E.; Glentarras, 1399, S.-E.

Appears March 22, J. T. J.

Salix aurita. *Linn.*

RECORDS : *Dfs.*—P. Gray, 1850. *Kcd.*—J. M'Andrew, 1882. *Wgt.* —G. C. Druce, 1883.

LOCALITIES : *Nithsdale*—Dumfries, P. Gr.; Nith, common, S.-E.; Sanquhar, Dr Dv. *Annandale*—Auchencat Burn, J. T. J.; Annan water, J. T. J. *Eskdale*—Ewesleesdowns, J. Rae; Tarras, S.-E.

VISITORS : Bombus, spp., Empis bilineata, very abundant, S.-E.

Salix phylicifolia. *Linn.*

RECORDS : Nigricans, *Dfs.*— Dr Davidson, 1886. *Wgt.* — G. C. Druce, 1883.
   Radicans (?) *Dfs.*—Mr Maughan, 1789.

Davalliana (?) *Dfs.*—Dr Davidson, 1886.
Tenuior (?) *Dfs.*—Dr Davidson, 1886.
Tetrapla (?) *Dfs.*—Dr Davidson, 1886.
Cotonifolia (?) *Kcd.*—Mr Maughan, 1789.

LOCALITIES: *Nithsdale*—Sanquhar, Dr Dv.; Glenglass, Dr Dv.; Euchan, Nith, Dr Dv.; Euchan Head, Dr Dv. *Annandale*—Auchencat Burn, Beerholm, J. T. J.; Black's Hope, J. T. J.

Appears May 17, J. T. J.

Salix repens. *Linn.*

RECORDS: *Dfs.*—Dr Davidson, 1886. *Kcd.*—Mr Maughan, 1789.
Wgt.—G. C. Druce, 1883.
Fusca (?) *Wgt.*—Dr Arnott, 1848.
Argentea (?) *Kcd.*—Dr Walker, 1762.

LOCALITIES: *Nithsdale*—Moniaive, J. Cr.; Tynron, J. Sh.; Sanquhar, Dr Dv.

Salix Lapponum. *Linn.*

RECORDS: 1407. Lapponum, *Dfs.*—Dr Walker, 1762.
a. arenaria, *Dfs.*—J. T. Johnstone, 1891.
1404. Arbuscula, *Dfs.*—Mr Maughan, 1789.

LOCALITIES: *Nithsdale*—Thornhill to Sanquhar, 1404, Mau. *Annandale*—Hartfell Foot, 1404, Mau.; 1407, north side of mountain to south of Loch Skene, Whitecoombe, Dr Wl., Dr Bl., J. Sd., J. T. J.; Spoonburn, E. F. L.

Appears May 16. On dry rocks of whinstone (2000 to 2400 feet), J. T. J., S.-E.

Salix myrsinites. *Linn.*

RECORD: *Dfs.*—Dr Walker, 1796.

LOCALITIES: *Nithsdale*—Euchan, Dr Wl. *Annandale*—600 feet above S. Lapponum, Whitecoombe, Dr Wl., Dr Bl., J. Sd.

Salix herbacea. *Linn.*

RECORD: *Dfs.*—W. Stevens, 1848.

LOCALITIES: *Annandale*—Black's Hope, J. T. J.; Whitecoombe, W. St., Dr Bl., J. T. J.; Loch Skene, Correifron, J. T. J.

Appears June 22, J. T. J.

Salix ambigua. *Ehr.* (aurita × repens).

RECORD: *Dfs.*—Dr Davidson, 1886.
LOCALITIES: (*Nithsdale*—Sanquhar, Dr Dv?) *Annandale*—Putts, Moffat, J. T. J.

Salix cinerea × nigricans.

RECORD: *Dfs.*—Dr Davidson, 1886.
LOCALITIES: *Nithsdale*—Sanquhar, Dr Dv. *Annandale*—Gudeshaw, J. T. J.

Salix aurita × phylicifolia.

RECORD: *Dfs.*—J. T. Johnstone, 1891.
LOCALITY: *Annandale*—Moffat, J. T. J.

The following names are given by Dr Walker about 1761, and appear to be Dumfriesshire willows, which probably exist, but I cannot place them under other names:—

Salix analifolia.—Mountain gullies; Annan and Nith (1200 to 1500 feet).
Salix evoniæ.—Dumfries to Moffat Road below Evan Bridge.
Salix linearis.—Caerlaverock.

Populus alba. *Linn.*

RECORDS: (Planted) *Dfs.*—Dr Davidson, 1886. *Kcd.*—J. M'Andrew, 1882. *Wgt.*—Rev. G. Wilson, 1893.
LOCALITIES: *Nithsdale*—Sanquhar, Dr Dv. *Annandale*—Moffat, J. T. J.
Prefers moist alluvium, A. M.

Populus tremula. *Linn.*

RECORDS: (Planted) *Dfs.*—Dr Davidson, 1886. *Kcd.*—J. M'Andrew, 1882. *Wgt.*—J. M'Andrew, 1889.

Populus nigra. *Linn.*

RECORDS: (Planted) *Dfs.*—Dr Davidson, 1886. *Kcd.*—J M'Andrew, 1882. *Wgt.*—G. C. Druce, 1883.
Prefers moist sunny places, W. Do.

Pinus sylvestris. *Linn.*

RECORDS : (Planted) *Dfs.*—Dr Davidson, 1886. *Kcd.*—J. M'Andrew, 1882. *Wgt.*—G. C. Druce, 1883.
Prefers dry, gravelly, moorish soil, A. M.

Juniperus communis. *Linn.*

RECORDS: *Dfs.*—Dr Singer, 1843. *Kcd.*—J. M'Andrew, 1882. *Wgt.*—J. M'Andrew, 1886.
LOCALITIES : *Nithsdale*—Mabie Hills, R. Sc.; Kello, Euchan, Dr Dv. *Annandale*—Moffat Hills, Raking Gill, J. T. J., common, Dr Sn., W. Bn. *Eskdale*—Castle O'er, J. Wn. and R. Bl.
Dry rocks; fully exposed.

Taxus baccata. *Linn.* (Yew).

RECORDS: (Planted) *Dfs.*—Dr Davidson, 1886. *Kcd.*—J. M'Andrew, 1882. *Wgt.*—G. C. Druce, 1883.
Prefers shady half-sheltered places, W. Do., A. M.

Typha latifolia. *Linn.* (Bulrush).

RECORDS : *Dfs.*—Dr Burgess, 1777. *Kcd.*—J. Fraser, 1843. *Wgt.*—J. M'Andrew, 1887.
LOCALITIES : *Nithsdale*—Black Loch at top of Tinwald Parish, Dr Br.; Blackwood Loch, S.-E.; Maxwelltown, Glencairn, J. Cr.; Closeburn Loch, Dr Br. *Annandale*—Kirk Loch, Lochmaben, G. N. Ll., S.-E., Hn.

Typha angustifolia. *Linn.*

RECORDS : *Dfs.*—Dr Singer, 1843. *Kcd.*—P. Gray, 1848.
LOCALITY : *Annandale*—Lochmaben, Dr Sn., J. Cru., F. W. G.

Sparganium ramosum. *Huds.* (Burr Reed).

RECORDS: *Dfs.* and *Kcd.*—P. Gray, 1850. *Wgt.*—G. C. Druce, 1883.
LOCALITIES : *Nithsdale*—Clarencefield, F. W. G.; Dumfries, P. Gr.; Maxwelltown Loch, P. Gr., Hn.; Cowhill Ad. and S. D. J.; Kingholm Merse, Hn.; Caitloch, J. Cr.; Thornhill, Cample, R. A.; Kelloside, Dr

Dv. *Annandale*—Lochmaben, Exc.; Evan Water, S.-E.; Moffat, common, J. T. J. *Eskdale*—Turner's Linn, Half-Morton, L. Sm.; Kirkandrews, E. Ty.; Glen Tarras, S.-E.

In ditches, ponds, etc.; sheltered from wind by other plants.

Sparganium simplex. *Huds.*

RECORDS: 1560. Simplex, *Dfs.*—J. Gray, 1850. *Kcd.*—P. Gray, 1846. *Wgt.*—Dr Balfour, 1843. 1561. Affine, *Dfs.* and *Kcd.*—G. N. Lloyd, 1837. *Wgt.* —G. C. Druce, 1883.

LOCALITIES: *Nithsdale*—Lochar Moss, 1561, G. N. Ll., Dr Gl.; Maxwelltown Loch, 1560, P. Gr., Hn.; Dawson's field, R. H. M.; Fingland Lane (1000 feet), 1561, J. Cr.; Nith above Saw Mill, Sanquhar, Dr Dv. *Annandale*—Earshaig Lakes, J. T. J.

Sparganium minimum. *Flr.*

RECORDS: *Dfs.*—Dr Burgess, 1777. *Kcd.*—G. N. Lloyd, 1837. *Wgt.*—G. C. Druce, 1883.

LOCALITIES: *Nithsdale*—Holywood, Dr Br.; Fingland Lane, Moniaive, J. Cr. *Annandale*—Castle Loch, Lochmaben, Myreside, Dr Br.; Earshaig, J. T. J.

Arum maculatum. *Linn.* (Cuckoo Pint).

RECORDS: (Escape) *Dfs.*—W. Stevens, 1848. *Kcd.*—Mrs Gilchrist Clark, 1867. *Wgt.*—Rev. G. Wilson, 1893.

LOCALITIES: *Nithsdale*—Jarbruck (374 feet), J. Cr.; Drumlanrig Woods, W. St.; Tibber's Castle, R. A.

Acorus calamus. *Linn.* (Sweet Mace).

RECORD: (Introduced) *Kcd.*—J. M'Andrew, 1882.

LOCALITY: Balmæ, J. M'A.

Lemna minor. *Linn.* (Duckweed).

RECORDS: *Dfs.* and *Kcd.*—P. Gray, 1850. *Wgt.*—G. C. Druce, 1883.

LOCALITIES: Very common in all the valleys in stagnant ponds, ditches, etc.

Zostera marina. *Linn.* (Grasswrack).

RECORDS : *Kcd.*—G. N. Lloyd, 1837. *Wgt.*—G. C. Druce, 1883.

LOCALITIES : *Along the shore*—Kirkcudbright Bay, G. N. Ll., P. Gr.; Dee and Urr mouth, J. M'A.; Auchencraig, J. C. W., between Arbigland and Southerness, J. Fr.

Zostera nana. *Roth.*

RECORD : *Kcd.*—Professor Oliver, 1887.

Ruppia maritima. *Linn.*

RECORDS : 1608. Spiralis *(fide* J. M'Andrew), *Kcd.*—G. N. Lloyd, 1837. *Wgt.*—Greville, Herbarium, 1836.

LOCALITIES : *Along the shore*—Lochryan, Stranraer, Grev.; Kirkcudbright, G. N. Ll.; Lot's Wife, Sandyhills, Douglas Hall, Gillfoot, P. Gr., J. M'A.

Potamogeton natans. *Linn.**

RECORDS: *Dfs.* and *Kcd.*—P. Gray, 1850. *Wgt.*—G. C. Druce, 1883.

LOCALITY : Very common in all the valleys.

Appears June 14. In ditches, low streams, often on wet mud.

Potamogeton polygonifolius. *Pourr.*

RECORDS : *Dfs.*—J. T. Johnstone, 1891. *Kcd.*—J. M'Andrew, 1882. *Wgt.*—Dr Arnott, 1848.

LOCALITIES : *Annandale*—Lochmaben, G. Bl., S.-E.; common at Moffat, J. T. J.

Much deeper water than preceding, usually stagnant ponds.

Potamogeton plantagineus. *Du Croy.*

RECORD : *Dfs.*—F. Buchanan White (*Scottish Naturalist*, vol. 2, New Series).

---

* In this genus I have simply followed the *London Catalogue*, as the integration of the sub-species has never to my knowledge been carefully done.

Potamogeton rufescens. *Schrad.*

RECORDS: *Dfs.*—J. Cruickshank, 1836. *Kcd.*—J. Fingland, 1890. *Wgt.*—G. C. Druce, 1883.

LOCALITIES: *Nithsdale*—Lochar Moss, J. Cru.; Fingland Lane, J. Fn. (Has been confirmed for three counties by Mr Bennet).

Potamogeton lanceolatus. *Sm.*

RECORDS: *Dfs.*—W. Stevens, 1848. *Kcd.*—P. Gray, 1850.

LOCALITIES: *Nithsdale*—Dumfries, P. Gr.; foot of Morton hills, Locherben, Auchenbainzie Loch, W. St.*

Potamogeton heterophyllus. *Schreb.*

RECORDS: *Dfs.*—Dr Davidson, 1886. *Kcd.*—J. Cruickshank, 1836. *Wgt.*—G. C. Druce, 1883.

LOCALITIES: *Nithsdale*—Lochrutton, F. W. G.; Maxwelltown Loch, J. Cru., P. Gr.; Guffockland Dam, Dr Dv. (Confirmed for three counties by Mr Bennet).

Potamogeton lucens. *Linn.*

RECORD: *b*, acuminatus, *Kcd.*—F. R. Coles, 1883.

LOCALITY: Tarff, F. R. C.*

Potamogeton zizii. *Roth.*

RECORDS: *Dfs.*—P. Gray, 1850. *Kcd.*—F. R. Coles, 1883. *Wgt.* —G. C. Druce, 1883.

LOCALITIES: *Nithsdale*—Dumfries, P. Gr.*

Potamogeton prælongus. *Wulf.*

RECORDS: *Kcd.*—F. R. Coles, 1883. *Wgt.*—G. C. Druce, 1883.

LOCALITY: Tarff, F. R. C. (Confirmed by Mr Bennet).

Potamogeton perfoliatus. *Linn.*

RECORDS: *Dfs.* and *Kcd.*—P. Gray, 1850.

LOCALITIES: *Nithsdale*—Dumfries, P. Gr.; Lincluden, Hn., S.-E.

* Require confirmation.

Potamogeton crispus. *R.*

RECORDS: *Dfs.* and *Kcd.*—P. Gray, 1850. *Wgt.*—G. C. Druce, 1883.

LOCALITIES: *Nithsdale*—Near Dumfries, P. Gr. (confirmed for first two counties by Mr Bennet); Crindau, Dumfries, S.-E. *Annandale*—Castle Loch, Lochmaben, F. W. G.; Annan, G. Bl.

Potamogeton densus. *Linn.*

RECORDS: *Dfs.* and *Kcd.*—P. Gray, 1850.
LOCALITY: *Nithsdale*—Dumfries, P. Gr.*

Potamogeton obtusifolius. *Mert and Koch.*

RECORDS: *Dfs.*—W. Stevens, 1848. *Kcd.*—J. M'Andrew, 1882.

LOCALITIES: *Nithsdale*—Loch by side of Edinburgh Road, about five miles from Dumfries, W. St.; common Loch, Thornhill, R. A.

Potamogeton compressus. *Sm.*

RECORD: *Kcd.*—P. Gray, 1850.

LOCALITIES: *Nithsdale*—Dumfries, P. Gr.; also Carlingwark Loch, F. R. C. (Confirmed by Mr Bennet.)

Potamogeton pusillus. *Linn.*

RECORDS: *Dfs.* and *Kcd.*—P. Gray, 1850. *Wgt.*—G. C. Druce, 1883.

LOCALITIES: *Nithsdale*—Dumfries, P. Gr.; Maxwelltown Loch, F. W. G.; Knockenhair, Auchengruith, Guffockland, Dr Dv. (Confirmed for Wigtown by Mr Bennet.)

Potamogeton pectinatus. *Linn.*

RECORD: *Wgt.*—J. M'Andrew, 1893.
LOCALITY: Ravenstone Loch.

Alisma Plantago. *Linn.* (Water Plantain).

RECORDS: *Dfs.* and *Kcd.*—P. Gray, 1850. *Wgt.*—G. C. Druce, 1883.

LOCALITIES: *Nithsdale*—Maxwelltown Loch, Hn., S.-E.; Nith,

* Requires confirmation.

Dumfries, Hn.; Cowhill, Ad. and S. D. J.; Kirkmahoe, S.-E. *Annandale*—Annan, S.-E.; Redpath Mill, S.-E.; Lochmaben, abundant, S.-E.

In ditches usually in water, but half wind-sheltered by reeds, ditch sides, etc.

VISITORS : Andrena albicans, S.-E. ; Helophilus pendulus ; Syrphus franiditarse, Hn ; luniger, Platychirius clypeatus, peltatus, albimana, Syritta pipiens, S.-E. ; Eristalis nemorum, Hyetodesia incana, Hn. ; Onesia sepulchralis, Siphona cristata, Anthomyia radicum, Hyelemyia strigosa, Vanessa Urticæ, S.-E.

Alisma ranunculoides. *Linn.*

RECORDS : *Dfs.*—Mr Maughan, 1789. *Kcd.*—P. Gray, 1846. *Wgt.*—Dr Arnott, 1848.

LOCALITIES : *Nithsdale*—Caerlaverock, Maxwelltown Loch, P. Gr. *Annandale*—Castle Loch, Lochmaben, Mau.*

Butomus umbellatus. *Linn.* (Flowering Rush).

RECORD : *Kcd.*—In Dumfries Herbarium, 1865.

LOCALITY : Kirkbean.

Triglochin palustre. *Linn.*

RECORDS: *Dfs.* and *Kcd.*—P. Gray, 1850. *Wgt.*—G. C. Druce, 1883.

LOCALITIES : *Nithsdale*—Maxwelltown Loch, F. W. G.; Townfoot Moor and Loch, R. H. M.; Holywood, Broomrigg, S.-E.; Glencairn, common, J. Cr.; Sanquhar, common, Dr Dv. *Annandale*—Common Dryfe, S.-E.; Moffat and (to 1200 feet) Correifron, common, J. T. J., S.-E. *Eskdale*—Liddel, Langholm, Bentpath, S.-E.; Castle O'er, J. Wn. and R. Bl.

In wet, peaty mosses, ditches, etc.; usually in long grass.

Triglochin maritimum. *Linn.*

RECORDS: *Dfs.* and *Kcd.*—P. Gray, 1850. *Wgt.*—G. C. Druce, 1883.

LOCALITIES : *Along the shore*—From a mile below Kingholm to Gretna. Very common in estuarine, wet mud ; exposed, F. W. G., S.-E.

* All require confirmation.

Malaxis paludosa. *Sn.*

RECORDS: *Dfs.*—J. Corrie, 1887. *Kcd.*—J. M'Andrew, 1882. *Wgt.*—Mr Winch, 1789.

LOCALITIES: *Nithsdale*—Girharrow Hill, J. Cr.; reported Keir, F. W. G.

Epipactis latifolia. *Auct.*

RECORDS: *Dfs.*—Dr Singer, 1843. *Kcd.*—P. Gray, 1850.

LOCALITIES: *Nithsdale*—Arbigland, Kirkbean, Fq.; Tinwald, Dr Sn.; Mavisgrove, P. Gr.; Kirkmahoe, Dr Gl.; Cowhill, Ad. and S. D. J.; Crawick Woods, Knockenhair Rifle Range, Dr Dv. *Eskdale*—Kirkandrews, E. Ty.

Listera ovata. *Br.* (Tway Blade).

RECORDS: *Dfs.*—F. A. Hope, 1881. *Kcd.*—J. Fraser, 1843. *Wgt.*—G. C. Druce, 1883.

LOCALITIES: *Nithsdale*—Arbigland, J. Fr.; Lochanhead, F. W. G.; Glen, Th.; Cowhill, Ad. and S. D. J.; Isle, S.-E.; Scaur, J. Sh.; Glencairn, abundant, J. Cr.; Longmire, Waterside, R. A.; Auchengruith, Euchan, Dr Dv. *Annandale*—Springkeld, S.-E.; Jardine Hall, Th.; Lochmaben, S.-E.; railway, Lockerbie, Murrayfield, G. Bl.; Kirkpatrick-Juxta, F. A. H.; Crooks, Commonside, Kerr, Breckonside, March Cottage, J. T. J.; Beld Craig, W. Ca., J. T. J. *Eskdale*—Woodslee, Penton, Glentarras, S.-E.; Irvine, C. Y.; Langholm, Meggat, S.-E.; Castle O'er, J. Wn. and R. Bl.

Appears June 20, J. T. J. On moist humus, holmlands; usually half-shaded and wind-sheltered by long grass or woods.

Listera cordata. *R. Br.*

RECORDS: *Dfs.*—Dr Singer, 1843. *Kcd.*—J. M'Andrew, 1882. *Wgt.*—J. M'Andrew, 1891.

LOCALITIES: *Nithsdale*—1700 feet, Criffel, A. B. H.; Carr, Clonrae Hill, J. Wl.; Scaur Water, J. Sh. *Annandale*—Garpol, J. Sd.; Wellburn, J. T. J., J. M'A.

Grows amongst heather roots, and is thus partly wind-sheltered, on whinstone soil.

Neottia nidus avis. *Rich.* (Bird's Nest).

RECORDS: *Dfs.*—G. Bell, 1893. *Kcd.*—P. Gray, 1848.

LOCALITIES: *Nithsdale*—West side vale of Dumfries, Hillhead, P. Gr.; * Capenoch Bridge, R A. *Annandale*—Three miles from Lockerbie, G. Bl., S.-E.

On a pretty dry bank on hazel roots; well sheltered from wind.

Orchis mascula. *Linn.* (Early Purple).

RECORDS: *Dfs.* and *Kcd.*—P. Gray, 1850. *Wgt.*—Rev. G. Wilson, 1893.

LOCALITIES: *Nithsdale*—Common Cargen, Cluden, Nith, Th., S.-E.; Cowhill, Ad. and S. D. J.; Sanquhar, Dr Dv. *Annandale*—Tundergarth, G. Bl.; Evan, S.-E.; New Mills, Wellburn, Grey Mare's Tail, J. T. J. *Eskdale*—Chapelknowe Moss, E. Ty.; Burnfoot (to 1200 feet), S.-E.

Appears April 23, G. Bl. On dry holms, whinstone soils, etc.; half-shaded and usually wind-sheltered.

Orchis maculata. *Linn.*

RECORDS: 1352. *Dfs.* and *Kcd.*—P. Gray, 1850. *Wgt.*—G. C. Druce, 1883.

LOCALITIES: *Nithsdale*—Very common Dumfries, P. Gr.; Sanquhar, Dr Dv. *Annandale*—Very common Queensberry, Correifron, etc., J. T. J., S.-E. *Eskdale*—Very common to Moodlaw Loch, Langholm, S.-E.; Castle O'er, J. Wn. and R. Bl.

Appears June 4 to 7, J. T. J. On marshy ground often peaty; and fully exposed to wind and sun or half-sheltered.

Orchis latifolia. *Linn.* (Bull's Dairy, J. Shaw).

RECORDS: *Dfs.* and *Kcd.*—P. Gray, 1850. *Wgt.*—G. C. Druce, 1883.

LOCALITIES: Very common in all the valleys to 700 feet.

Appears June 7 to 12, J. T. J. On wet marshy holms, boulder clay, whinstone soils; usually in part wind-sheltered by rushes, etc.

VISITORS: Empis vitripennis, Eristalis pertinax, S.-E.

* I think this has been supposed to be in Dumfriesshire. Mr Wilson finds no specimen from Dumfries in the Edinburgh Herbarium.

Orchis pyramidalis. Linn.

RECORDS : *Dfs.*— G. F. Scott-Elliot, 1891. *Kcd.*— Edinburgh Herbarium, see J. M'Andrew, 1882. *Wgt.*—Dr Graham, 1836.

LOCALITIES : *Eskdale*—Merrylaw, Whitehope, S.-E.

Orchis conopsea. Linn.

RECORDS : *Dfs.*—W. Keddie, 1854. *Kcd.*—P. Gray, 1846. *Wgt.* —Dr Balfour, 1843.

LOCALITIES : *Nithsdale*—Dalskairth, P. Gr.; Scaur, Glen, J. Sh., Hn., R. A.; Templand, R. A.; Euchan, Dr Dv.   *Annandale*—Beld Craig, Kd., J. T. J.; New Edinburgh Road, Wellburn, Riddingsfield, Peter's Moss, Langside, J. T. J.   *Eskdale*—Ewesleesdowns, Stennies, S.-E.

Appears June 16 to 29, J. T. J. On moist whinstone soils ; slightly wind-sheltered by rushes, etc.

VISITORS : Argynnis Aglaia, S.-E.; Syrphus ribesii, Hn.

Orchis incarnata. Linn.

RECORDS : *Dfs.*—E. F. Linton and J. T. Johnstone, 1894. *Wgt.*— G. C. Druce, 1883.

LOCALITY : *Annandale*—Correifron, E. F. L., J. T. J.   July 7, 1894.

Habenaria bifolia. R. Br. (Butterfly).

RECORDS : 1468. Bifolia, *Dfs.*—W. Keddie, 1854. *Kcd.*—P. Gray, 1848. *Wgt.*—Rev. J. Gorrie, 1891.
1469. Chloroleuca, *Dfs.*—W. Stevens, 1848. *Kcd.*—P. Gray, 1846. *Wgt.*—Dr Graham, 1835.

LOCALITIES : *Nithsdale*—Lochanhead, F. W. G.; Hillhead, 1468, P. Gr., C. E. M.; Glen, Terregles, Maxwelltown, 1469, Th., P. Gr.; Cowhill, Ad. and S. D. J.; Drumlanrig, 1469, W. St.; Sanquhar (also 1469), Dr Dv.; Templand, R. A.; Scaur, Hn.   *Annandale*— Murrayfield, Corrie, G. Bl.; Beld Craig, J. Kd.; fourth milestone Dumfries Road, Swinefoot, 1468, J. T. J.; Gardenholm, Breckonside, Peter's Moss, 1469, J. T. J.   *Eskdale*—Kirkandrews, E. Ty.; Glentarras, S.-E.

Appears June 12 to July 7, J. T. J. On moist holms, leaf-mould, whinstone soils, etc.; usually in part wind-sheltered.

VISITORS : Large brown and sulphur moths, R. A.

Habenaria albida. R. Br.

RECORDS : *Dfs.*—W. Stevens, 1848. *Kcd.*—P. Gray, 1848. *Wgt.*—Rev. J. Gorrie, 1893.

LOCALITIES : *Nithsdale*—West side vale of Dumfries, P. Gr.; Merkland, Exc.; Penpont, W. St.; Tynron, J. Wl.; Euchan Waterfall, Dr Dv. *Annandale*—Queensberry, W. St.; Beld Craig, Kd.; Wellburn, J. T. J.; Garpol, J. T. J.

Appears June 28, J. T. J. On dry moors and whinstone soils; fully exposed.

Habenaria viridis. R. Br. (Frog).

RECORDS : *Dfs.*—S. W. Carruthers, 1888. *Kcd.*—P. Gray, 1846. *Wgt.*—G. C. Druce, 1883.

LOCALITIES : *Nithsdale*—Dalskairth, P. Gr.; Lochanhead, F. W. G.; Glencairn, J. Cr.; Sanquhar, Dr Dv.; Scaur, R. A. *Annandale*—Ericstane, Mrs We.; Commonside, Selcoth, at foot of Linn, J. T. J.; Greygillhead, S. W. Ca. *Eskdale*—Irvine House, E. Ty.

Appears July 7 to 26, J. T. J. On dry hill pastures, chiefly on whinstone soils ; fully exposed, or part wind-sheltered by grass, etc.

Iris pseudacorus. Linn. (Yellow Flag).

RECORDS : *Dfs.*—P. Gray, 1850. *Kcd.*—J. M'Andrew, 1882. *Wgt.*—G. C. Druce, 1883.

LOCALITIES : *Nithsdale*—Arbigland, Southerness, C. E. M.; Caerlaverock, S.-E.; very common Cargen, etc., near Dumfries, P. Gr., Hn., S.-E.; Glencairn, J. Cr.; Crawick, Spango, Laggries, Dr Dv. *Annandale*—Craigies Corrie, G. Bl.; Moffat, common, J. T. J. *Eskdale*—Tarras, Wauchope, S.-E.; Castle O'er, J. Wn. and R. Bl.

Appears July 9, J. T. J. On wet, marshy places, in mud ; exposed to sun and wind, or in shade.

Iris fœtidissima. Linn.

RECORDS: (Introduced) *Kcd.*—" Natural History Society Trans." *Wgt.*—Sir H. Maxwell, 1889.

LOCALITY : Rascarrel, Rerrick, Exc.

Crocus nudiflorus. Sm.

RECORD : (Escape) *Dfs.*—Miss Ethel Taylor, 1890.
LOCALITY : *Eskdale*—Stuart's Wood, Canobie, E. Ty.

Galanthus nivalis. *Linn.*

RECORD : (Escape) *Kcd.*—F. R. Coles, 1883.
LOCALITY : Spout Glen.

Narcissus pseudonarcissus. *Linn.*

RECORDS : (Escape) *Dfs.*—Dr Davidson, 1886.   *Wgt.*—Sir H. Maxwell, 1889.
LOCALITY : *Nithsdale*—Sawmills, Sanquhar, Dr Dv.

Paris quadrifolia. *Linn.* (Herb Paris).

RECORDS : (Escape) *Dfs.*—Dr Singer, 1843.   *Kcd.*—Rev. J. Fraser, 1882.
LOCALITIES : *Nithsdale*—Druidhall Mill, Penpont, J. Sh., R. A.; Kettleton, R. A.; Tinwald, Dr Sn.; Kirkconnel Lea, Miss° Ro.; Enterkinfoot, Dr Gr.   *Annandale*—Tundergarth Linn, very abundant, G. Bl. *Eskdale*—Bilholm and Lyneholm, Ramage, R. Bl., S.-E.
In shade and shelter of linns, on humus.

Polygonatum multiflorum. *All.* (Solomon's Seal).

RECORD : (Escape) *Dfs.*—Dr Burgess, 1789.
LOCALITIES : *Nithsdale*—Tibbers Castle Woods, Dr Br., R. A. *Annandale*—Kirkmichael, R. H. M.

Convallaria majalis. *Linn.* (Lilly of the Valley).

RECORD : (Escape) *Kcd.*—Rev. J. Fraser, 1848.
LOCALITY : Woods near Urr between Colvend and Dalbeattie, J. Fr.

Asparagus officinalis. *Linn.* (Asparagus).

RECORD : (Escape) *Kcd.*—Field Club Excursion No. 3, 1893.
LOCALITY : Seashore, Arbigland, apparently quite established.

Ruscus aculeatus. *Linn.* (Butcher's Broom).

RECORD : (Escape) *Dfs.*—G. F. Scott-Elliot, 1890.   *Kcd.*—J. M'Andrew, 1882.
LOCALITIES : *Nithsdale*—Cowhill, S.-E.   *Annandale*—Hydropathic, J. T. J.

Gagea Lutea. *Ker.*

RECORD: (Introduced) *Kcd.*—J. Cruickshank, 1836.
LOCALITY: *Nithsdale*—The Grove, J. Cru., P. Gr., Dr Gl., R. R., Th.
In a hazel copse.

Ornithogallum umbellatum. *Linn.*

RECORDS: (Escape) *Dfs.*—G. F. Scott-Elliot, 1890. *Wgt.*—Miss Hannay, 1893.
LOCALITIES: *Nithsdale*—Nethermills, Kirkbean, Hn.; Isle orchard, S.-E. *Annandale*—Hedges near Annan, Hn., S.-E. *Eskdale*—Bilholm, J. Wn. and R. Bl.; Woodslee, S.-E. (quite established and appearing annually).

Scilla verna. *Huds.*

RECORDS: *Kcd.*—J. T. Syme, 1836. *Wgt.*—Dr Arnott, 1848.
LOCALITIES: *Along the shore*—Portpatrick, Arn.; Cruggleton, Hn.; Borness, T. Bl.; Falboque Bay, Borgue, J. M'A.; Brighouse, F. R. C.; Ross, J. T. S., G. C.

Scilla nutans. *Sm.* (Wild Hyacinth).

RECORDS: *Dfs.* and *Kcd.*—P. Gray, 1850 and 1848. *Wgt.*—G. C. Druce, 1883.
LOCALITIES: Very common in all the valleys, and reaching 1100 feet, Beeftub.
Appears May 4 to 26, J. T. J. Moist or dry leaf mould of shady woods; wind-sheltered.
VISITORS: Syrphus cinctellus, Rhingia rostrata, Platychirius albimana, Siphona geniculata, S.-E.

Allium scorodoprasum. *Linn.*

RECORDS: *Kcd.*—Dr Macnab, 1837. *Wgt.*—Miss Hannay, 1893.
LOCALITIES: St. Mary's Isle, J. Fr.; Kirkcudbright, Dr M'N.

Allium oleraceum. *Linn.*

RECORD: *Dfs.*—J. Shaw, 1882 (sown?).
LOCALITIES: *Nithsdale*—By Nith, Dumfries, Hn.; Airdwood, Tynron, J. Sh.; Roadside, Kirkland, S.-E.; Closeburn, Mr Watson. *Annandale*—Dinwoodie Lodge, by roadside, J. Wi.

Allium vineale. *Linn.* (Crow Garlic).

RECORDS : *Dfs.*—J. Fingland (1887 ?). *Kcd.*—P. Gray, 1848. *Wgt.*—J. M'Andrew, 1889.

LOCALITIES : *Along the shore*—Cruggleton Castle to Port Allan, J. M'A.; Ross, G. C.; Port o' Warren, Glenstocking, Lincluden Abbey, P. Gr., F. W. G.; Torduff Point, J. Fn.

On rocks or river banks.

Allium carinatum. *Linn.*

RECORD : (Escape) *Kcd.*—F. R. Coles, 1884.

LOCALITY : Lake shore, Isle, J. M'A.

Allium ursinum. *Linn.*
(Garlic, Ramsons, Ramps, J. Sh.)

RECORDS : *Dfs.*—Dr Burgess, 1789. *Kcd.*—P. Gray, 1850. *Wgt.*—G. C. Druce, 1883.

LOCALITIES : Very common in all the valleys as high as the level of deciduous woods.

Appears April 29, G. Bl.: May 5 to 27, J. T. J. On wet or dry bare humus of woods; shaded and sheltered.

VISITORS : Apis, very abundant; Anthomyidæ, Dolichopodidæ, S.-E.

Narthecium ossifragum. *Huds.* (Bog Asphodel).

RECORDS : *Dfs.*—Dr Little, 1834. *Kcd.*—P. Gray, 1850. *Wgt.*—G. C. Druce, 1883.

LOCALITIES : *Nithsdale* -Lochanhead, F. W. G.; Dumfries, common, P. Gr.; Lochar Moss, Hn.; Cowhill, Ad. and S. D. J.; Closeburn, Dr Gl.; Sanquhar, Dr Dv. *Annandale*—Johnstone, Dr Lt.; Wellburn Carr., Breconside, Bold Craig, etc., J. T. J. *Eskdale*—Castle O'er, J. Wn. and R. Bl.

Appears June 25 to July 8, J. T. J. In wet marshes or peat mosses; quite exposed.

VISITORS : Apis, abundant, S.-E.; Bombus lucorum, abundant, Hn., S.-E.; hortorum, Hn.

Tofieldia palustris. *Huds.*

RECORD : *Dfs.*—Dr Singer, 1843.

LOCALITY : Moffat, Dr Sn.*

* Requires confirmation.

FLORA OF DUMFRIESSHIRE. 173

Juncus communis. *Mey.* (Rush).

RECORDS: 1535. Effusus, *Dfs.*—Dr Davidson, 1886. *Kcd.*—J. M'Andrew, 1882. *Wgt.*—G. C. Druce, 1883.
1536. Conglomeratus, *Dfs.* and *Kcd.*—P. Gray, 1850. *Wgt.*—G. C. Druce, 1883.

LOCALITIES: *Nithsdale*—Very common Dumfries, 1536, P. Gr.; Thornhill, 1535, R. A.; Lochanhead, 1536, F. W. G.; Sanquhar, 1535, Dr Dv. *Annandale*—Hartfell, 1535, J. T. J. *Eskdale*—Very common, S.-E.

VISITOR: Vespa, sp.; regularly sucking, Hn.

Juncus glaucus. *Ehrh.*

RECORDS: *Dfs.*—G. Gordon, 1836. *Kcd.*—J. M'Andrew, 1882. *Wgt.*—G. C. Druce, 1883.

LOCALITIES: *Along the shore*—Near Dumfries, G. Go., P. Gr.; Solway shore (*sic*), J. Fn.

Juncus balticus. *Willd.*

RECORD: *Kcd.*— Rev. J. Fraser, 1843.
LOCALITY: Gillfoot, J. Fr.*

Juncus tenuis. *Willd.*

RECORD: *Kcd.*—J. M'Andrew, 1886.
LOCALITY: Roadside, New Galloway, J. M'A. (Sept. 7).

Juncus articulatus. *Linn.* (Spret, J. Shaw).

RECORDS: 1539. Supinus, *Kcd.*—J. M'Andrew, 1882. *Wgt.*—G. C. Druce, 1883.
    Var. Fluitans? *Dfs.*—Dr Davidson, 1886. *Wgt.*— J. M'Andrew.
    Var. Uliginosus? *Dfs.*—Dr Davidson, 1886. *Kcd.* —F. R. Coles, 1883.
1542. Lamprocarpus, *Dfs.* and *Kcd.*—P. Gray, 1850. *Wgt.*—G. C. Druce, 1883.
    b. nigritellus, *Wgt.*—C. C. Bailey, 1883.
1543. Acutiflorus, *Dfs.*—Dr Davidson, 1886. *Kcd.*— J. M'Andrew, 1882. *Wgt.*—G. C. Druce, 1883.

* Requires confirmation.

LOCALITIES: *Nithsdale*—Loch Kindar, 1539, R. H. M.; Aucheness-nane, T. Br.; Lochanhead, 1543, F. W. G.; Auchengruith Mill Dam, 1539, Dr Dv.; Thornhill, 1543, R. A.; Sanquhar, 1539, 1543, Dr Dv. *Annandale*—Ecclefechan, S.-E.; Well Hill, 1539, J. T. J.; Moffat, 1543, J. T. J. *Eskdale*—Meggat, 1539, S.-E.; Meikledale, Whitehope (to 1400 feet), 1539, S.-E.; Liddel, 1542, S.-E.

Marshy spots on hills.

Juncus obtusiflorus. *Ehrh.*

RECORDS: *Kcd.*—G. N. Lloyd, 1837. *Wgt.*—Dr Balfour, 1843.

LOCALITIES: *Along the shore*—Glenluce, Dr Bl.; Creetown to Carsluith (Trans. 1841-4), Ross, Balmae, G. N. Ll.; Colvend, J. M'A.; Glencaple, F. W. G.

Juncus compressus. *Jacq.*

RECORDS: Gerardi, *Dfs.*—J. Fingland, 1886. *Kcd.*—J. M'Andrew, 1882. *Wgt.*—Dr Balfour, 1843.

LOCALITIES: *Along the shore*—Glenluce, Dr Bl.; common in Kirkcudbright, J. M'A.; Glencaple, F. W. G; Annan Waterfoot, J. Fn., Hn.; Annan to Gretna, occasionally, Sark mouth, Kirtle mouth, S.-E.

On wet estuarine mud along the shore; abundant.

Juncus Squarrosus. *Linn.* (Stoolbent, J. Shaw).

RECORDS: *Dfs.*—P. Gray, 1850. *Kcd.*—J. M'Andrew, 1882. *Wgt.* —G. C. Druce, 1883.

LOCALITIES: *Nithsdale*—Dumfries, P. Gr.; Sanquhar, Dr Dv. *Annandale*—Grey Mare's Tail, common Moffat, E. F. L., J. T. J. *Eskdale*—Whitchope, Causey Grain (to 1500 feet), S.-E.

On wet places, on peat and whinstone soils, chiefly above 1000 feet.

Juncus bufonius. *Linn.*

RECORDS: *Dfs.*—Dr Burgess, 1789. *Kcd.*—J. M'Andrew, 1882. *Wgt.*—G. C. Druce, 1883.

LOCALITIES: *Nithsdale*—Brow to Stank, Dr Br., S.-E.; Glencaple, F. W. G.; Dumfries, common, P. Gr., S.-E.; Sanquhar, Dr Dv. *Annandale*—Glenkillburn, Dr Br.; Evan, S.-E.; Moffat, very common, J. T. J. *Eskdale*—Very common, S.-E.; Castle O'er, J. Wn. and R. Bl.

On bare, wet, often peaty, mud by roadsides, margins of lochs, etc., to 1700 feet.

Juncus maritimus. *Lam.*

RECORDS: *Dfs.*—G. Gordon, 1835. *Kcd.*—G. N. Lloyd, 1837. *Wgt.*—Dr Graham, 1835.

LOCALITIES: *Along the shore*—Glenluce, West Tarbet, Dr Bl., Dr Gr.; Southwick, J. Fr.; Kirkcudbright Bay, G. N. Ll.; Urr mouth, J. Fr., P. Gr.; Glencaple, Hn., R. H. M., R. A.; Caerlaverock, G. Go., W. St., R. A.

In salt estuarine marshes.

VISITOR: Vespa vulgaris, Hn.

Juncus trifidus. *Linn.*

RECORD: *Dfs.*—J. Sadler, 1858.

LOCALITY: *Annandale*—Whitecoombe, J. Sd.*

Juncus castaneus. *Sm.*

RECORD: *Dfs.*—Dr Singer, 1843.

LOCALITY: *Annandale*—Moffat Hills, Dr Sn.*

Juncus biglumis. *Linn.*

RECORD: Triglumis, *Dfs.*—Dr Singer, 1843.

LOCALITY: *Annandale*—Moffat Hills, Dr Sn.*

Luzula pilosa. *Willd.*

RECORDS: 1550. Vernalis, *Dfs.* and *Kcd.*—P. Gray, 1850. *Wgt.*—G. C. Druce, 1883.

LOCALITIES: *Nithsdale*—Very common Ruthwell, Dr Gl.; Routen Brig, Scaur, S.-E.; Blackwood Linn, Exc.; Sanquhar, Dr Dv. *Annandale*—Tundergarth, G. Bl.; Beld Craig, J. T. J.; Lochanburn, S.-E. *Eskdale*—Byreburn, Tarras, Esk to Langholm, Glencorfe, Ewes, S.-E.

Appears April 30, J T. J. On wet or rather dry humus, chiefly over whinstone; quite shaded and sheltered.

Luzula sylvatica. *Bichen.*

RECORDS: *Dfs.* and *Kcd.*—P. Gray, 1850 *Wgt.*—G. C. Druce, 1883.

* Require confirmation.

LOCALITIES : Very common in all the valleys in linns at low altitudes, e.g., Scaur, S.-E.; Beld Craig, J. T. J.; Penton, S.-E.; or on rocks at 2000 feet, Black's Hope, S.-E.; 1500 feet Archie Hill, S. E.

On moist leaf-mould ; in shade or shelter or quite exposed (above 1000 feet), on whinstone rocks.

Luzula campestris. Br.

RECORDS : Aggregate, *Dfs.* and *Kcd.*—P. Gray, 1850. *Wgt.*—G. C. Druce, 1883.
SUBSPECIES—1554. Campestris, *Wgt.*—G. C. Druce, 1883.
1555. Erecta, *Dfs.*—G. F. Scott-Elliot, 1892. *Wgt.*—G. C. Druce, 1883.
b. congesta, *Dfs.* and *Kcd.*—P. Gray, 1850. *Wgt.*—G. C. Druce, 1883.

LOCALITIES : Very common in all the valleys (var. b. Dumfries, P. Gr.; Sanquhar, Dr Dv.; Hartfell, Correifron, J. T. J.), to 2000 feet Moffat, and 1600 feet Pikethow, S.-E.

On peat moors, gravelly sand, boulder clay, whinstone soils, etc.; fully exposed.

VISITORS : Strenia clathrata, abundant ; Volucella bombylans, S.-E.

Luzula albida.

RECORD : (Escape) *Dfs.*—J. T. Johnstone, 1889.

LOCALITIES : *Nithsdale*—Riddings, R. A.   *Annandale*—Hillside, Lockerbie, G. Bl.; Garpol, J. T. J.

On dry humus under beech trees.

Schoenus nigricans. Linn.

RECORDS : *Dfs.*—G. F. Scott-Elliot, 1891. *Kcd.*—J. T. Syme, (1835 ?) *Wgt.*—G. C. Druce, 1883.

LOCALITIES : *Along the shore*—Port William, S.-E. ; Garliestown, Hn.; Ross, J. T. G.; Kippford, P. Gr.; common, J. M'A.; Torduff, S.-E.

In marshes by the sea.

Cladium mariscus. Br.

RECORDS : *Kcd.*—J. M'Andrew, 1882. *Wgt.*—Mr M'Kie, before 1835.

LOCALITIES : Ravenston Loch, Whithorn, Dr M'N.; Boreland, F. R. C.; Barnhourie Loch, J. Fr., Fq.; Twynholm, J. M'A.; Trostrie, T. Bl.

Rhynchospora alba. *Vahl.*

RECORDS : *Dfs.* and *Kcd.*—G. N. Lloyd, 1831. *Wgt.*—J. T. Syme, 1836.

LOCALITIES : *Nithsdale*—Loch Kindar, E. M. C.; Kirkconnel Moss, G. N. Ll., Dr Gl. *Eskdale*—T. B. Bl.

Rhynchospora fusca. *Roem and Schlecht.*

RECORD : *Kcd.*—J. M'Andrew, 1882.

LOCALITY : Auchencairn Moor, J. M'A.

Blysmus compressus. *Panz.*

RECORDS : *Dfs.*—Dr F. W. Grierson, 1882. *Kcd.*—G. Gordon, 1835.

LOCALITIES : *Annandale*—Glencaple, F. W. G.; Torduff Point, J. Fn., S.-E. *Eskdale*—Bentpath, Bilholm, S.-E.

On wet boulder clay or whinstone soils ; exposed.

Blysmus rufus. *Link.*

RECORDS : *Dfs.*—Dr F. W. Grierson, 1882. *Kcd.*—P. Gray, 1848. *Wgt.*—J. M'Andrew, 1887.

LOCALITIES : *Along the shore*—Portwilliam, J. M'A.; Castle Hill to Glenluffin, P. Gr.; Glencaple, F. W. G., R. H. M., S.-E.; Sark to Kirtle mouth, S.-E.

On estuarine mud.

Scirpus acicularis. *Linn.*

RECORDS : *Dfs.*—G. N. Lloyd, 1831. *Kcd.*—(Reported) J. Fraser, 1843. *Wgt.*—G. C. Druce, 1883.

LOCALITIES : *Annandale*—Castle Loch, Lochmaben, G. N. Ll.*

Scirpus palustris. *Linn.*

RECORDS : *Dfs.*—Dr Davidson, 1886. *Kcd.*—J. M'Andrew, 1882. *Wgt.*—G. C. Druce, 1883.

LOCALITIES : *Nithsdale*—Nith backwater, Martindale Bridge, Broomrigg, S.-E. ; Auchenessnane, T. Br. ; Nithside, Sanquhar, Dr Dv. *Annandale*—shore, Annan, S.-E.; Castle Loch, Lochmaben, F. W. G., S.-E.

On wet mud of lochs or river sides.

* All require confirmation.

Scirpus multicaulis. *Sm.*

RECORDS : *Dfs.*—P. Gray, 1850. *Kcd.*—J. M'Andrew, 1882. *Wgt.*—Dr Balfour, 1843.

LOCALITIES : *Nithsdale*—near Dumfries, P. Gr. *Annandale*—Moffat, J. T. J.

Damp stony places on hills, J. T. J.

Scirpus pauciflorus. *Lightf.*

RECORDS : *Dfs.*—P. Gray, 1846. *Kcd.*—J. M'Andrew, 1882. *Wgt.*—Dr Balfour, 1843.

LOCALITIES : *Nithsdale*—Criffel, A. B. Hall ; Lochar Moss, Maxwelltown Loch, P. Gr.

Scirpus cæspitosus. *Linn.* (Deer's Hair).

RECORDS : *Dfs.*—Dr Davidson, 1886. *Kcd.*—P. Gray, 1850˙ *Wgt.*—G. C. Druce, 1883.

LOCALITIES : Very common on all the moors to the highest summits. On wet peat ; fully exposed.

Scirpus fluitans. *Br.*

RECORDS : *Kcd.*—P. Gray, 1850. *Wgt.*—J. M'Andrew, 1890.

LOCALITY : *Nithsdale*—Dumfries, P. Gr. (common J. M'A.)

Scirpus setaceus. *Linn.*

RECORDS : *Dfs.*—Dr Davidson, 1886. *Kcd.*—P. Gray, 1850. *Wgt.*—G. C. Druce, 1883.

LOCALITIES : *Nithsdale*—Dumfries, P. Gr.; Auchenessnane, T. Br.; Sanquhar, Dr Dv. *Annandale*—Moffat, J. T. J.; Beld Craig, F. W. G. *Eskdale*—Wauchope, S.-E.

On damp, muddy roadsides, Dr Dv., J. T. J.

Scirpus riparius. *Spreng.*

RECORD : *Wgt.*—Dr Graham, 1835.

LOCALITIES : East Tarbet, Glenluce, Mull, Dr Bl.; roadside, Drummore, Dr Gr.; Port Logan, Portpatrick, Arn.

Scirpus lacustris. Linn.

RECORDS: 1533. Lacustris, *Dfs.* and *Kcd.*—P. Gray, 1850. *Wgt.*
—J. M'Andrew, 1887?
1534. Tabernæmontani, *Kcd.* (reported), J. M'Andrew, 1882.

LOCALITIES: *Nithsdale*—Arbigland, 1534, J. M'A.; Dumfries, P. Gr. *Annandale*—Lochmaben, S.-E.
In lochs, part sheltered by reeds, etc.

Scirpus maritimus. Linn.

RECORDS: *Dfs.*—G. Gordon, 1837. *Kcd.*—Mr Farquharson, 1873. *Wgt.*—Dr Arnott, 1848.

LOCALITIES: *Along the shore*—Cowans, Portlogan, Arn.; Creetown, north of Tarff, F. R. C.; Colvend, Fq.; Lochar, G. Go.; Kingholm, S.-E.; Solway Bridge to Torduff, Annan town to mouth, occasionally to and along Kirtle, S.-E. *Eskdale*—Moodlaw Loch, S.-E.
In estuarine mud; fully exposed.

Scirpus sylvaticus. Linn.

RECORDS: *Dfs.*—Dr Burgess, 1789. *Kcd.*—J. M'Andrew, 1882.

LOCALITIES: *Nithsdale*—Broomrigg, S.-E.; Sanquhar, by Nith, Dr Dv.; Thornhill, R. A. *Annandale*—Cumburn, opposite Lochrighead, Kirkmichael, Dr Br.; three miles below Beattock, Nether Murthat, S.-E., J. T. J. *Eskdale*—Kirkandrews, E. Ty.

Eriophorum vaginatum. Linn.

RECORDS: *Dfs.* and *Kcd.*—P. Gray, 1850. *Wgt.*—G. C. Druce, 1883.

LOCALITIES: Very common in all the valleys to 2000 feet.
On wet peat moors and haggs.

Eriophorum polystachyum. Linn. (Draw Moss).

RECORDS: 1644. Angustifolium, *Dfs.*—Dr Davidson, 1886. *Kcd.*
—J. M'Andrew, 1882. *Wgt.*—G. C. Druce, 1883.
b. elatius, *Dfs.*—Dr Davidson, 1890.
1645. Latifolium, *Kcd.*—A. Croall, 1840. *Wgt.*—G. C. Druce, 1883.

LOCALITIES: *Nithsdale*—Very common, Dumfries, S.-E.; Lochanhead, F. W. G.; Maxwelltown Loch, 1645, A. Cro., P. Gr., J. Cru.; Sanquhar, Dr Dv.; Knockenstob, Glenmaddie, 1644, *b.*, Dr Dv.
On marshy, often peatty, soils.

Carex dioica. *Linn.*

RECORDS: *Dfs.*—Dr Davidson, 1886. *Kcd.*—P. Gray, 1846. *Wgt.*—G. C. Druce, 1883.

LOCALITIES: *Nithsdale*—Maxwelltown Loch, Irongray, P. Gr.; Girharrow, Glencairn, J. Cr.; Auchenessnane, T. Br.; Cample Cleugh, R. A.; Braeheads, Sanquhar, Dr Dv. *Annandale*—Frenchland Burn, J. T. J.
Wet places.

Carex pulicaris. *Linn.* (Flea Carex).

RECORDS: *Dfs.*—Dr Davidson, 1886. *Kcd.*—P. Gray, 1850. *Wgt.*—G. C. Druce, 1883.

LOCALITIES: *Nithsdale*—Very common Lochanhead, F. W. G.; Dumfries, P. Gr.; Glencairn, J. Cr. *Annandale*—Moffat, F. W. G., J. T. J. *Eskdale*—Meggat Mosspaul, Meikledale, S.-E.
In wet places on peat, mud of small rills, etc.

Carex rupestris. *All.*

RECORD: *Dfs.*—J. Sadler, 1858.
LOCALITY: *Annandale*—Moffat, J. Sd.*

Carex pauciflora. *Lightf.*

RECORDS: *Dfs.*—W. Stevens, 1848. *Kcd.*—J. M'Andrew, 1882.
LOCALITY: *Nithsdale*—Dalveen Pass Lowthers, Drumlanrig, W. St.
Wet places on hills.

Carex leporina. *Linn.* (=C. ovalis).

RECORDS: *Dfs.* and *Kcd.*—P. Gray, 1850. *Wgt.*—G. C. Druce, 1883.

LOCALITIES: *Nithsdale*—Dumfries, P. Gr.; Glencairn, common, J. Cr.; Thornhill, Dr Gl.; Sanquhar, common, Dr Dv. *Annandale*—Gallows Hill, Frenchland Burn, J. T. J. *Eskdale*—Meikledale, Meggat, Bentpath, S.-E.

* Requires confirmation.

Carex elongata. Linn.

RECORDS: *Dfs.*—T. Brown, 1893. *Kcd.*—J. M'Andrew, 1887.

LOCALITIES: Kenmure Holms, J. M'A. Auchenessnane, Penpont, T. Br.

Carex stellulata. Good.

RECORDS: *Dfs.* and *Kcd.*—P. Gray, 1850. *Wgt.*—G. C. Druce, 1883.

LOCALITIES: *Nithsdale*—Maxwelltown Loch, Lochar Moss, F.W.G., J. Cr.; Sanquhar, Dr Dv. *Annandale*—Reaching 2400 feet, Loch Skene, S.-E. *Eskdale*—Langholm, etc., to 1500 feet, S.-E.

On wet peat or mud on the moors.

Carex canescens. Linn. (=C. Curta.)

RECORDS: *Dfs.* and *Kcd.*—P. Gray, 1850. *Wgt.*—G. C. Druce, 1883.

LOCALITIES: *Nithsdale*—Broomrigg, S.-E.; Dumfries, P. Gr.; Glencairn, J. Cr.; New Loch, R. A.; Barr Bank, Farthing, Mollock, Sanquhar, Dr Dv. *Annandale*—Lochmaben, S.-E.; Annan Water, Nethermill, J. T. J. *Eskdale*—Blackburn, Moodlaw Loch, Causey Grain, 1400 feet, S.-E.

In marshy grass by burns.

Carex remota. Linn.

RECORDS: *Dfs.*—J. Cruickshank, 1836. *Kcd.*—Dr Balfour, 1843. *Wgt.*—J. M'Andrew, 1892.

LOCALITIES: *Nithsdale*—Brownhall, J. Cru.; Dalmakerran (450 feet), J. Cr.; Nithbank, R. A.; Sanquhar, Dr Dv. *Annandale*—Selkirk Road, J. T. J.; Beld Craig, S.-E. *Eskdale*—Woodslee, Penton, Glentarras, Burnfoot, Langholm, S.-E.

Appears July 6. On damp leaf-mould or holmlands; in shade and shelter.

Carex axillaris. Good.

RECORDS: *Dfs.*—G. Bell, 1893. *Kcd.*—T. Bell, 1882.

LOCALITIES: Trostrie Loch, T. Bl.; Tundergarth, G. Bl.*

* Require confirmation.

Carex paniculata. Linn.

RECORDS: 1662. Teretiuscula, *Dfs.*—J. Shaw, 1882. *Kcd.*—P. Gray, 1850. *Wgt.*—J. M'Andrew, 1889.
1664. Paniculata, *Dfs.*—Dr Davidson, 1886. *Kcd.*—J. M'Andrew, 1882. *Wgt.*—J. M'Andrew, 1889.

LOCALITIES: *Nithsdale*—Maxwelltown, R. H. M.; Tynron, 1662, J. Sh.; Auchenessnane, T. Br.; Thornhill, 1662, J. Wall.; Kirkbog, Morton Mains, Blackhill, R. A.; East of Morton, Carron, 1664, J. Wall.

Carex vulpina. Linn.

RECORDS: *Dfs.*—J. Shaw, 1882. *Kcd.*—Dumfries Herbarium, 1862. *Wgt.*—G. C. Druce, 1883.

LOCALITY: *Eskdale*—Twiglees, Eskdalemuir, J. Sn.

Carex muricata. Linn.

RECORDS: *Dfs.*—J. Cruickshank, 1839. *Kcd.*—P. Gray, 1846. *Wgt.*—G. C. Druce, 1883.

LOCALITIES: *Nithsdale*—Caerlaverock, J. Cru.; Lincluden, P. Gr.; roadside, Dunscore, Moniaive, J. Cr.; Auchenessnane, T. Br.; Sanquhar, Dr Dv.; Trigony, R. A. *Annandale*—Selkirk Road at Craigieburn, E. F. L., J. T. J.

Chiefly by shore.

Carex arenaria. Linn.

RECORDS: 1660. Disticha, *Dfs.*—J. Fingland, 1888. *Kcd.*—P. Gray, 1850. *Wgt.*—J. M'Andrew, 1889.
1661. Arenaria, *Kcd.*—J. M'Andrew, 1882. *Wgt.*—G. C. Druce, 1883.

LOCALITIES: *Nithsdale*—Dumfries, P. Gr.; Townhead farm, R. H. M.; Powfoot to Newbie, 1661, S.-E.*

Carex saxatilis. Linn. (=C. pulla).

RECORD: *Dfs.*—J. Fingland, 1891?

Carex cæspitosa. Linn.

RECORDS: 1684. Rigida, *Dfs.*—W. Stevens, 1848.
1685. Aquatilis.
b. Watsoni, *Dfs.*—Dr Davidson, 1886. *Kcd.*—J. M'Andrew, 1882.
e. minor, *Dfs.*—J. T. Johnstone, 1890.

* Requires confirmation.

1687. Goodenowii, *Dfs.*—Dr Davidson, 1886. *Kcd.*—J. M'Andrew, 1882. *Wgt.*—G. C. Druce, 1883.

LOCALITIES: *Nithsdale*—Lochar Moss, 1687, F. W. G.; Friars' Carse, S.-E.; Glencairn, J. Cr.; Sanquhar, 1658, *b.*, and 1687, Dr Dv.; Kirkbog, 1685, *b.* and *e.*, Thornhill, R. A. *Annandale*—Moffat Hills, 1684, Dr Bl.; Loch Skene, Correifron, 1685, *e.*, J. T. J. *Eskdale*— Archie Hill, Moodlaw Loch, White Hope (1500 feet), S.-E.

1684 chiefly on rocks; 1685 and 1687 in marshes or mud of small rills.

Carex acuta. *Linn.*

RECORDS: *Kcd.*—J. M'Andrew, 1882. *Wgt.*—Rev. J. Gorrie, 1893.

LOCALITIES: Loch Ken, etc., J. M'A.

Carex atrata. *Linn.*

RECORD: *Dfs.*—W. Stevens, 1848.

LOCALITIES: *Annandale*—Hill near Hartfell, W. St.; Loch Skene and Midlaw Burn, E. F. L., J. T. J.; Beld Craig, Kd.

On rock ledges; fully exposed.

Carex præcox. *Jacq.*

RECORDS: *Dfs.* and *Kcd.*—P. Gray, 1850. *Wgt.*—G. C. Druce, 1883.

LOCALITIES: *Nithsdale*—Very common, Dumfries, P. Gr.; Sanquhar, Dr Dv. *Annandale*—Whitecoombe, to the top, S.-E.; Moffat, very common, J. T. J. *Eskdale*—Very common by Esk and Tarras, S.-E.

In dry or rarely wet places, usually in short turf, on boulder clay, whinstone soils, sandy soils, etc.; fully exposed.

Carex flava. *Linn.*

RECORDS: Flava, *Dfs.* and *Kcd.*—P. Gray, 1850. *Wgt.*—G. C. Druce, 1883.
    b. minor, *Dfs.*—Dr Davidson, 1886. *Wgt.*—G. C. Druce, 1883.
    Oederi, *Dfs.*—J. Fingland, 1887. *Kcd.*—P. Gray, 1846. *Wgt.*—J. M'Andrew, 1889.

LOCALITIES: *Nithsdale*—Maxwelltown Loch, P. Gr., F. W. G.; Townfoot, R. H. M.; Auldgirth, J. Fn.; Glencairn, J. Cr.; Sanquhar, common, Dr Dv. *Annandale*—Annan to Kirtle, S.-E.; Loch Skene

(2200 feet), S.-E.; Correifron, E. F. L.; Moffat, common, J. T. J. *Eskdale*—Pikethow and Mosspaul (to 1400 feet), S.-E.

On wet mud of rills, ditches, and common amongst the hills. (Common in a dried up pond, J. Fn.)

Carex distans. Linn.

RECORDS: 1710. Lævigata, *Dfs.*—Mr Winch, 1837. *Ked.*—J. M'Andrew, 1887. *Wgt.*—J. M'Andrew, 1892.
1711. Binervis, *Dfs.*—Dr Davidson, 1886. *Ked.*—P. Gray, 1850. *Wgt.*—Dr Balfour, 1843.
1712. Distans, *Dfs.*—Dr Davidson, 1886. *Ked.*—J. M'Andrew, 1882. *Wgt.*—Dr Balfour, (1845?).
1714. Fulva, *Dfs.*—P. Gray, 1850. *Ked.*—J. Cruickshank, 1839. *Wgt.*—J. M'Andrew, 1887.
b. Hornschuchiana, *Dfs.*—Dr Davidson, 1886. *Wgt.*—G. C. Druce, 1883.

LOCALITIES: *Nithsdale*—Arbigland, G. Go.; Glencaple, 1712, F.W.G.; Maxwelltown Loch, 1714, J. Cru., P. Gr.; Longwood, 1714, R. H. M.; Auchenessnane, 1714, T. Br.; Newark Wood and Matthew's Folly, 1716; Dr Dv.; Cample Gilchristland, 1710, R. A.; Thornhill, 1711, 1714, R. A.; Sanquhar, 1711, 1714, and 1714, *b.*, Dr Dv. *Annandale*—Craigboar, 1711, S. E.; Garpol, Beattock, 1710, J. Sd.; Cornal Tower wood, 1714, F. W. G.; Beld Craig, 1710, 1714, Kd.; Moffat, 1710, Win.; Breconside, 1710, S.-E.; Black's Hope, 1711, J. T. J.; Correifron, 1711, J. T. J.; 1714, (1714, *c.*?), S.-E.; Loch Skene, 1711, E. F. L., to 2200 feet. *Eskdale*—Wisp, Causeway Grain, 1711, S.-E.; Langholm, 1710, S.-E.

Of these 1711 grows on wet rock ledges, 1712 rather dry places by the shore, 1714 on pretty dry whinstone rocks, S.-E.; 1714, *b.*, moist clayey banks, Dr Dv.

Carex pilulifera. Linn.

RECORDS: *Dfs.* and *Ked.*—P. Gray, 1850. *Wgt.*—G. C. Druce, 1883.

LOCALITIES: *Nithsdale*—Irongray, P. Gr.; Dalwhat, J. Cr.; Auchenessnane, T. Br.; Sanquhar, Dr Dv. *Annandale*—Craigboar, S.-E.; Whitecoombe, Dr Bl.; Moffat, common, J. T. J.

Dry rock ledges.

Carex filiformis. Linn.

RECORDS: *Dfs.*—Dr Davidson, 1886. *Ked.*—P. Gray, 1846. *Wgt.* —Rev. J. Gorrie, 1893.

LOCALITIES: *Nithsdale*—Maxwelltown Loch, J. Cru., P. Gr.; Girharrow, Moniaive, J. Cr.; Auchenessnane, T. Br.; Sanquhar, Dr. Dv.

Carex hirta. Linn.

RECORDS: *Dfs.*—Dr Davidson, 1886. *Kcd.*—P. Gray, 1850. *Wgt.*—G. C. Druce, 1883.

LOCALITIES: *Nithsdale*—Dumfries, P. Gr.; Moniaive, J. Cr.; Cample, Templand Bridge, R. A.; Sawmills, Sanquhar, Dr Dv. *Annandale*—Moffat, J. T. J. *Eskdale*—Glentarras, S.-E.
On moist loam.

Carex pallescens. Linn.

RECORDS: *Dfs.*—W. Keddie, 1854. *Kcd.*—J. M'Andrew, 1882. *Wgt.*—G. C. Druce, 1883.

LOCALITIES: *Nithsdale*—Moniaive, J. Cr.; Gilchristland, R. A.; Auchenessnane, T. Br.; Sanquhar, Dr Dv. *Annandale*—Beeftub road, S.-E.; Cornal Tower, F. W. G.; Beattock, Kd.: Beld Craig, S.-E.; Selkirk road, Beeftub, J. T. J. *Eskdale*—Byreburn, Causeway Grain (1400 feet), S.-E.
Wet places amongst rocks; half-shaded.

Carex extensa. Good.

RECORDS: *Dfs.*—W. Stevens, 1848. *Kcd.*—J. T. Syme (1836?). *Wgt.*—Dr Graham, 1835.

LOCALITIES: *Along the shore*—Common; Arbigland, W. Stables; Glencaple, R. A.; Sark to Annan (*b.* pumila?), S.-E. (*Inland*—Durisdeer, Edinburgh road, Grey Mare's Tail, W. St.?)
In estuarine mud, shingles, etc.

Carex punctata. Good.

RECORDS: *Kcd.*—J. Fraser, 1882. *Wgt.*—J. M'Andrew, 1887.
LOCALITY: On moist rock ledges at foot of bank, Colvend, Fq.

Carex panicea. Linn.

RECORDS: 1701. Panicea, *Dfs.* and *Kcd.*—P. Gray, 1850. *Wgt.*—G. C. Druce, 1883.
1702. Vaginata, *Dfs.*—J. T. Johnstone, 1890.

LOCALITIES: *Nithsdale* — Dumfries, P. Gr.; Glencairn, J. Cr. *Annandale*—Riddings, J. T. J.; Cornal Tower, F. W.. G.; Loch Skene, 1702, J. T. J.; Beld Craig, 1701, S.-E.
In long grass by springs (1701), S.-E.

Carex capillaris. Linn.

RECORD: *Dfs.*—W. Stevens, 1848.

LOCALITIES: *Annandale*—Hartfell, W. St.; Loch Skene, J. Fr., E. F. L.; Midlawburn, E. F. L., J. T. J.

In wet mud of springs.

Carex limosa. Linn.

RECORDS: 1689. Magellanica, *Dfs.*—W. Stevens, 1848.
1690. Limosa, *Dfs.*—J. Corrie, 1892. *Kcd.*—P. Gray, 1882.

LOCALITIES: *Nithsdale*—Terregles, 1690, J. Cru.; Maxwelltown Loch, 1690, P. Gr.; Girharrow, 1689, J. Cr.; Stroanshalloch, 1690, J. Cr.; Morton Castle, 1689, W. St.

Carex glauca. Scop.

RECORDS: *Dfs.*—Dr Davidson, 1886. *Kcd.*—J. M'Andrew, 1882. *Wgt.*—G. C. Druce, 1883.
c. stictocarpa, *Dfs.*—J. T. Johnstone, 1891.

LOCALITIES: *Nithsdale*—Lochanhead, F. W. G.; Holywood, S.-E.; Glencairn, J. Cr.; Townhead farm, R. A.; Sanquhar, Dr Dv. *Annandale*—Annan to Kirtle, Lochmaben, S.-E.; Grey Mare's Tail, E. F. L.; Black's Hope (var. *c.*), J. T. J.; Whitecoombe, S.-E. *Eskdale*—Meikledale, Tarras, S.-E.

In wet peat or mud, by streams, loch margins, etc.; usually in long grass.

Carex sylvatica. Huds.

RECORDS: *Dfs.*—W. Keddie, 1854. *Kcd.*—J. M'Andrew, 1882. *Wgt.*—J. M'Andrew, 1892.

LOCALITIES: *Nithsdale*—Dumfries, Dr Bl.; Jarbruck (370 feet), J. Cr.; Caitloch, Tynron (450 feet), J. Cr.; Drumlanrig, R. A.; Euchan, Dr Dv. *Annandale*—Beld Craig, Kd., J. Sd., S.-E.; Garpol, J. Sd. *Eskdale*—Penton, Tarras, S.-E.

On wet humus; in shade and shelter.

Carex pendula. Linn.

RECORDS: *Dfs.*—Dr Davidson, 1886. *Kcd.*—J. M'Andrew, 1882.[*]
Wgt.—J. M'Andrew, 1892.

LOCALITIES: *Nithsdale* — Carron, Nithbank, Drumlanrig Bridge, Craighope Linn, R. A., Dr Dv. *Eskdale*—Byreburn, Gilnockie, S.-E.

[*] Specimens illustrating Tongland Flora in Dumfries Herbarium, marked C. pendula, are, according to Mr Bennet, c. aquatilis, var. virescens.

Carex ampullacea. Good.

RECORDS: *Dfs.*—Dr Davidson, 1886. *Kcd.*—J. M'Andrew, 1882. *Wgt.*—G. C. Druce, 1883.

LOCALITIES: *Nithsdale*—Broomrigg, S.-E.; Glencairn, J. Cr.; Kirkbog, R. A.; Sanquhar, Dr Dv. *Annandale*—Lochmaben, F. W. G.; Wellhill, J. T. J.; Queensberry, Loch Skene, S.-E. *Eskdale*—Moodlaw, Kirkburn, Causeway Grain, S.-E.

In ditches and pools, in water.

Carex vesicaria. Linn.

RECORDS: *Dfs*—J. Fingland, 1888? *Kcd.*—P. Gray, 1850. *Wgt.* —G. C. Druce, 1883.

LOCALITIES: *Nithsdale*—Maxwelltown, R. H. M.; Broomrigg, S.-E.; Dalgoner Cairn, J. Cr.; Auchenknight Loch, R. A. *Annandale*— Moffat, common, J. T. J. *Eskdale*—Causeway Grain, S.-E.

Carex paludosa. Good.

RECORDS: 1720. Acutiformis, *Dfs.*—J. Wallace, 1882. *Kcd.*—Rev. J. Fraser, 1882. *Wgt.*—J. M'Andrew, 1887. 1721. Riparia, *Dfs.*—J. Cruickshank, 1839. *Kcd.*—J. M'Andrew, 1887.

LOCALITIES: *Nithsdale*—Caerlaverock, R. A., J. Fr.; Moniaive, J. Cr.; Logie Kirk, 1721, J. Cru. *Annandale*—Moffat, 1720, J. T. J. *Eskdale*—Twiglees, Eskdalemuir, J. Wall.; Wauchope, S.-E.

Carex stricta. Good.

RECORD: *Kcd.*—P. Gray, 1850?

Carex fulva × flava.

RECORD: *Dfs.*—E. F. Linton, 1889.

LOCALITY: *Annandale*—Correifron, E. F. L.

Milium effusum. Linn. (Millet).

RECORDS: *Dfs.*—Dr Singer, 1843. *Kcd.*—P. Gray, 1846.

LOCALITIES: *Nithsdale*—Cluden Craigs, Mavisgrove, P. Gr.; Sanquhar, Dr Dv.; Tinwald, Dr Sn. *Annandale*—Garpol, Moffat, J. Sd.; *Eskdale*—Penton Linn, S.-E.

In moist leaf mould of shady linns; wind-sheltered.

Anthoxanthum odoratum. *Linn.* (Vernal).

RECORDS: *Dfs.*—Dr Davidson, 1886. *Kcd.*—P. Gray, 1850. *Wgt.*—G. C. Druce, 1883.

LOCALITIES: Very common in all the valleys to 2400 feet.

On pretty dry boulder clay, peat mixed with mould, roadsides, etc.; fully exposed.

VISITOR: Hyclemyia strigosa, abundant, S.-E.

Anthoxanthum puellii. *Lecoq.*

RECORD: *Kcd.*—J. M'Andrew, 1887.

LOCALITY: Burnfoot.

Phalaris canariensis. *Linn.* (Canary).

RECORDS: (Escape) *Dfs.*—Dr Singer, 1843. *Kcd.* and *Wgt.*—J. M'Andrew, 1882 and 1890.

LOCALITIES: *Nithsdale*—Tinwald, Dr Sn.; Dumfries, P. Gr.; Templand Bridge, R. A.; Crawick wood, Dr Dv. *Annandale*—Waste ground, J. T. J.

Digraphis arundinacea. *Trin.* (Reed Canary).

RECORDS: *Dfs.* and *Kcd.*—P. Gray, 1850. *Wgt.*—G. C. Druce, 1883.

LOCALITIES: *Nithsdale*—Maxwelltown Loch, Cluden Mills, common by Nith to Sanquhar, S.-E., Dr Dv. *Annandale*—Annan mouth, Springkeld, Lochmaben, Milke, Evan, Beld Craig, S.-E. *Eskdale*—Woodslee, Penton, Liddel, Irvine, Wauchope, S.-E.

In water, river backwaters, and along holms, banks of rivers, more rarely in stagnant ponds; usually wind-sheltered.

Phleum pratense. *Linn.* (Timothy).

RECORDS: *Dfs.* and *Kcd.*—P. Gray, 1850. *Wgt.*—G. C. Druce, 1883.

LOCALITIES: Very common, Dumfries, P. G., F. W. G., S.-E.; Sanquhar, Dr Dv.; Meggat, S.-E.

Prefers moist holms, boulder clay, etc.

Phleum arenarium. *Linn.* (Sea Cat's Tail).

RECORDS : *Dfs.*—Dr Davidson, 1890. *Kcd.*—J. M'Andrew, 1882.
LOCALITIES : *Nithsdale*—Braeheads, Sanquhar, Dr Dv.

Alopecurus geniculatus. *Linn.*

RECORDS : *Dfs.*—P. Gray, 1850. *Kcd.*—J. Fraser, 1843. *Wgt.*—
G. C. Druce, 1883.
LOCALITIES : *Nithsdale*—Southerness, J. Fr.; Dumfries, P. Gr.;
Auchenessnane, T. Br.; Sanquhar, Dr Dv. *Annandale*—Kirtle mouth
to Annan, S.-E.; Moffat, J. T. J.
Wet places, estuarine mud, ditches, by roads, etc.

Alopecurus agrestis. *Linn.* (Foxtail).

RECORD : *Dfs.*—J. Fingland, 1887.
LOCALITY : Annan Water Foot, J. Fn.

Alopecurus pratensis. *Linn.*

RECORDS : *Dfs.* and *Kcd.*—P. Gray, 1850. *Wgt.*—G. C. Druce, 1883.
LOCALITY : Very common in all the valleys.
Pretty dry holmlands, well manured arable ground, etc.

Agrostis alba. *Linn.* (Marsh Bent).

RECORDS : 1756. Palustris, *Dfs.* and *Kcd.*—P. Gray, 1850. *Wgt.*—
C. C. Bailey, 1883.
1757. Vulgaris, *Dfs.* and *Kcd.*—P. Gray, 1850. *Wgt.*—
G. C. Druce, 1883.
Var. pumila, *Dfs.*—J. T. Johnstone, 1890. *Wgt.*
—G. C. Druce, 1883.
Var. sylvatica, *Dfs.*—T. Brown, 1893.
LOCALITY : Common in all the valleys.
1757, roadsides, J. M'A.; 1756, wet ditches, marshes, on boulder
clay and whinstone soils, J. M'A., Dr Dv., S.-E.; 1756, var. pumila,
Correifron, J. T. J.

Agrostis canina. *Linn.* (Brown Bent).

RECORDS: *Dfs.*—Dr Davidson, 1886. *Kcd.*—J. M'Andrew, 1882. *Wgt.*—J. M'Andrew, 1893.

LOCALITIES: *Nithsdale*—Auchenessnane, T. Br.; Sanquhar, Dr Dv. *Annandale*—Correifron, E. F. L.

On peat moors, cinders of railway stations.

Psamma arenaria. *Beauv.* (Maram).

RECORDS: *Dfs.*—Dr Burgess, 1789. *Kcd.*—P. Gray, 1846. *Wgt.*—Dr Arnott, 1848.

LOCALITIES: *Along the shore*—Killiness, Arn.; Port-William, J. M'A.; Almorness Point, J. M'A.; Carsethorn, P. Gr.; Newby, Dr Br., S.-E.; Powfoot, J. Fr., S.-E.

Calamagrostis lanceolata. *Roth.*

RECORD: *Kcd.*—J. M'Andrew, 1883.

LOCALITY: Kenmure holms, J. M'A.

Aira cæspitosa. *Linn.*

RECORD: *Dfs.* and *Kcd.*—P. Gray, 1850. *Wgt.*—G. C. Druce, 1883.

LOCALITIES: *Nithsdale*—Dumfries, P. Gray; Auchenessnane, T. Br.; Closeburn, Dr Gl.; Sanquhar, Dr Dv. *Annandale*—Springkeld, Kirtlebridge, Ecclefechan, Evan water, S.-E. *Eskdale*—Meggat, Tarras, Gilnockie, Burnfoot, Eskdalemuir Kirk, S.-E.

On moist humus, boulder clay, holms, roadsides, peat, etc.; usually in shade and wind-sheltered.

Aira flexuosa. *Linn.*

RECORDS: *Dfs.* and *Kcd.*—P. Gray, 1846 and 1850. *Wgt.*—G. C. Druce, 1883.

LOCALITIES: *Nithsdale*—Glen, Dr. Gl.; Lochanhead, F. W. G.; Auchenessnane, T. Br.; Closeburn, Dr Gl.; Dumfries, common, P. Gr.; Sanquhar, Dr Dv. *Annandale*—Templand, Crofthead, S.-E. *Eskdale* Common by Esk, Meggat, S.-E.

On dry peat moors, boulder clay, mud-stones.

Aira præcox. *Linn.* (Early Hair).

RECORDS: *Dfs.*—Dr Davidson, 1886. *Kcd.*—P. Gray, 1850. *Wgt.*—G. C. Druce, 1883.

LOCALITIES: *Nithsdale*—Common, Dalawoodie, Exc., J. Fn.; Sanquhar, Dr. Dv. *Eskdale*—Penton Linn, S.-E.

Dry sloping banks in short turf, often under beeches.

Aira caryophyllea. *Linn.* (Silvery Hair).

RECORDS: *Dfs.* and *Kcd.*—P. Gray, 1850. *Wgt.*—G. C. Druce, 1883.

LOCALITIES: *Nithsdale*—Dumfries, P. Gr.; Routen Brig, Racks, S.-E.; Holywood, Hn.; Sanquhar, Dr Dv. *Annandale*—Moffat, J. M'A.; *Eskdale*—Langholm railway, S.-E.

On cinders of railways.

Avena fatua. *Linn.* (Wild Oat).

RECORDS: *Dfs.*—J. Shaw, 1882. *Wgt.*—G. C. Druce, 1883.
LOCALITY: *Nithsdale*—Tynron, J. Sh.

Avena strigosa. *Schreb.*

RECORDS: *Dfs.*—W. Stevens, 1848. *Kcd.*—P. Gray, 1850. *Wgt.*—G. C. Druce, 1883.
LOCALITY: Near Dumfries, W. St., P. Gr.

Avena pratensis. *Linn.* (Perennial Oat).

RECORDS: 1780. Pubescens, *Dfs.*—P. Gray, 1846. *Wgt.*—J. M'Andrew, 1893.
1781. Pratensis, *Dfs.* and *Kcd.*—P. Gray, 1850.
b. alpina, *Dfs.*—E. F. Linton, 1890.

LOCALITIES: *Nithsdale*—Terregles by Nith, 1781, P. Gray; Craigs, 1780, P. Gr.; Auchenessnane, 1780, T. Br.; Nith holms, Sanquhar, 1781, Dr Dv. *Annandale*—Caledonian Railway, Lockerbie, 1780, S.-E.; Craigbeck, 1781, J. T. J.; Craigboar, Penbreck, 1781, J. T. J.; Hartfell, Black's Hope, Correifron, Grey Mare's Tail, Whitecoombe (all 1781 *b.*), E. F. L., J. T. J. *Eskdale*—Whitehope, S.-E.

1780, on dry, exposed, cindery railway embankments; 1781, holms by rivers; var. *b.*, on wet rock ledges, half-shaded and sheltered.

Avena flavescens. *Beauv.* (Yellow Oat).

RECORDS: *Dfs.*—Dr Davidson, 1886. *Kcd.*—J. M'Andrew, 1893.

LOCALITIES: *Nithsdale*—Roadside, Crawick Bridge, Bankhead, Dr Dv.

Arrhenathrum avenaceum. *Beauv.* (False Oat).

RECORDS: *Dfs.* and *Kcd.*—P. Gray, 1850. *Wgt.*—G. C. Druce, 1883.
    b. nodosum, *Dfs.*—Dr Davidson, 1886. *Kcd.* and *Wgt.*—J. M'Andrew, 1882.

LOCALITIES: *Nithsdale*—Ruthwell, Dr Gl.; Dumfries, P. Gray, Auchenessnane, T. Br.; Sanquhar (also var. *b.*), Dr Dv. *Annandale*—Ecclefechan, Garpol, S.-E. *Eskdale*—Penton, Burnfoot, Whitehope, Mosspaul, S.-E.

On pretty dry roadsides, railways, humus, boulder clay, usually half-shaded.

Holcus lanatus. *Linn.* (Soft Grass).

RECORDS: *Dfs.* and *Kcd.*—P. Gray, 1850. *Wgt.*—G. C. Druce, 1883.

LOCALITIES: Very common in all the valleys.

On pretty dry holms, boulder clay, whinstone soils, granite, etc.; fully exposed.

Holcus mollis. *Linn.*

RECORDS: *Dfs.* and *Kcd.*—P. Gray, 1850. *Wgt.*—G. C. Druce, 1883.

LOCALITIES: Very common in all the valleys.

On moist humus, shingles, boulder clay, whinstone soils; usually half-shaded, S.-E.

Lepturus filiformis. *Trin.*

RECORDS: *b.* incurvatus, *Dfs.*—J. Cruickshank, 1839. *Kcd.*—Dr Lightfoot, 1789.

LOCALITIES: *Along the shore*—Southerness, J. Fr.; Arbigland, Lg.; Caerlaverock, W. St., P. Gr., J. Cru.; Powfoot to Newbie, J. Fr., S.-E.

In wet estuarine marshes.

FLORA OF DUMFRIESSHIRE. 193

Nardus stricta. *Linn.* (Matgrass).

RECORDS: *Dfs.* and *Kcd.*—P. Gray, 1850. *Wyt.*—G. C. Druce, 1883.

LOCALITIES: Very common all over peat moors (to 2000 feet). On dry peat, boulder clay, stony ground; exposed.

Elymus arenarius. *Linn.* (Lyme).

RECORD: *Dfs.*—Dr Singer, 1843.
LOCALITY: *Annandale*—Newbie, Dr Sn.*

Agropyrum repens. *Beauv.* (Couch Arrow).

RECORDS: 1851. Repens, *Dfs.* and *Kcd.*—P. Gray, 1850. *Wgt.*—G. C. Druce, 1883.
1853 Acutum, *Kcd.* and *Wgt.*—G. C. Druce, 1883.
1854. Junceum, *Dfs.*—J. Fingland, 1887. *Kcd.*—J. M'Andrew, 1882. *Wgt.*—Dr Arnott, 1848.

LOCALITIES: *Nithsdale*—Lochanhead, F. W. G.; Maxwelltown Loch, P. Gr.; Kingholm, S.-E., R. A.; Sanquhar, Dr Dv. *Annandale*—Powfoot, S.-E. *Eskdale*—Railway, Liddel Bridge, S.-E.

Rather wet meadows, railway embankments, 1851; sandy sea shores, 1854.

Agropyrum caninum. *Beauv.*

RECORDS: *Dfs.*—Dr Davidson, 1886. *Kcd.*—J. M'Andrew, 1882. *Wgt.*—G. C. Druce, 1883.

LOCALITIES: *Nithsdale*—Holywood Station, S.-E.; Auchenessnane, T. Br.; Sanquhar, Dr Dv. *Annandale*—Mouth of Kirtle, Annan Station, Garpol, Beld Craig, S.-E. *Eskdale*—Penton, S.-E.

On wet humus, boulder clay, or dry cinders of railways.

Lolium perenne. *Linn.* (Rye Grass).

RECORDS: *Dfs.* and *Kcd.*—P. Gray, 1850. *Wgt.*—G. C. Druce, 1883.
Var. *e.*, Italicum, *Dfs.*—Dr Davidson, 1886.

LOCALITIES: Very common in all the valleys to 900 feet. (Var. *e.*, Sanquhar, Dr Dv.)

* Requires confirmation.

Lolium temulentum. *Linn.* (Darnel).

RECORD: *Dfs.*—Dr Balfour, 1843.
  b. arvense, *Dfs.*—Dr Davidson, 1886.

LOCALITIES: *Nithsdale*—Dumfries, Dr Bl.; Kirkconnel, by Nith, Dr Dv.

Brachypodium sylvaticum. *R. and S.* (False Brome).

RECORDS: *Dfs.* and *Kcd.*—P. Gray, 1850. *Wgt.*—G. C. Druce, 1883.

LOCALITIES: *Nithsdale*—Craigs, by Nith, P. Gray; Sanquhar, Dr Dv.; Auchenessnane, T. Br. *Annandale*—Moffat, E. F. L. *Eskdale*—Byreburn, Liddel, Burnfoot, S.-E.

On humus or boulder clay; half-shaded.

Bromus erectus. *Huds.*

RECORDS: *Dfs.*—Dr Balfour, 1843. *Kcd.*—P. Gray, 1846.

LOCALITIES: *Nithsdale*—Newabbey Churchyard, P. Gray; Dumfries, Dr Bl. *Annandale*—Lockerbie, S.-E.

Bromus asper. *Linn.*

RECORD: *Dfs.*—Dr Burgess, 1789. *Kcd.*—J. M'Andrew, 1882. *Wgt.*—J. M'Andrew, 1890.

LOCALITIES: *Nithsdale*—Dumfries, P. Gr.; Auchenessnane, T. Br.; Sanquhar, Dr Dv. *Annandale*—Crofthead, Auchenbraith Linn, Kirkmichael, S.-E.; Lamonby Mill, G. Bl. *Eskdale*—Very common Penton, Langholm Lodge, Tarras, S.-E.

In wet humus of woods; shaded or half-shaded.

Bromus sterilis. *Linn.*

RECORDS: *Dfs.*—Dr Davidson, 1886. *Kcd.*—J. M'Andrew, 1882. *Wgt.*—J. M'Andrew, 1893.

LOCALITIES: *Nithsdale*—Old Brick-field, Dumfries, S.-E.; Auldgirth Station, S.-E.; Sanquhar, Dr Dv. *Annandale*—? *Eskdale*—Gretna Green Station, Langholm, Bentpath, S.-E.

On dry cinders of railways, old brick-fields, etc.

Bromus arvensis. Linn.

RECORDS : 1840. Secalinus, *Dfs.* and *Kcd.*—P. Gray, 1850.
Var. velutinus, *Wgt.*—Dr Graham, 1836.
1841. Racemosus, *Dfs.*—Dr Davidson, 1886. *Kcd.*—
J. M'Andrew, 1882. *Wgt.*—G. C. Druce, 1883.
1842. Commutatus, *Dfs.*—J. T. Johnstone, 1893. *Kcd.*
—P. Gray, 1846. *Wgt.*—G. C. Druce, 1883.
1843. Mollis, *Dfs.* and *Kcd.*—P. Gray, 1850. *Wgt.*—
G. C. Druce, 1883.

LOCALITIES : *Nithsdale*—Near Dumfries, 1840 and 1841, P. Gray; banks of Nith and Terregles, 1842, P. Gr.; Auldgirth Station, S.-E.; Auchenessnane, 1841, T. Br.; Sanquhar, 1841 and 1843, Dr Dv. *Annandale*—Newbie, 1843, S.-E.; Ecclefechan, S.-E.; Moffat, 1843, J. M'A., 1840, J. T. J.

Bromus giganteus. Linn.

RECORDS : *Dfs.*—T. Brown, 1883. *Kcd.*—G. Gordon, 1836. *Wgt.* —G. C. Druce, 1883.

LOCALITIES : *Nithsdale*—Auchenessnane, T. Br.; Sanquhar, Dr Dv. *Annandale*—Milke, Dryfe, G. Bl.; Moffat, J. M'A.

Festuca ovina. Linn. (Sheep's Fescue).

RECORDS : 1826. Ovina, *Dfs.* and *Kcd.*—P. Gray, 1850. *Wgt.*—
G. C. Druce, 1883.
b. capillata, *Wgt.*—G. C. Druce, 1883.
1827. Rubra, *Dfs.*—P. Gray, 1850. *Kcd.*—Dr Burgess,
1789. *Wgt.*—G. C. Druce, 1883.
b. arenaria, *Wgt.*—G. C. Druce, 1883.

LOCALITIES : *Nithsdale*—Arbigland, 1827, Dr Br.; Dumfries, 1827, P. Gr.; Friars' Carse, 1827, S.-E.; Sanquhar, Dr Dv. *Annandale*— Very common 1826, J. T. J.; Moffat, 1827, J. T. J.; Grey Mare's Tail, Dr Bl. *Eskdale*—Whitehope Edge, Causeway Grain (to 1500 feet), Gilnockie, Meikledale, 1827, S.-E.

In rather wet places on peat, whinstone soils, etc., very common 1826; dry whinstone rocks, rather sandy soil, carboniferous sandstone, 1827.

Festuca elatior. Linn.

RECORDS : 1831. Elatior, *Linn, Dfs.*—Dr Davidson, 1886. *Kcd.*—
J. M'Andrew, 1882. *Wgt.*—G. C. Druce, 1883.

b. pseudololiacea, *Dfs.*—G. F. Scott-Elliot, 1892.
c. loliacea, *Dfs.*—Dr Davidson, 1890.
d. pratensis, *Dfs.* and *Kcd.*—P. Gray, 1850.   *Wgt.*
    G. C. Druce, 1883.
1832. Arundinacea, *Dfs.*—T. Brown, 1893.

LOCALITIES : *Nithsdale*—Dumfries, var. *d.*, P. Gr.; Kingholm, S.-E.; Auchenessnane, 1832, T. Br.; Sanquhar, var. *c.*, Dr Dv.   *Annandale*— Moffat, 1831, J. M'A.; 1832, S.-E.; Annan, Kirtle, 1831, var. *b.*, and 1832, S.-E.; Gasworks, Moffat, 1831, var. *c.*, J. M'A.   *Eskdale*—Kirkburn, Bentpath, 1832, S.-E.; Byreburn, 1832, S.-E.

Moist holms, roadsides, etc.; estuarine mud, 1832.

Festuca gigantea. *Vill.*

RECORDS : *Dfs.*—Dr Singer, 1843.   *Kcd.*—J. M'Andrew, 1882.

LOCALITIES : *Nithsdale*—Tinwald, Dr Sn.; Tynron, J. Sh.; Thornhill, J. M'A.

Festuca sylvatica. *Vill.*

RECORDS : *Dfs.* and *Kcd.*—T. Brown, 1882.

LOCALITIES : *Nithsdale*—Kirkbean, Tynron, T. Br.   *Eskdale*—Byreburn, S.-E.

Festuca myurus. *Linn.*

RECORDS : 1824. Myurus, *Dfs.*—Dr Singer, 1843.   *Kcd.*—T. Brown,
    1882?
1825. Sciuroides, *Dfs.*—P. Gray, 1850.   *Wgt.*—G. C.
    Druce, 1883.

LOCALITIES : *Nithsdale*—Tinwald, 1824, Dr Sn.; Dumfries, 1825, P. Gr.; Auchenessnane, 1825, T. Br.; Sanquhar, Dr Dv.   *Annandale*— Moffat, 1824, J. T. J. and J. M'A.

Dactylis glomerata. *Linn.* (Cocksfoot).

RECORDS : *Dfs.* and *Kcd.*—P. Gray, 1850.   *Wgt.*—G. C. Druce, 1883.

LOCALITIES : Very common in all the valleys to 900 feet.

On moist or dry holms, roadsides, humus, etc.; often in complete shade and shelter.

Cynosurus cristatus. *Linn.* (Dogs' Tail).

RECORDS: *Dfs.* and *Kcd.*—P. Gray, 1850. *Wgt.*—G. C. Druce, 1883.

LOCALITIES: Very common in all the valleys to 2000 feet. On roadsides, peat, etc.; almost every soil, and fully exposed.

Briza media. *Linn.* (Quaking Grass).

RECORDS: *Dfs.* and *Kcd.*—P. Gray, 1850. *Wgt.*—G. C. Druce, 1883.

LOCALITIES: *Nithsdale.*—Lochanhead, F. W. G.; Dumfries, P. Gr.; Morton Mains, Townhead, R. A.; Sanquhar, Dr Dv. *Annandale*—Evan, S.-E.; Beeftub, Kd.; Moffat, common, J. T. J. *Eskdale*—Tarras, Bilholm, Mosspaul, Whitehope, 1400 feet, S.-E.

On moist whinstone soils, boulder clay, etc.; fully exposed.

Poa plicata. *Fr.*

RECORD: *Dfs.*—Dumfries Herbarium, 1863.
LOCALITY: Ruthwell.

Poa aquatica. *Linn.*

RECORD: *Dfs.*—P. Gray, 1846. *Kcd.*—J. Fraser, 1843.

LOCALITIES: *Nithsdale.*—Caerlaverock, P. Gr., R. H. M. *Annandale*—Kirtle Bridge, Beld Craig, Garpol, S.-E. *Eskdale*—Tarras water, Bexburn Linn, near Burnfoot, S.-E.

In wet mud by ponds and burns.

Poa fluitans. *Scop.*

RECORDS: *Dfs.* and *Kcd.*—P. Gray, 1850. *Wgt.*—G. C. Druce, 1883.

LOCALITIES: *Nithsdale*—Ruthwell, Dr Gl.; Routen Brig, S.-E.; Dumfries, P. Gr.; Auchenessnane, T. Br.; Sanquhar, Dr Dv.; Grey Mare's Tail, R. A. *Annandale*—Kirtle, S.-E.; Ecclefechan, S.-E.; common, Moffat, J. T. J., S.-E. *Eskdale*—Penton, along Esk at Burnfoot, Wauchope, etc., S.-E.

In water of ditches; usually in long grass.

Poa maritima. *Huds.*

RECORD: *Dfs.*—P. Gray, 1846. *Kcd.* and *Wgt.*—G. C. Druce, 1883.

LOCALITY: *Along the shore*—Caerlaverock, P. Gr.

Poa loliacea. *Huds.*

RECORDS: *Kcd.*—J. M'Andrew, 1882. *Wgt*—Dr Graham, 1836.

LOCALITIES: East of Drummore, Dr Gr.; South Creetown, J. M'Andrew.

Poa annua. *Linn.*

RECORDS: *Dfs.* and *Kcd.*—P. Gray, 1850. *Wgt.*—G. C. Druce, 1883.

LOCALITIES: Very common in all the valleys.
By roadsides, waste ground, etc. (not on peat).

Poa distans. *Linn.*

RECORD: *Kcd.*—J. M'Andrew, 1883.

Poa sudetica. *Hœnke.*

RECORD: *Dfs.*—T. Brown, 1893.
LOCALITY: Auchenessnane.

Poa pratensis. *Linn.*

RECORDS: *Dfs.* and *Kcd.*—P. Gray, 1850. *Wgt.*—G. C. Druce, 1883.
    b. subcærulea, *Dfs.*—G. F. Scott-Elliot, 1893. *Wgt.*—G. C. Druce, 1883.

LOCALITIES: Very common in all the valleys, var. *b.* Langholm, S.-E.

On pretty dry holms, sandy stony soils, roadsides, etc.; fully exposed in short turf.

Poa balfourii. *Bab.*

RECORD: *Dfs.*—Rev. E. F. Linton, 1889.
LOCALITY: Midlaw Burn.

Poa trivialis. *Linn.*

RECORDS : *Dfs.* and *Kcd.*—P. Gray, 1850. *Wgt.*—G. C. Druce, 1883.
LOCALITIES : Very common in all the valleys.
On moist or dry holms, etc.; fully exposed.

Poa nemoralis. *Linn.*

RECORDS : *Dfs.*—T. Brown, 1882.
c. glaucantha, *Dfs.*—Dr Davidson, 1886.
LOCALITIES : *Nithsdale*—Wood near Drumlanrig, T. Br.; rocks by Nith, Sanquhar, Dr Dv.; rocks on Kellowater, var *c.*, Dr Dv. *Annandale*—Moffat, J. T. J.
In shade.

Catabrosa aquatica. *Beauv.*

RECORDS : *Dfs.*—Dr Singer, 1843. *Wgt.*—Dr Arnott, 1848.
LOCALITIES : *Along the shore*—Portlogan to Port Gill, Arn.; Merse, J. M'A.; Tinwald, Dr Sn.

Molinia cærulea. *Mœnch.*

RECORDS: *Dfs.*—Dr Davidson, 1886. *Kcd.*—J. M'Andrew, 1882. *Wgt.*—G. C. Druce, 1883.
LOCALITIES: *Nithsdale*—Very common, Lochar Moss, F. W. G.; Auchenessnane, T. Br.; Sanquhar, Dr Dv. *Annandale*—Very common Moffat Hills, J. T. J., S.-E. *Eskdale*—Common Whitehope (to 1500 feet), and by Esk, S.-E.
Usually marshes, on peat or boulder clay, very common.

Melica nutans. *Linn.*

RECORDS: *Dfs.*—J. Sadler, 1858. *Kcd.*—J. M'Andrew, 1882.
LOCALITIES: *Nithsdale*—Scaur, J. Sh., T. Br.; Nith Linns, R. A.; Crawick, Euchan, Kello, Dr Dv. *Annandale*—Beld Craig, Garpol, J. Sd.; Wellburn, Craiks Wood, J. T. J.
In glens at a higher altitude than the following.

Melica uniflora. Linn.

RECORDS: *Dfs.* and *Kcd.*—P. Gray, 1850 and 1846. *Wgt.*—G. C. Druce, 1883.

LOCALITIES: *Nithsdale*—Kirkbean, Hn.; Mavisgrove, Glen, Craigs, P. Gr., S.-E.; Routen Brig, S.-E.; Scaur, Exc., T. Br.; Nithbank, R. A.; Sanquhar, common, Dr Dv. *Annandale*—Tundergarth, S.-E.; Garpol, Beld Craig, S.-E.; Craiks Wood, Wellburn, J. T. J. *Eskdale*—Penton, Langholm, Tarras, Bexburn, S.-E.

On moist leaf mould; in shady and sheltered linns.

Triodia decumbens. Beauv.

RECORDS: *Dfs.*—Dr Davidson, 1886. *Kcd.*—P. Gray, 1846. *Wgt.*—G. C. Druce, 1883.

LOCALITIES: *Nithsdale*—Maxwelltown Loch, P. Gr.; Hellcleugh, R. A.; Sanquhar, Dr Dv. *Annandale*—Torduff Point, S.-E.

Dry whinstone pastures, boulder clay.

Kœleria cristata. Pers.

RECORDS: *Dfs.*—Dr Davidson, 1890. *Kcd.*—J. M'Andrew, 1882. *Wgt.*—Dr Graham, 1835.

LOCALITY: *Nithsdale*—Spango, Dr Dv.

Dry pastures, Dr Dv.

Arundo phragmites. Linn. (Reed).

RECORDS: *Dfs.*—Dr Singer, 1843. *Kcd.*—P. Gray, 1850. *Wgt.*—G. C. Druce, 1883.

LOCALITIES: *Nithsdale*—Dumfries, P. Gr.; Scaur, J. Wl. *Annandale*—Lochmaben Lochs, Dr Sn., F. W. G., S.-E.

Isoetes lacustris. Linn. (Quillwort).

RECORDS: *Dfs.* — W. Keddie,* 1854. *Kcd.* — J. Fraser, 1864. *Wgt.*—C. C. Bailey, 1883.

LOCALITY: *Annandale*—Loch Skene, W. Kd.

Pilularia globifera. Linn.

RECORDS: *Kcd.*—J. Cruickshank, 1843? *Wgt.*—G. C. Druce, 1883.

LOCALITY: Lincluden House Pond, Dumfries, J. Cru., P. Gray.

* Requires confirmation.

Lycopodium clavatum. Linn.

RECORDS: *Dfs.* and *Kcd.*—P. Gray, 1850. *Wgt.*—Rev. G Wilson, 1891.

LOCALITIES: *Nithsdale*—Dalskairth, Lochar Moss, P. Gray; Loch Urr, Girharrow, J. Cr.; Craigdarroch, Glenmaddie, Pamphalinn, Dr Dv. *Annandale*—Well Burn, Beld Craig Burn, J. T. J.; Breckonside, W. Ca.; Loch Skene, S.-E. *Eskdale*—Langholm Hill, Bea.

Lycopodium annotinum. Linn.

RECORDS: *Dfs.*—Dr Singer, 1843.

LOCALITIES: *Nithsdale*—Barn Hills, Morton, J. Cr. *Annandale*—Moffat, Dr Sn.

Lycopodium alpinum. Linn.

RECORDS: *Dfs.*—J. Sadler, 1862. *Kcd.*—P. Gray, 1848.

LOCALITIES: *Nithsdale*—Dalskairth House (to north), P. Gr.; Eastern Slope of Lochanhead Hills, P. Gr.; near marsh, on Lowthers, Dr Dv.; Townhead, R. A. *Annandale*—Hartfell, Hindhill, W. Ca.; Garpol, J. Sd.; Black's Hope, Whitecoomb, S.-E.

Lycopodium selago. Linn.

RECORDS: *Dfs*—W. Carruthers, 1882. *Kcd.*—P. Gray, 1846. *Wgt.*—J. M'Andrew, 1890.

LOCALITIES: *Nithsdale*—Dalskairth Hills, P. Gr.; Caitloch, J. Cr.; Uplands, Sanquhar, Dr Dv. *Annandale*—Auchencat Burns, S.-E.; Hartfell and Craigs, W. Ca., S.-E.; Grey Mare's Tail, J. T. J.; Crofthead, S.-E.

On dry places on hills, mudstones, etc.

Lycopodium selaginoides. Linn.

RECORDS: *Dfs.*—W. Carruthers, 1882. *Kcd.*—P. Gray, 1846. *Wgt.*—J. M'Andrew, 1890.

LOCALITIES: *Nithsdale*—Dalskairth, Criffel, P. Gr.; general, Glencairn, J. Cr.; Cample, R. A.; common Sanquhar, Dr Dv. *Annandale*—Beld Craig Burn, Correifron, S.-E.; Well Hill Marshes, J. T. J.; Grey Mare's Tail, W. Ca. *Eskdale*—Castle O'er, S.-E.

Equisetum maximum. *Lam.*

RECORDS : *Kcd.*—J. M'Andrew, 1888. *Wgt.*—Dr Arnott, 1848.

LOCALITIES : Dunskey Glen, J. M'A.; south of Portwilliam, Monreith, G. C. Dr.; Stoneykirk, Arn.

Equisetum pratense. *Linn.*

RECORD : *Dfs.*—Dr Davidson, 1886.

LOCALITY : *Nithsdale*—Holmwood, Nith above Sawmill, Sanquhar, Dr Dv.

Equisetum arvense. *Linn.*

RECORDS : *Dfs.*—P. Gray, 1850. *Kcd.*—J. M'Andrew, 1882. *Wgt.*—G. C. Druce, 1883.
Var. *b.* Alpestre? *Dfs.*—J. T. Johnstone, 1893.

LOCALITIES : Very common in all the valleys, reaching 2000 feet, Black's Hope (var. alpestre), J. T. J.

On dry sandy soils, cinders of railways ; a common garden weed.

Equisetum sylvaticum. *Linn.*

RECORDS: *Dfs.* and *Kcd.*—P. Gray, 1850. *Wgt.*—G. C. Druce, 1883.

LOCALITIES : *Nithsdale*—Dumfries, P. Gr.; Redpaths, R. A.; common Sanquhar, Dr Dv. *Annandale*—Catch Hall Loaning, Dryfe, Caledonian Railway, G. Bl.; Kirkpatrick-Juxta, F. A. H.; New Edinburgh Road, J. T. J.

Rather wet places ; often in shade.

Equisetum limosum. *Sm.*

RECORDS: *Dfs.* and *Kcd.*—P. Gray, 1850. *Wgt.*—G. C. Druce, 1883.
b. fluviatile, *Dfs.*—Dr Singer, 1843.

LOCALITIES : *Nithsdale*—Dumfries, P. Gr.; Maxwelltown Loch, R. H. M.; Newlands, R. A.; common Sanquhar, Dr Dv. *Annandale*—Lochmaben (also var. *b.*), Dr Sn., S.-E.; Kirkpatrick-Juxta, F. A. H.; Reddings, J. T. J.

Equisetum palustre. *Linn.*

RECORDS : *Dfs.* and *Kcd.*—P. Gray, 1850. *Wgt.*—G. C. Druce, 1883.
b. polystachyum, *Dfs.*—R. Armstrong, 1893.

LOCALITIES : *Nithsdale*—Dumfries, P. Gr.; Nethercog, Kirkconnel Reservoir, Dr Dv. *Annandale*—Dryfe, G. Bl. *Eskdale*—Wauchope, S.-E.

Equisetum hyemale. *Linn.*

RECORDS : *Dfs.*—J. M'Andrew, 1882. *Kcd.*—P. Gray, 1848.

LOCALITIES : *Nithsdale*—Scaur, Exc.; Thornhill, J. M'A. *Annandale*—Balgray, Dryfe, G. Bl., S.-E.

Hymenophyllum tunbridgense. *Sm.*

RECORD : *Dfs.*—Dr Burgess, 1789.

LOCALITIES : *Nithsdale*—Drumlanrig, J. T. S.; Auchenessnane, T. Br. *Annandale*—Rocks on south side of Water of Æ a little below Hollas, foot of sandy holm on Glenkillburn, Kirkmichael, Dr Br.*

Hymenophyllum unilaterale. *Bory.*

RECORDS : *Dfs.*—W. Stevens, 1848. *Kcd.*—J. M'Andrew, 1882.

LOCALITIES : *Nithsdale*—Dalveen Pass, Scaur, W. St., T. Br.; Glenjaun (1000 feet), Glencrosh (700 feet), Benbuie (700 feet), J. Cr.; Kello and Mennock sources, Dr Dv. *Annandale*—Garpol, Dr Bl.; Wellburn, Kd.; Cornal Tower Burn, Midlaw Burn, J. T. J.; Grey Mare's Tail, Dr Dv. *Eskdale*—1800 feet at Midlaw Burn, J. T. J.; Tarras, Arkleton, Bea.

On moist sandstone, or whinstone rocks, quite shaded, W. St., Bea.

Pteris aquilina. *Linn.* (Bracken).

RECORDS : *Dfs.*—P. Gray, 1850. *Kcd.*—J. M'Andrew, 1882. *Wgt.*—G. C. Druce, 1883.

LOCALITIES : Very common all over the hills.

On boulder clay, whinstone soils, sandy poor soil (not on peat or rich alluvium).

---

* I suspect this to be H. unilaterale, but as I have never found Dr Burgess wrong, I leave the record.

Cryptogramme crispa. *R. Br.* (Parsley).
(ALLOSORUS CRISPUS).

RECORDS: *Dfs.*—Dr Burgess, 1789. *Kcd.*—Mr Maughan, 1789. *Wgt.*—Rev. J. Gorrie, 1889.

LOCALITIES: *Nithsdale*—Newabbey, Mau.; Craigs, P. Gr.; Auldgirth, Th.; Auchenstrowan (1000 feet), Caitloch (1000 feet), Castlehill (700 feet), J. Cr.; Morton Range, W. St., R. A. *Annandale*—Very common Æ Water (at 500 feet), Bremner, Queensberry, S.-E.; Beeftub, Kd., S.-E.; Hartfell, S.-E.; Black's Hope, Correifron, Grey Mare's Tail, Earshaig Burn, J. T. J., S.-E.; Beld Craig Burn, Kd., J. Sd. *Eskdale*—Langholm Hill, Bea.

On whinstone fragments at foot of cliffs, old walls; fully exposed.

Lomaria spicant. *Dew.* (Hard Fern).
(BLECHNUM BOREALE).

RECORDS: *Dfs.* and *Kcd.*—P. Gray, 1850. *Wgt.*—G. C. Druce, 1883.

LOCALITIES: Very common in all the valleys.

On moist or pretty dry leaf mould, in shady sheltered woods, Bea., S.-E.

Asplenium adiantum nigrum. *Linn.* (Black Maidenhair).

RECORDS: *Dfs.*—W. Keddie, 1854. *Kcd.*—P. Gray, 1848. *Wgt.*—G. C. Druce, 1883.

LOCALITIES: *Nithsdale*—Auldgirth Bridge, S.-E.; Minnygrile (650 feet), Crechan (500 feet), J. Cr.; Black Linn, Morton, R. A.; Sanquhar Castle, Ryehill Railway Wall, Dr Dv. *Annandale*—Tundergarth, G. Bl.; Wamphray Glen, Duff, Kinnel, J. T. J.; Garpol, W. Ca.; Beeftub, S.-E.; Beld Craig, Kd., W. Ca.; Cornal Tower Burn, W. Ca. *Eskdale*—Tarras Hill, Bea.

On dry whinstone rocks, old walls.

Asplenium marinum. *Linn.*

RECORDS: *Dfs.*—(Nith estuary, P. Gray, 1848.) *Kcd.*—G. N. Lloyd, 1837. *Wgt.*—Dr Balfour, 1843.

LOCALITIES: *Along the shore*—Common.

Asplenium viride. *Huds.* (Green Maidenhair).

RECORDS: *Dfs.*—W. Stevens, 1848. *Kcd.*—J. M'Andrew, 1882.

LOCALITIES: *Nithsdale*—Euchan Glen, J. Sh., Dr Dv.; Townhead Hill, James Smith, R. A. *Annandale*—Hartfell Craigs, J. T. J., S.-E.; Grey Mare's Tail, W. St., W. Ca.; Beld Craig, Kd., J. Sd., W. Ca. *Eskdale*—Tarras Water, Ewes Hills, Bea.

On moist whinstone or mudstone; partly shaded and wind-sheltered.

Asplenium trichomanes. *Linn.* (Maidenhair).

RECORDS: *Dfs.* and *Kcd.*—P. Gray, 1850. *Wgt.*—G. C. Druce, 1883.

LOCALITIES: *Nithsdale*—Glen Bridge, S.-E.; Dumfries, common, P. Gr.; Glencairn, common, J. Cr.; Cleugh, R. A.; Sanquhar, common, Dr Dv. *Annandale*—Very common Beld Craig, Kd., W. Ca., J. T. J.; Loch Skene (2000 feet), S.-E. *Eskdale*—Common Langholm Bridge, S.-E., Bea.

On dry rarely wet whinstone rocks, old walls, bridges, &c.

Asplenium rutamuraria. *Linn.* (Wall Rue).

RECORDS: *Dfs.*—P. Gray, 1850. *Kcd.*—J. M'Andrew, 1882. *Wgt.*—J. M'Andrew, 1889.

LOCALITIES: *Nithsdale*—Routen Brig, S.-E.; Craigdarroch, Jarbruck, J. Cr.; Auldgirth Bridge, R. A., S.-E.; Morton, Wallace Hall, R. A.; Euchan Bridge, Sanquhar Castle, Dr Dv. *Annandale*—Annan Bridge at Moffat, W. Ca.; Beeftub, J. T. J.; Milke Bridge, S.-E.; Beld Craig, Kd.; Auchencas, W. Ca. *Eskdale*—Langholm Bridge, Bentpath Bridge, S.-E.

On dry rocks, bridges, walls of whinstone or limestone.

Athyrium filix fœmina. *Roth.* (Lady Fern).

RECORDS: *Dfs.* and *Kcd.*—P. Gray, 1850. *Wgt.*—G. C. Druce, 1883.

Var. Fieldiæ-cristatum, Mr Harper, 1892.

LOCALITIES: Very common in all the valleys (var., Kirkconnel, Harper).

In moist leaf mould of shady woods.

Ceterach officinarum. *Willd.* (Scaly Fern).

RECORDS : *Dfs.*—W. Stevens, 1848. *Kcd.*—P. Gray, 1848. *Wgt.*—J. M'Andrew, about 1876.

LOCALITIES : *Nithsdale*—Walls, Drumlanrig, W. St. *Annandale*—Manse garden, Kirkpatrick-Juxta, Mr Brodie ; old tower Dryfe water, F. W. G.

Scolopendrium vulgare. *Symons.* (Hart's-tongue).

RECORDS : *Dfs.* and *Kcd.*—P. Gray, 1850. *Wgt.*—J. M'Andrew, 1887.

LOCALITIES : *Nithsdale*—Cluden Craigs, Glen, P. Gr.; two stations, Glencairn, J. Cr.; Black Linn, R. A.; Euchan, Dr Dv. *Annandale*—Scotbrig, Kirtle, Gimmonbie, G. Bl.; Archbank, Beld Craig, Wamphray, W. Ca. *Eskdale*—Middlebie burn, Westerkirk, Bea.

Woodsia ilvensis. *R. Br.*

RECORDS : *Dfs.*—W. Stevens, 1848.

LOCALITIES : *Annandale*—Beeftub (D. Oliver) ; Black's Hope (P. N. Frazer), J. T. J.; Correifron, P. N. Frazer, W. Ca.; Whitecoombe, Loch Skene, W. St., W. Ca.*

In rocky ravines ; completely shaded.

Crystopteris fragilis. *Bernh.* (Bladder Fern).

RECORDS : *Dfs.*—W. Keddie, 1854. *Kcd.*—G. N. Lloyd, 1837. *Wgt.*—J. M'Andrew, 1890.

LOCALITIES : *Nithsdale*—Kirkconnel House, G. N. Ll., R. R.; Cluden Craigs, P. Gr.; Glenjaun (1400 feet), J. Cr.; Cample Cleugh, Carron Glen, Enoch Bridge, R. A.; Kello, Orchard Burn, Dr Dv. *Annandale*—Duff, Kinnel, J. T. J.; Tundergarth Linn, G. Bl.; Garpol, Kd.; Hartfell, W. Ca., S.-E.; Moffat Well, Loch Skene, W. Ca., S.-E.; *Eskdale*—Tarras, Esk, Bea.

Moist rocks, tops of walls ; in shade or exposed, usually wind-sheltered.

Polystichum lonchitis. *Roth.* (Holly).

RECORD : *Dfs.*—Simon Halliday, 1872.

LOCALITY : *Annandale*—Black's Hope, J. T. J.

---

* Of the hundreds of plants described as existing in 1856, there are now only two remaining, all the rest having been exterminated mainly by the ravages of the " Innerleithen Alpine Club."

Polystichum lobatum. *Presl.*

RECORDS: *a.* Genuinum, *Dfs.*—Mr Yalden, 1789. *Kcd.* and *Wgt.*
—J. M'Andrew, 1882 and 1890.
b. Aculeatum, *Dfs.*—Dr Burgess, 1789. *Kcd.*—P. Gray, 1850. *Wgt.*—J. M'Andrew, 1890.

LOCALITIES: *Nithsdale*—Glencairn *(b.)*, abundant, J. Cr.; Blackwood Linn, Exc.; Drumlanrig *(b.)*, Dr Br.; Carronbridge, Sanquhar, Dr Dv. *Annandale*—Raehills *(a.)*, Dr Sn.; Moffat Well *(a.)*, Yalden, W. Ca.; Hartfell, Cornal Tower Burn, Garpol, Brackenside Burn, W. Ca.; Grey Mare's Tail *(a.)*, J. Sd.; Beld Craig *(a.)*, Dr Sn., *(b.?)*, Kd., W. Ca. *Eskdale*—Wauchope *(b.)*, Bea.; Bentpath, S.-E.; Ewes water *(a.)*, Bea.
On rocks, usually whinstone; in shade and shelter.

Lastrea oreopteris. *Presl.* (Mountain Shield).

RECORDS: *Dfs.*—W. Carruthers, 1863. *Kcd.*—J. M'Andrew, 1882. *Wgt.*—Confirmed by J. M'Andrew, in 1892.

LOCALITIES: *Nithsdale*—Glencairn, abundant, J. Cr.; Sanquhar, common, Dr Dv. *Annandale*—Common on hills, Garpol, Beld Craig, W. Ca. *Eskdale*—Deanbanks, Bea.
Sloping banks in shade, Bea.

Lastrea filix-mas. *Presl.* (Male Fern).

RECORDS: *Dfs.* and *Kcd.*—P. Gray, 1850. *Wgt.*—G. C. Druce, 1883.
Var. *e.*, abbreviata, *Dfs.*—E. F. Linton, 1889.

LOCALITIES: Very common in all the valleys, to 2200 feet Loch Skene; var. *e.*, Grey Mare's Tail, E. F. Linton.
On boulder clay, whinstones, etc.

Lastrea spinulosa. *Presl.*

RECORDS: *Dfs.*—J. M'Andrew, 1893. *Kcd.*—J. M'Andrew, 1887. *Wgt.*—G. C. Druce, 1883.

LOCALITIES: *Nithsdale*—Comlongan, J. M'A.; Manse Wood, R. A. *Annandale*—Sallow Hill, Gardenholm Wood, etc., J. T. J.

Lastrea dilatata. *Presl.* (Broad Buckler).

RECORDS: *Dfs.*—W. Keddie, 1854. *Kcd.*—J. M'Andrew, 1882. *Wgt.*—G. C. Druce, 1883.

LOCALITIES : *Nithsdale*—Glencairn, P. Gr.; Craighope Linn, R. A.; Sanquhar, Dr Dv. *Annandale*—Gallows Hill, Anchencas, W. Ca.; Beld Craig, Kd., W. Ca. *Eskdale*—Common in woods, Bea.
Moist shady woods.

Polypodium vulgare. *Linn.* (Polypody).

RECORDS: *Dfs.* and *Kcd.*—P. Gray, 1850. *Wgt.*—G. C. Druce, 1883.

LOCALITIES : *Nithsdale*—Very common Glencairn, J. Cr.; Sanquhar, Dr Dv. *Annandale*—Very common Garpol, W. Ca.; Hartfell, S.-E. *Eskdale*—Very common Dean banks, Wauchope, Bea.
Moist or dry whinstone rocks, old walls, old trees ; chiefly in shade.

Phegopteris dryopteris. *Fee.* (Oak Fern).

RECORDS: *Dfs.*—Dr Burgess, 1777. *Kcd.*—P. Gray, 1846. *Wgt.*—J. M'Andrew, 1890.

LOCALITIES : *Nithsdale*—Cluden, Dalskairth, P Gr.; Blackwood, abundant, S.-E.; Glencairn, J. Cr.; Craighope Linn, Hn., R. A. *Annandale*—Gimmenbie, G. Bl.; Raehills, F. A. H.; Gardenholm Linn, Wellburn, Black's Hope, Correifron, etc., J. T. J.; Beld Craig, Kd., W. Ca., etc. *Eskdale*—Dean Banks, Dr Br., Bea.; Broomholm, Dr Br.
Common on moist leaf-mould ; in shade and shelter.

Phegopteris polypodioides. *Fee.* (Beech Fern).

RECORDS : *Dfs.*—Dr Burgess, 1777. *Kcd.*—P. Gray, 1846. *Wgt.*—J. M'Andrew, 1890.

LOCALITIES : *Nithsdale*—Mabie, Dalskairth, P. Gr.; Glencairn, J. Cr.; Carron Bridge, Blackwood, M'Glashan ; Craighope Linn, White Quarry Scaur, R. A.; Sanquhar, Dr Dv. *Annandale*—Queensberry, S.-E.; Gimmenbie, G. Bl.; Kirkpatrick-Juxta, Raehills, F. A. H.; Garpol, Cornal Tower Burn, Loch Skene (to 2200 feet), W. Ca., S.-E. *Eskdale* —Penton, Dr Br., S.-E.; Broomholm, Dr Br.; Langholm, Dr Br., Bea.
On moist leaf-mould ; in shade and shelter.

Osmunda regalis. *Linn.* (Royal Fern).

RECORDS : *Dfs.*—G. N. Lloyd, 1837. *Kcd.*—P. Gray, 1848. *Wgt.* —Dr Arnott, 1843.

LOCALITY : *Nithsdale*—Lochar Moss, G. N. Ll.

Ophioglossum vulgatum. *Linn.* (Adder's Tongue).

RECORDS: *Dfs.*—W. Keddie, 1854. *Kcd.*—P. Gray, 1848. *Wgt.*—Rev. J. Gorrie, 1890.

LOCALITIES: *Nithsdale*—Kirkbean, P. Gr.; Twomerkland (400 feet), Caitloch (600 feet), Dalmakerran (750 feet), J. Cr.; New Loch, R. A.; shooting range, Sanquhar, Dr Dv. *Annandale*—Whitestane hill, Tundergarth Linn, Poolhouses, Whiteknowe, G. Bl.; old camp, Kirkmichael, Bremner; Shieldhill, S.-E.; Beld Craig, Kd.; Beattock hill, W. Ca. *Eskdale*—Langholm hills, Bea.

In pastures on whinstone soil.

Botrychium lunaria. *Sw.* (Moonwort).

RECORDS: *Dfs.*—Dr Balfour, 1856. *Kcd.*—P. Gray, 1846. *Wgt.*—J. M'Andrew, 1886.

LOCALITIES: *Nithsdale*—Glen, Barrhill (300-800 feet), P. Gr.; Glencairn, common, J. Cr.; Thornhill, common, and Moss, R. A.; Ulzieside, Sanquhar, Dr Dv. *Annandale*—Milke, G. Bl.; Hunterheck hill, W. Ca.; Black's Hope, Midlaw burn, Dumcrief, Annan water, J. T. J.; Beattock hill, Carr.; Dobb's Linn to Grey Mare's Tale, Dr Bl., W. Ca. *Eskdale*—Langholm hill, Bea.

On wet or dry pastures on whinstone soils, mudstones or boulder clay.

Received too late for insertion in the Flora:—

Elatine hexandra. *D. C.*

RECORD: *Dfs*—Lockerbie, G. Bell, 1896.

INDEX OF INSECTS AND HOST PLANTS.

Ablyteles cerinthius, Carduus heterophyllus.
Adrastus limbatus, Pyrola minor.
Ægynnis aglaia, Orchis conopsea.
Allantus nothi, Ranunculus acris, Trollius, Lathyrus pratensis, Spiræa ulmaria, Poterium sanguisorba, Carum verticillatum, Pimpinella saxifraga, Myrrhis Odorata, Conopodium denudatum, Conium maculatum, Scabiosa succisa, Chrysanthemum segetum, Matricaria inodora, Achillea millefolium, Carduus arvensis, Crepis virens, paludosa, Euphrasia officinalis, Polygonum Bistorta.
Allantus scrophulariæ, Scrophularia officinalis.
Andrena albicans, Berberis, Hypericum quadratum, Geranium pratense, Alchemilla vulgaris, Lapsana communis, Alisma plantago, Cerastium triviale, Taraxicum.
Andrena bicolor, Capsella bursapastoris, Lotus corniculatus, Crepis virens
Andrena coitana, Barbarea vulgaris, Rosa canina, Œnanthe crocata, Crepis virens, Euphrasia officinalis.
Andrena denticulata, Senecio jacobæa.
Andrena furcata, Crepis virens.
Andrena Gwynana, Campanula rotundifolia.
Andrena parvula, Geranium molle.
Andrena Trimmeriana, Corydalis claviculata.
Andrena Wilkella, Taraxacum.
Anthidium manicatum, Arctium lappa.
Antholobium triviale, Cardamine amara.
Anthomyia pluvialis, Draba verna, Fragaria vesca, Heracleum spondylium, Crepis virens.
Anthomyia radicum, Ranunculus lingua, Glaucium luteum, Barbarea vulgaris, Nasturtium palustre, Cardamine hirsuta, Sisymbrium alliaria, Brassica monensis, Sinapis, Cochlearia, Capsella, Arenaria peploides, Spergula, Hypericum perforatum, quadrangulum, Geranium sanguineum, pratense, Erodium cicutiarum, Oxalis acetosella, Geum urbanum, Rubus Chamæmorus, Fragaria, Potentilla tormentilla, maculata, anserina, Alchemilla vulgaris, Pyrus Aucuparia, Epilobium montanum, Sanicula, Cicuta, Ægopodium, Carum verticillatum, Œnanthe crocata, Myrrhis, Sambucus Ebulus, Galium palustre, Aparine, Aster Tripolium, Bellis, Gnaphalium silvaticum, Chrysanthemum segetum, Senecio aquaticus, Carduus heterophyllus, Tragopogon pratensis, Leontodon hirtus, Hypochæris, Taraxicum, Crepis virens, Hieracium murorum, Lapsana, Jasione, Campanula rotundifolia, Veronica serpyllifolia, montana, Chamædrys, Armeria, Alisma plantago.
Anthomyia sulciventris, Ranunculus lingua, sceleratus, Papaver dubium, Veronica chamædrys, Cardamine hirsuta.
Apathus campestris, Scabiosa succisa.
Apathus quadricolor Corydalis claviculata, Solidago virgaurea, Senecio jacobea, Centaurea nigra, Convolvulus sepium, Mentha aquatica, Thymus serpyllum.
Apathus vestalis, Prunus cerasus, Rubus Idæus.
Apion germari, Cardamine hirsuta.
Apion pavidum, Ranunculus sceleratus, Trollius Europæus.

Apis mellifica, Caltha palustris, Nymphea, Corydalis claviculata, Barbarea vulgaris, Capsella, Lepidium Smithii, Raphanus, Spergularia, Hypericum perforatum, pulchrum, Malva moschata, Geranium sanguineum, sylvaticum, pratense, Ulex Europæus, Cytisus Laburnum, Trifolium medium, repens, Lotus corniculatus, Lathyrus pratensis, Prunus cerasus, Spiræa ulmaria, Rubus idæus, Potentilla comarum, Rosa canina, Cratægus oxyacantha, Epilobium angustifolium, Lythrum Salicaria, Saxifraga stellaris, Heracleum spondylium, Valeriana officinalis, Solidago, Leontodon hirtus, Erica vulgaris, cinerea, Menyantnes trifoliatus, Anchusa arvensis, Linaria vulgaris, Thymus serpyllum, Plantago lanceolata, Mercurialis, Allium ursinum, Narthecium ossifragum.

Ascia podagrica, Ulex Europæus, Fragaria vesca, Potentilla reptans, Saxifraga stellaris, Veronica chamædrys.

Athalia spinarum, Alchemilla vulgaris.

Bombus Derhamellus, Corydalis claviculata, Hypericum quadrangulum, Malva moschata, sylvestris, Rubus fruticosus.

Bombus hortorum, Corydalis lutea, Trifolium pratense, Vicia sylvatica, sepium, sativum, Geum rivale, Lythrum salicaria, Ænanthe crocata, Lonicera caprifolium, Carduus lanceolatus, palustris, Centaurea nigra, Primula vulgaris, Convolvulus sepium, Vaccinium myrtillus, Linaria vulgaris, Melampyrum pratense, Scrophularia officinalis, Digitalis, Euphrasia, Scutellaria galericulata, Stachys, Lamium album, Betonica, Palustris, Galeopsis tetrahit, Teucrium scorodonia, Narthecium ossifragum.

Bombus lapidarius, Trifolium pratense, Lathyrus pratensis, Lonicera caprifolium, Arctium lappa, Centaurea nigra, Rhinanthuscristagalli.

Bombus lucorum, Caltha, Berberis, Nymphea? Lychnis diurna, Floscuculli, Hypericum perforatum, quadrangulum, hirsutum, Viola canina, Malva sylvestris, Ulex Europæus, Genista tinctoria, Cytisus Scoparius, Laburnum, Ononis arvensis, Trifolium arvense, pratense, medium, Lotus corniculatus, Vicia cracca, sepium, Lathyrus pratensis, macrorrhizus, Spiræa ulmaria, Rubus idæus, Potentilla Tormentilla, comarum, Rosa canina, Pyrus aucuparia, Epilobium angustifolium, Heracleum spondylium, Conium maculatum, Sambucus Ebulus, Valeriana officinalis, Scabiosa succisa, Aster Trifolium, Arctium Lappa, Carduus lanceolatus, heterophyllus, Jasione montana, Campanula rotundifolia, Erica cinerea, tetralix, Symphytum officinale, Linaria vulgaris, Veronica officinalis, Euphrasia, Rhinanthus cristagalli, Pedicularis palustris, silvatica, Thymus serpyllum, Stachys palustris, silvatica, Galeopsis tetrahit, Narthecium ossifragum.

Bombus muscorum, Corydalis claviculata, Viola canina, Lychnis floccuculli, Hypericum quadrangulum, Geranium sanguineum, sylvaticum, pratense, Robertianum, Genista tinctoria, Cystisus scoparius, Trifolium pratense, Lotus corniculatus, Anthyllis, Vicia cracca, sylvaticum, sepium, sativum, Lathyrus pratensis, macrorrhizus, Geum rivale, Rubus idæus, fruticosus, Potentilla comarum, Rosa canina, Epilobium angustifolium, Scabiosa succisa, Arctium Lappa, Carduus heterophyllus, Centaurea nigra, Vaccinium myrtillus, Symphytum officinale, Hyoscyamus niger, Anchusa, Scrophularia officinalis, Melampyrum pratense, Mentha aquatica, Calamintha clinopodium, Ajuga reptans, Nepeta Glechoma, Prunella vulgaris, Lamium album, Scutellaria galericulata, Stachys Betonica, sylvatica, palustris, Galeopsis tetrahit, Lanium purpureum, Teucrium scorodonia.

FLORA OF DUMFRIESSHIRE.

Bombus pratorum, Berberis, Brassica Sinapis, Lychnis vespertina, Hypericum perforatum, quadrangulum, hirsutum, Malva moschata, Geranium sylvaticum, pratense, Robertianum, Trifolium medium, Rubus Idæus, fruticosus, Rosa Canina, Epilobium angustifolium, Lythrum Salicaria, Sedum acre, Lonicera Caprifolium, Scabiosa succisa, Carduus palustris arvensis, Campanula latifolia, Symphytum officinale, Veronica Chamædrys, Bartsia, Odontites, Rhinanthus, Cristagalli, Pedicularis sylvaticus, Melampyrum pratense, Stachys Betonica, Sylvatica, Teucrium Scorodonia, Armeria.
Bombus terrestris, Lamium album.
Borborus equinus, Lysimachia nemorum, Veronica montana.
Brachycentrum sabulosum, Caltha palustris.
Byturus Rosae, Nasturtium officinale.
Cænonympha pamphylius, Medicago lupulina.
Calliphora erythrocephala, Pyrus Aucuparia, Chrysanthemum leucanthemum, Thymus serpyllum.
Caricea tigrina, Nasturtium officinale.
Charæas graminis, Sambucus ebulus, Scabiosa succisa.
Chilosia peltata, Cardamine amara, pratensis.
Chlorosia formosa, Brassica Sinapis, Galium Mollugo.
Chrysogaster cemetorum, Œnanthe crocata.
Chrysogaster metallina, Crepis paludosa.
Chrysogaster nigrinus, Heracleum spondylium.
Colletes fodica, Senecio Jacobea.
Colletes succincta, Erica cinerea.
Criorhina oxyacantha, Scabiosa succisa.
Cynomyia mortuorum, Helianthemum vulgare, Sanguisorba officinalis, Sedum Rhodiola, Chrysanthemum leucanthemum, Thymus Serpyllum.
Cyrtonema stabularis, Epilobium angustifolium.
Dolichopus febrilis, Pyrus aucuparia, Cratægus oxyacanthæ.
Drymeia hamata, Hypericum pulchrum, Geranium sanguineum, Potentilla Tormentilla.
Empis bilineata, Ranunculus Ficaria, Arabis hirsuta, Viola canina, Pyrus aucuparia, Saxifraga hypnoides, stellaris, Hieracium murorum, Veronica Beccabunga.
Empis ignota, Arenaria trinervis.
Empis lividia, Nasturtium officinale, Alliaria officinalis, Lychnis vespertina, Stellaria graminea, Hypericum perforatum, Vicia cracca, Carduus lanceolatus, palustris, Erythræa Centaurium, Convolvulus sepium.
Empis opaca, Myrrhis odorata, Taraxicum dens-leonis.
Empis pennata, Stellaria holostea, Geranium sylvaticum, Robertianum.
Empis punctata, Ranunculus Ficaria, Brassica Sinapis.
Empis tessellata, Geranium sylvaticum, Robertianum, Pyrus Aucuparia, Polygonum Bistorta.
Empis trigramma, Alchemilla vulgaris, Myrrhis odorata.
Empis vitripennis, Silene inflata, Spergula, Hypericum hirsutum, Linum catharticum, Geranium sylvaticum, Orchis latifolia.
Epuræa æstiva, Pyrus aucuparia.
Eristalis arbustorum, Nasturtium officinale, Œnanthe crocata, Chærophyllum temulum, Valeriana officinalis, Scabiosa succisa, Aster Tripolium, Chrysanthemum segetum, Achillea ptarmica, Senecio Jacobea.
Eristalis asnea, Trollius Europæus.
Eristalis intricarius, Carduus lanceolata, Erythræa Centaurium.
Eristalis nemorum, Matricaria inodora, Crepis paludosa, Alisma plantago.

Eristalis pertinax, Brassica Sinapis, Silene inflata, Trifolium repens, Rubus fruticosus, Rosa canina, Œnanthe crocata, Viburnum opulus, Galium cruciatum, Valeriana officinalis, Aster, Tripolium, Bellis, Hieracium murorum, Erica cinera, Orchis latifolia.
Eristalis sepulchralis, Nasturtium officinale.
Eristalis tenax, Brassica Sinapis, Œnanthe crocata, Valeriana officinalis, Scabiosa, Mentha aquatica.
Eristalis tumidata, Brassica Sinapis.
Escophanes occupator, Alchemilla vulgaris.
Formica cunicularis, Brassica monensis.
Hæmatophora pluvialis, Achillea millefolium.
Halictus albipes, Geranium sanguineum, Epilobium hirsutum.
Halictus cylindricus, Geranium sylvaticum, Sambucus Ebulus, Carduus arvensis.
Halictus mori, Barbarea vulgaris, Brassica monensis, Veronica Beccabunga, Polygonum Persicaria.
Halictus rubicundus, Senecio Jacobea, Taraxicum.
Halictus subfasciatus, Senecio Jacobea.
Halictus villosulus, Taraxicum.
Haltica oleracea, Alliaria officinalis, Brassica monensis.
Hedrerigaster Urticæ, Epilobium montanum.
Helophilus frutetorum, Potentilla Tormentilla, Galium cruciatum, Senecio Jacobea, Crepis paludosa, Myosotis palustris.
Helophilus pendulus, Nasturtium officinale, Cardamine pratensis, Achillea millefolium, Euphrasia officinalis, Alisma plantago.
Hilara maura, Armeria vulgaris.
Hydrellia griseola, Cerastium vulgatum, Erodium Cicutarium.
Hydrœcia nictitans, Scabiosa succisa.
Hydrotea dentipes, Thalictrum minus, Ranunculus aquatilis, lingua, acris, Papaver dubium, Glaucium luteum, Capsella bursapastoris, Hypericum elodes, Cytisus Laburnum, Trifolium arvense, Geum urbanum, Rubus Chamæmorus, Potentilla reptans, maculata, anserina, Saxifraga stellaris, Sanicula Europea, Carum verticillatum, Myrrhis odorata, Galium verum, Solidago virgaurea, Chrysanthemum segetum, Matricaria inodora, Achillea ptarmica, millefolium, Senecio aquaticus, Jacobea, Carduus arvensis, Hypochœris radicata, Sonchus arvensis, Crepis virens, Myosotis palustris, Veronica Beccabunga, Mentha arvensis.
Hyelemyia strigosa, Ranunculus lingua, Nasturtium officinale, palustris, Alliaria officinalis, Geum urbanum, Myrrhis odorata, Conium maculatum, Achillea millefolium, Lapsana communis, Campanula rotundifolia, Veronica montana, Alisma plantago, Anthoxanthum odoaratum.
Hyctodesia basalis, Ranunculus ficaria, Hypericum pulchrum, Chrysanthemum leucanthum, segetum, Senecio aquaticus, Carduus lanceolatus.
Hyetodesia incana, Ranunculus acris, Medicago lupulina, Potentilla comarum, anserina, Alchemilla vulgaris, Sanguisorba officinalis, Cratægus oxyacantha, Cicuta virosa, Chœrophyllum temulum, Sambucus ebulus, Galium cruciatum, Mollugo, Scabiosa succisa, Solidago virgaurea, Senecio aquaticus, Alisma plantago.
Hyetodesia jucana, Taraxicum densleonis.
Kmetes sidaris, Ranunculus acris.
Latridium porcatus, Draba verna.
Leptis tringaria, Hypericum quadrangulum.
Leucogonia lucorum, Potentilla tormentilla.

Licus ferrugineus, Geranium sanguineum.
Lonchoptera lutea, Ranunculus acris.
Lophius albomarginatus, Mentha arvensus.
Lucilia Cæsar, Nasturtium officinale, Crambe, Helianthemum vulgare,
　　　　　　Spergularia rubra, Sanguisorba officinalis, Cicuta virosa,
　　　　　　Œnanthe crocata, Heracleum spondylium, Chrysanthemum
　　　　　　leucanthum, segetum, Matricaria inodora.
Lucilia cornicina, Crambe maritima.
Mantua napi, Ranunculus ficaria, Cardamine amara, pratensis, Brassica
　　　　　　Sinapis.
Megachile Willughbiella, Carduus lanceolatus.
Melanostoma gracilis, Anemone nemorosa.
Melanostoma mellina. Anemone nemorosa, Ranunculus ficaria, Helianthe-
　　　　　　mum vulgare, Geranium lucidum, Trifolium procumbens,
　　　　　　Spiræa ulmaria, Rosa canina, Achillea millefolium, Euphrasia
　　　　　　officinalis.
Meligethes æneus, Ranunculus acris, Papaver dubium, Glaucium luteum,
　　　　　　Nasturtium palustre, Brassica monensis, Cochlearia officinalis,
　　　　　　Cakile maritima, Crambe maritima, Raphanus raphanistrum,
　　　　　　Stellaria media, nemorum, holostea, Geranium Robertianum,
　　　　　　Pyrus acuparia.
Meligethes brassicæ, Ranunculus ficaria.
Melegethes viridescens, Conopodium denudatum.
Micropalpus vulpinus, Erica tetralix, Thymus serphyllum.
Micropteryx calthella, Caltha palustris, Barbarea vulgaris, Cardamine
　　　　　　amara.
Mimesa Dahlbomi, Cicuta virosa.
Morellia hortorum, Spiræa ulmaria, Scabiosa succisa, Solidago virgaurea,
　　　　　　Achillea millefolium, Senecio Jacobea, Carduus arvensis,
　　　　　　Centaurea nigra, Crepis virens, Sonchus arvensis, oleraceus.
Musca corvina, Galium verum, Matricaria inodora.
Nemapoda stercoraria, Spergularia rubra.
Nemoteles notatris, Spergularia rubra, Sedum anglicum.
Nomada lateralis, Geranium sylvaticum.
Notiphila cinera, Spiræa ulmaria, Aster Tripolium.
Notiphila riparia, Ranunculus aquatilis.
Nysson dimidiatus, Senecio Jacobæa.
Odynerus spinipes, Ægopodium podagraria.
Onesia sepulchralis, Sanguisorba officinalis, Cicuta virosa, Solidago vir-
　　　　　　gaurea, Carduus arvensis, Crepis virens, Alisma plantago.
Opomyza germinationis, Capsella bursapastoris, Cerastium vulgatum.
Orchisa minor, Conopodium denudatum, Veronica hederifolia.
Ortalis emissa, Heracleum spondylium.
Panipla vulgaris, Cicuta virosa.
Parhydra aquila, Ranunculus aquatilis.
Perineima uassata, Galium Mollugo, Polygonum Bistorta.
Phyllobium calcaratum, Cardamine amara.
Pieris brassica, Fumaria officinalis, Scabiosa succisa, Sambucus ebulus.
Pieris napi, Rubrus fruticosus, Epilobium tetragonum, Scabiosa succisa.
Pieris rapi, Scabiosa succisa.
Platychirius albimana, Ranunculus ficaria, Trollius Europæus, Papaver
　　　　　　dubium, Barbarea vulgaris, Arabis hirsuta, Alliaria officinalis,
　　　　　　Brassica Sinapis, Spergula arvensis, Geranium molle, Ornith-
　　　　　　opus perpusillus, Potentilla anserina, Rosa canina, Epilobium
　　　　　　tetragonum, Circæa lutetiana, Saxifraga stellaris, Viburnum
　　　　　　opulus, Galium cruciatum, Saxatile, Jasione montana, Myo-
　　　　　　sotis palustris, Veronica, Beccabunga, montana, Ajuga reptans,
　　　　　　Alisma Plantago, Scilla nutans

Platychirius clypeatus, Ranunculus lingua, Nasturtium officinalis, Arabis
    Thaliana, Helianthemum vulgare, Viola canina, Ulex Europæus,
    Medicago lupulina, Trifolium arvense, procumbens, Vicia
    cracca, Epilobium montanum, tetragonum, Saxifraga stellaris,
    Galium cruciatum, Chrysanthemum segetum, Carduus lance-
    olatus, arvensis, Sonchus oleraceus, Crepis paludosa, Jasione
    montana, Myosotis palustris, Sylvatica, Veronica Beccabunga,
    Euphrasia officinalis, Alisma Plantago.
Platychirius manicatus, Lychnis diurna, Vespertina, Silene inflata, Ceras-
    tium vulgatum, Stellaria holostea, Spergula arvensis, Linum
    catharticum, Malva sylvestris, Geranium sylvaticum, pratense,
    lucidum, dissectum, Conopodium denudatum, Crepis virens,
    Euphrasia officinalis.
Platychirius peltatus, Berberis, Glaucium luteum, Nasturtium officinale,
    Lychnis vespertina, Stellaria graminea, Geranium sylvaticum,
    Rosa canina, Lythrum Salicaria, Galium palustre, Jasione
    montana, Erica tetralix, Veronica Beccabunga, Stachys palus-
    tris, Galeopsis tetrahit, Alisma Plantago.
Platychirius scutatus, Arenaria trinervis, Spergula arvensis.
Pollenia rudis, Nasturtium officinale, Carum verticillatum, Sambucus
    Ebulus, Aster Tripolium, Centaurea nigra, Crepis paludosa.
Polyommatus icarus, Medicago lupulina, Plantago lanceolata.
Polyommatus phlæas, Scabiosa succisa.
Prosopis hyalinata, Geranium sanguineum, Potentilla Tormentilla.
Psila fumitara, Myrrhis odorata.
Pyrophona rosarum, Nasturtium officinale.
Rhamphomyia albosegmentata, Ranunculus acris.
Rhingia rostrata, Trollius Europæus, Cardamine amara, Lychnis floscuculii,
    Spergula arvensis, Hypericum perforatum, Geranium sangu-
    ineum, Geum rivale, Œnanthe crocata, Carduus heterophyllus,
    Centaurea nigra, Veronica Beccabunga, Ajuga reptans, Scilla
    nutans.
Rhyphus fenestralis, Ulex Europeus.
Sarcophaga carnaria, Aster Tripolium.
Satyrus janira, Sambucus ebulus, Centaurea nigra.
Scatophaga inquinita, Hypericum hirsutum.
Scatophaga litorea, Nymphea, Spergularia rubra, Alchemilla vulgaris.
Scatophaga lutaria, Conopodium denudatum.
Scatophaga stercoraria, Nasturtium officinale, Ranunculus sceleratus,
    Spergularia rubra, Spergula arvensis, Hypericum hirsutum,
    Vicia hirsuta, Alchemilla vulgaris, Sanguisorba officinalis,
    Heracleum spondylium, Saxifraga hypnoides, Chrysosplenium
    oppositifolium, Mentha arvensis.
Scæna ribesii, Cardamine amara, pratensis.
Sepsis cynipsea, Papaver dubium, Hypericum quadrangulum, Chærophyllum
    temulum, Myosotis versicolor.
Sericomyia borealis, Hypericum pulchrum, Rubus fruticosus, Carduus
    lanceolatus, Euphrasia officinalis.
Sericomyia lapponum, Rosa canina.
Siphona cristata, Barbarea vulgaris, Stellaria graminea, Hypericum pulch-
    rum, Rubus chamæmorus, Poterium sanguisorba, Pyrus
    Aucuparia, Epilobium montanum, Chrysosplenium oppositi-
    folium, Carum verticillatum, Myrrhis odorata, Chœrophyllum
    temulum Galium verum, palustre, Matricaria inodora, Achillea
    millefolium, Senecio vulgaris, Taraxicum densleonsis, Cam-
    panula rotundifolia, Vaccinium oxycoccos, Mentha arvensis,
    Polygonum Persicaria, Alisma Plantago.

Siphona geniculata, Capsella bursapastoris, Viola palustris, Geum urbanum-
Saxifraga stellaris, Chrysosplenium oppositifolium, Ægopo-
dium podagraria, Myrrhis odorata, Conopodium denudatum,
Chœrophyllum temulum, Valeriana pyrenaica, Scabiosa succisa,
Aster Tripolium, Solidago virgaurea, Bellis, Senecio sylvati-
cus, aquaticus, Carduus arvensis, Sonchus oleraceus, Lapsana
communis, Jasione montana, Vaccinium oxycoccos, Erythrea
Centaurium. Veronica Beccabunga, Scilla nutans.
Sphærophoria Menthrastii, Matricaria inodora.
Sphegina clunipes, Arenaria trinervis.
Stremia clathrata. Luzula campestris.
Syritta pipiens, Barbarea vulgaris, Nasturtium officinalis, Arabis hirsuta,
Sisymbrium officinalis, Brassica monensis, Sinapis, Capsella,.
bursapastoris, Arenaria serpyllifolia, Stellaria graminea,
Spergula arvensis, Malva sylvestris, Epilobium montanum,
Œnanthe crocata. Conopodium denudatum, Conium macu-
latum, Galium cruciatum, verum, Mollugo, Matricaria inodora,
Senecio vulgaris, Alisma plantago.
Syrphus albostriatus, Hieracium murorum.
Syrphus arcuatus, Cerastium vulgatum, Hypericum hirsutum,
Syrphus balteatus, Ranunculus repens, Hypericum perforatum, Rosa canina
Centaurea nigra, Galeopsis tetrahit.
Syrphus bifasciatus, Brassica Sinapis, Cytisus Laburnum.
Syrphus cinctellus, Geranium lucidum, Rubus Idæus, Sanicula Europea,
Conopodium denudatum, Galium cruciatum, saxatile, Scilla
nutans.
Syrphus corollæ, Capsella bursapastoris, Trifolium procumbens, Lotus
corniculatus, Heracleum spondylium, Aster Tripolium.
Syrphus franiditarse, Alisma plantago.
Syrphus luniger, Alisma plantago.
Syrphus nemorum, Heracleum spondylium.
Syrphus ribesii, Berberis, Capsella bursapastoris, Potentilla reptans, Cytisus
Laburnum, Rubus idæus, Heracleum Spondylium, Galium
cruciatum, verum, palustre. saxatile, Valeriana officinalis.
Matricaria inodora, Achillea millefolium, Senecio aquaticus,
Carduus lanceolatus, Sonchus oleraceus, Lapsana communis,
Veronica montana, Orchis conopsea.
Syrphus umbellatarum, Alliaria officinalis.
Telephorus bicolor, Nasturtium officinale.
Telephorus discoideus, Alchemilla vulgaris.
Telephorus fulvus, Lotus corniculatus, Conium maculatum, Senecio Jacobea,
Crepis virens, Mentha arvensis.
Telephorus rusticus, Nasturtium officinale.
Tenthredo ater, Achillea millefolium.
Thyomis morcida, Nasturtium officinale.
Tipbia minuta, Ægopodium podagraria.
Tryphon vulgaris, Ægopodium podagraria, Carum verticillatum.
Vanessa Atalanta, Sambucus ebulus, Scabiosa succisa, Carduus lanceolatus,
Alisma plantago.
Vanessa Urticæ, Lythrum Salicaria, Sambucus ebulus.
Vespa rufa, Sanicula Europea, Œnanthe crocata.
Vespa sylvestris, Epilobium angustifolium, Cicuta virosa, Galium aparine,
Scrophularia nodosa.
Vespa vulgaris, Scabiosa succisa, Juncus maritimus.
Volucella bombylans, Nasturtium officinale, Polygonum Bistorta, Luzula
campestris.
Xylophasia polyodon, Sambucus ebulus.
Zostenophorus tæniatus, Cardamine pratensis.

INDEX TO THE GENERIC NAMES OF PLANTS.

	Page.		Page.		Page.
Acer	41	Avena	191	Conopodium	81
Achillea	95	Barbarea	10	Convallaria	170
Aconitum	7	Bartsia	132	Convolvulus	120
Acorus	161	Bellis	91	Coriandrum	79
Adoxa	83	Berberis	7	Cornus	83
Ægopodium	76	Beta	145	Coronilla	48
Æthusa	78	Betula	155	Corydalis	10
Agrimonia	61	Bidens	93	Corylus	155
Agropyron	193	Blysmus	177	Cotyledon	68
Agrostis	189	Borago	124	Crambe	19
Aira	190	Botrychium	209	Cratægus	64
Ajuga	141	Brachypodium	194	Crepis	104
Alchemilla	59	Brassica	14	Crithmum	79
Alisma	164	Briza	197	Crocus	169
Alliaria	14	Bromus	194	Cryptogramme	204
Allium	171	Buda	30	Cuscuta	120
Alnus	155	Butomus	165	Cynosurus	197
Alopecurus	189	Cakile	19	Cystopteris	206
Althea	36	Calamagrostis	190	Cytisus	43
Anagallis	116	Calamintha	136	Dactylis	196
Anchusa	123	Callitriche	153	Daphne	151
Andromeda	112	Caltha	6	Datura	125
Anemone	2	Camelina	16	Daucus	82
Angelica	79	Campanula	109	Dianthus	22
Antennaria	92	Capsella	18	Digitalis	128
Anthemis	95	Cardamine	12	Digraphis	188
Anthoxanthum	188	Carduus	99	Dipsacus	89
Anthyllis	47	Carex	180	Doronicum	98
Antirrhinum	126	Carlina	101	Draba	17
Apium	75	Carpinus	155	Drosera	73
Aquilegia	6	Carum	76	Echium	124
Arabis	12	Castanea	155	Elymus	193
Arctium	99	Catabrosa	199	Empetrum	152
Arctostaphylos	112	Caucalis	82	Epilobium	64
Aremonia	64	Centaurea	101	Epipactis	166
Arenaria	26	Centunculus	116	Equisetum	202
Armeria	142	Cerastium	27	Erica	112
Arrhenatherum	192	Ceterach	206	Eriophorum	179
Artemisia	96	Chærophyllum	81	Erodium	39
Arum	161	Cheiranthus	10	Erophila	17
Arundo	200	Chelidonium	8	Eryngium	75
Asparagus	170	Chenopodium	145	Erysimum	14
Asperula	87	Chrysanthemum	93	Erythræa	118
Asplenium	204	Chrysosplenium	72	Euonymus	41
Aster	90	Cichorium	105	Eupatorium	90
Astragalus	48	Cicuta	75	Euphorbia	151
Astrantia	75	Circæa	67	Euphrasia	132
Athyrium	205	Cladium	176	Fagus	155
Atriplex	145	Cochlearia	16	Festuca	195
Atropa	125	Conium	28	Filago	91

	Page.		Page.		Page.
Fragaria	56	Littorella	143	Parietaria	154
Fraxinus	118	Lobelia	109	Paris	170
Fumaria	9	Lolium	193	Parnassia	72
Gagea	171	Lomaria	204	Pedicularis	133
Galanthus	170	Lonicera	84	Peplis	68
Galeopsis	139	Lotus	47	Petasites	97
Galium	84	Luzula	175	Peucedanum	79
Genista	42	Lychnis	23	Phalaris	188
Gentiana	119	Lycopodium	201	Phegopteris	208
Geranium	37	Lycopsis	123	Phleum	188
Geum	54	Lycopus	134	Pilularia	200
Glaucium	9	Lysimachia	114	Pimpinella	77
Glaux	145	Lythrum	68	Pinguicula	117
Gnaphalium	92	Malaxis	166	Pinus	160
Habenaria	168	Malva	35	Plantago	142
Hedera	83	Marrubium	138	Platystemon	8
Helianthemum	19	Matricaria	94	Poa	197
Helleborus	7	Meconopsis	8	Polemonium	120
Heracleum	80	Medicago	44	Polygala	22
Hesperis	16	Melampyrum	133	Polygonatum	170
Hieracium	105	Melica	200	Polygonum	148
Hippophae	151	Melilotus	44	Polypodium	208
Hippuris	74	Mentha	134	Polystichum	206
Holcus	192	Menyanthes	119	Populus	159
Humulus	154	Mercurialis	152	Potamogeton	162
Hutchinsia	18	Mertensia	121	Potentilla	56
Hydrocotyle	74	Meum	78	Poterium	61
Hymenophyllum	203	Milium	187	Primula	114
Hyoscyamus	124	Mimulus	128	Prunella	137
Hypericum	31	Molinia	199	Prunus	52
Hypochaeris	103	Montia	31	Psamma	190
Iberis	18	Myosotis	121	Pteris	203
Ilex	41	Myrica	154	Pulicaria	93
Impatiens	40	Myriophyllum	73	Pulmonaria	121
Inula	93	Myrrhis	80	Pyrola	113
Iris	169	Narcissus	170	Pyrus	64
Isatis	19	Nardus	193	Quercus	156
Isoetes	200	Narthecium	172	Radiola	35
Jasione	109	Nasturtium	11	Ranunculus	2
Juncus	173	Neottia	167	Raphanus	19
Juniperus	160	Nepeta	137	Reseda	20
Koeleria	200	Nuphar	7	Rhamnus	41
Lamium	140	Nymphea	7	Rhinanthus	132
Lapsana	108	Œnanthe	77	Ribes	70
Lastroea	207	Œnothera	67	Rosa	61
Lathraea	126	Ononis	43	Rubus	55
Lathyrus	51	Onopordon	101	Rumex	146
Lavatera	35	Ophioglossum	209	Ruppia	162
Lemna	161	Orchis	167	Ruscus	170
Leontodon	102	Origanum	136	Rhynchospora	177
Lepidium	18	Ornithogalum	171	Sagina	25
Lepturus	192	Ornithopus	48	Salicornia	144
Ligusticum	78	Orobanche	125	Salix	156
Ligustrum	118	Osmunda	208	Salsola	144
Linaria	126	Oxalis	40	Sambucus	83
Linum	34	Oxyria	148	Samolus	116
Listera	166	Oxytropis	48	Sanguisorba	60
Lithospermum	121	Papaver	8	Sanicula	74

	Page.		Page.		Page.
Saussurea	99	Sonchus	103	Trientalis	145
Saxifraga	70	Sparganium	161	Trifolium	45
Scabiosa	89	Spergula	31	Triglochin	165
Scandix	80	Spergularia	30	Trigonella	45
Schoenus	176	Spiraea	53	Triodia	200
Scilla	171	Stachys	138	Trollius	6
Scirpus	177	Statice	142	Tussilago	96
Scleranthus	143	Stellaria	28	Typha	160
Scolopendrium	206	Suaeda	144	Ulex	42
Scrophularia	127	Subularia	17	Ulmus	154
Scutellaria	137	Symphoricarpos	83	Urtica	153
Sedum	68	Symphytum	123	Utricularia	117
Sempervivum	70	Tanacetum	96	Vaccinium	110
Senebiera	19	Taraxicum	104	Valeriana	88
Senecio	97	Taxus	160	Valerianella	89
Serratula	99	Teesdalia	17	Verbascum	126
Sherardia	88	Teucrium	141	Veronica	128
Sibbaldia	59	Thalictrum	1	Viburnum	84
Silene	23	Thlaspi	17	Vicia	48
Sisymbrium	14	Thymus	136	Vinca	118
Sium	77	Tilia	36	Viola	19
Solanum	125	Tofieldia	172	Woodsia	206
Solidago	91	Tragopogon	102	Zostera	162